2011 International Symposium on System on Chip

(SoC 2011)

Tampere, Finland
31 October – 2 November 2011

IEEE Catalog Number: CFP11554-PRT
ISBN: 978-1-4577-0671-4

**Copyright © 2011 by the Institute of Electrical and Electronic Engineers, Inc
All Rights Reserved**

Copyright and Reprint Permissions: Abstracting is permitted with credit to the source. Libraries are permitted to photocopy beyond the limit of U.S. copyright law for private use of patrons those articles in this volume that carry a code at the bottom of the first page, provided the per-copy fee indicated in the code is paid through Copyright Clearance Center, 222 Rosewood Drive, Danvers, MA 01923.

For other copying, reprint or republication permission, write to IEEE Copyrights Manager, IEEE Service Center, 445 Hoes Lane, Piscataway, NJ 08854. All rights reserved.

***This publication is a representation of what appears in the IEEE Digital Libraries. Some format issues inherent in the e-media version may also appear in this print version.**

IEEE Catalog Number: CFP11554-PRT
ISBN 13:   978-1-4577-0671-4

Additional Copies of This Publication Are Available From:

Curran Associates, Inc
57 Morehouse Lane
Red Hook, NY 12571 USA
Phone: (845) 758-0400
Fax: (845) 758-2633
E-mail: curran@proceedings.com
Web: www.proceedings.com

2011 International Symposium on System on Chip (SoC 2011)

Tampere, Finland
31 October - 2 November 2011

IEEE Catalog Number: CFP11554-POD
ISBN: 978-1-45770-671-4

Table of Contents

Analyzing Transport and MAC Layer in System-Level Performance Simulation   1

Subayal Khan[1], Jukka Saastamoinen[1], Mikko Majanen[1], Jyrki Huusko[1], Jari Nurmi[2]

[1]VTT Technical research Center of Finland, [2]Tampere University of Technology. Department of Computer Systems

Moonrake Chip - GALS Demonstrator in 40 nm CMOS Technology   9

Miloš Krstić[1], Xin Fan[1], Eckhard Grass[1], Luca Benini[2], M. R. Kakoee[2], Christoph Heer[3], Birgit Sanders[3], Alessandro Strano[4], Davide Bertozzi[4]

[1]IHP, [2]University of Bologna, [3]Intel Mobile Communications, [4]University of Ferr

Analyzing Synchronous Dataflow Scenarios for Dynamic Software-defined Radio Applications 14

Firew Siyoum[1], Marc Geilen[1], Orlando Moreira[2], Rick Nas[2], Henk Corporaal[1]

Eindhoven University of Technology, [2]2ST-Ericsson Eindhoven

A Hybrid Model of Speculative Execution and Scout Threading for Auto-Memoization Processor   22

Tomoki IKEGAYA[1], Ryosuke ODA[1], Tatsuhiro YAMADA[1], Tomoaki TSUMURA[1], Hiroshi MATSUO[1], Yasuhiko NAKASHIMA[2]

[1]Nagoya Inst. of Tech., [2]Nara Inst. of Sci. and Tech.

Customizable Datapath Integrated Lock Unit   29

Pekka Jääskeläinen, Erno Salminen, Otto Esko, Jarmo Takala

Tampere University of Technology

Exploring Instruction caching strategies for tightly-coupled shared-memory clusters   34

Daniele Bortolotti, Francesco Paterna, Christian Pinto, Andrea Marongiu, Martino Ruggiero, Luca Benini

University of Bologna - DEIS

Static Analysis Method for Deadlock Detection in SystemC Designs   42

Mikhail Moiseev[1], Alexey Zakharov[1], Ilya Klotchkov[2], Sergey Salishev[2]

[1]Saint-Petersburg State Polytechnical University, [2]Intel Labs

SAMOSA: Scratchpad Aware Mapping Of Streaming Applications   48

Zubair Wadood Bhatti[1], Davy Preuveneers[1], Narasinga Rao Miniskar[2], Roel Wuyts[2], Yolande Berbers[1]

[1]K.U.Leuven, [2]IMEC

A System Level Power Consumption Estimation for MPSoC   56

Anthosh Kumar Rethinagir[1], Rabie ben Atitallah[2], Jean-Luc Dekeyser[1]

[1]Univ.Valenciennes, [2]Univ. Lille

OpenCL Implementation of Cholesky Matrix Decomposition   62

Claudio Brunelli, Eero Aho, Heikki Berg

Nokia

Low-power Arithmetic Unit for DSP Applications   68

Mehdi Modaressi[12], Hossein Nikonia[1], Amir-Hossein Jahangir[1]

[1]Sharif University of Technology, [2]Tehran University

Co-designs of Parallel Rijndael   72

Issam W. Damaj

American University of Kuwait

A Set of Traffic Models for Network-on-Chip Benchmarking   78

Esko Pekkarinen, Lasse Lehtonen, Erno Salminen, Timo D. Hämäläinen

TUT

Effects of Loop Unrolling and Use of Instruction Buffer on Processor Energy Consumption 82

Vladimír Guzma, Teemu Pitkänen, Jarmo Takala

Tampere University of Technology

Applying IP-XACT in Product Data Management 86

Erno Salminen, Timo D. Hämäläinen, Marko Hännikäinen
TUT

Increasing Energy Efficiency of Automotive E/E-Architectures with Intelligent Communication Controllers for FlexRay 92

Christoph Schmutzler[1], Abdallah Lakhtel[1], Martin Simons[1], Jürgen Becker[2]

[1]Daimler AG, [2]Karlsruhe Institute of Technology, ITIV

AN AUTOMATIC EXPERIMENTAL SET-UP FOR ROBUSTNESS ANALYSIS OF DESIGNS IMPLEMENTED ON SRAM FPGAS 96

Uli Kretzschmar, Armando Astarloa, Jesús Lazaro, Jaime Jimeńez, Aitzol Zuloaga
Universidad del Pais Vasco UPV/EHU

A Coarse-Grained Reconfigurable Protocol Processor 102

Mohammad Badawi, Ahmed Hemani
Royal Institute of Technology, Stockholm, Sweden

moviTest: A Test Environment Dedicated to Multi-Core Embedded Architectures 108

Teodor Tite[1], Adelina Vig[1], Nicolae Olteanu[2], Cristian Cuna[2]

[1]University "Politehnica" of Timisoara, Department of Computer and Software Engineering, Timisoara, Romania, [2]Movidius SRL, Tools Department, Timisoara, Romania

Mismatch Characterization of High-Speed NoC Links using Asynchronous Sub-sampling 112

Sebastian Höppner, Dennis Walter, Georg Ellguth, René Schüffny
TU Dresden

Impact of Proactive Temperature Management on Performance of Networks-on-Chip 116

Tim Wegner, Martin Gag, Dirk Timmermann
University of Rostock

Bringing Network-on-Chip Links to 45nm 122

Marco Ferraresi, Giuseppina Gobbo, Daniele Ludovici, Davide Bertozzi
University of Ferrara

Synchronizing Distributed State Machines In A Coarse Grain Reconfigurable Architecture 128

Omer Malik, Ahmed Hemani
Royal Institute of Technology - KTH

Automatic Calibration of Streaming Applications for Software Mapping Exploration 136

Weihua Sheng, Stefan Schürmans, Maximilian Odendahl, Rainer Leupers, Gerd Ascheid
RWTH Aachen University

Building a RTOS for MPSoC Dataflow Programming 143

Yaset Oliva[1], Maxime Pelcat[1], Jean-Francois Nezan[1], Jean-Christophe Prevotet[1], Aridhi Slaheddine[2]

[1]IETR/INSA Rennes, [2]Texas Instruments

2011 International Symposium on System on Chip (SoC)

Tampere, Finland, October 31 - November 2, 2011

Welcome

Welcome to the System-on-Chip event. SoC is an annual symposium held at Tampere, Finland - The SoC City. It builds on the tradition of a series of SoC events organized annually since 1999. The mission of SoC is to provide a forum that is fully and comprehensively dedicated to SoC issues.

The event is based on a balanced mixture of

- world-class invited presentations
- exhibition of state-of-the-art commercial technology
- contributed paper track (since 2003)
- industrial paper track
- panel discussion
- tutorial course (since 2001)

The main organizer for the event is Tampere University of Technology Department of Computer Systems in cooperation with timely research activities of the field. Since 2003 the event has technical co-sponsorship by the Circuits and Systems Society of IEEE.

On these pages you will find information on SoC 2011.
For more information, please see SoC homepage or email Prof. Jari Nurmi (Jari.Nurmi@tut.fi).

Foreword

International Symposium on System-on-Chip 2011 is the 13th annual SoC event in Tampere, it builds on a tradition started back in 1999. In addition to the invited lectures and exhibit, the conference is open for contributions from researchers on this broad but focused field. The symposium also features a panel discussion on topics of high interest to the SoC community. This will reflect the theme of the year, System-Level Design Challenges. The mission of SoC 2011 is to provide a forum that is fully and comprehensively dedicated to SoC. We enjoy the privilege to have IEEE Circuits and Systems Society as our technical co-sponsor.

The event was the first to use solely "SoC" as its name and focus. Later on, many counterparts have emerged worldwide, adopting this magnificent abbreviation in their names. Still, it is the major international SoC event in the Northern Europe, equally appreciated by the companies and academics in Europe but also increasingly in Americas and Far East. This is also reflected in the spectrum of countries where the papers presented in SoC 2011 originate from, they come from 16 countries all over the world (first author). We think that a very interesting thing is that the six top countries are not very surprisingly Finland, Germany, France, Sweden, Japan and Italy, but also countries like Iran, Kuwait, Romania and Russia are represented. Thanks to all contributors for their submissions, whether or not exceeding the publication threshold this time.

We would like to acknowledge the sponsorship received from Nokia Corporation, Renesas Mobile Corporation, and IEEE Finland Section. Even more than that, we are especially pleased about the presence of numerous Nokia and Renesas representatives in the event, which has also become a tradition.

We are grateful to the technical program committee members and other reviewers of the submitted papers, with their help we could provide valuable feedback to the authors to improve the quality of the Proceedings. Last but definitely not least, we extend our thanks to the invited speakers of this year. Traditionally, the backbone of the event has been formed by the invited talks. We believe that with the selected six distinguished people from the academy and industry all over the world, the event will be in a good shape. They all approach the theme of the year from different viewpoints. This year we share part of the invited talks with another co-located event, the Conference on Design and Architectures for Signal and Image Processing (DASIP).

Thanks also to our steering committee comprising Prof. Jan Rabaey, Prof. Heinrich Meyr, Prof. Hannu Tenhunen, and Dr. Fabio Campi, chaired by the permanent general chair Prof. Jari Nurmi. Welcome to Tampere, the SoC City!

Jari Nurmi
General Chair

Jarmo Takala
Program Chair

Olli Vainio, Jussi Raasakka
Proceedings Chairs

Steering Committee

- Jari Nurmi, TUT, Finland (chairman)
- Jan Rabaey, UC Berkeley, USA
- Heinrich Meyr, RWTH Aachen / CoWare, Germany
- Hannu Tenhunen, KTH, Sweden / UTU, Finland / INPG, France / Fudan University, China
- Fabio Campi, ST Microelectronics, Italy

Retired Steering Committee Members

- Mika Kuulusa, Nokia, Finland

General Chair

- Prof. Jari Nurmi, Tampere University of Technology, Finland

Proceedings Co-chairs

- Prof. Olli Vainio, Tampere University of Technology, Finland
- Jussi Raasakka, Tampere University of Technology, Finland

Technical Program Committee Chair

- Prof. Jarmo Takala, Tampere University of Technology, Finland

Technical Program Committee

- Andrea Acquaviva, University of Verona, Italy
- Brian Bailey, independent consultant, USA
- Heikki Berg, Nokia, Finland
- Davide Bertozzi, University of Ferrara, Italy
- Shuvra S. Bhattacharyya, University of Maryland, USA
- Abdelhafid Bouhraoua, KFUPM, Saudi-Arabia
- Claudio Brunelli, Nokia, Finland
- Jean-Luc Dekeyser, LIFL, France
- Peeter Ellervee, TU Tallinn, Estonia
- M. A. Al Faruque, Siemens, USA
- William Fornaciari, Politecnico di Milano, Italy
- Kees Goossens, TU Eindhoven, The Netherlands
- Lasse Harju, ST-Ericsson, Finland
- Hannu Heusala, University of Oulu, Finland
- Heikki Hurskainen, Microteam, Finland
- Jouni Isoaho, University of Turku, Finland
- Tariq Jamil, Sultan Qaboos University, Oman
- Murali Jayapala, IMEC, Belgium
- Tuomas Järvinen, ST-Ericsson, Finland
- Kimmo Kuusilinna, Nokia, Finland
- Vesa Lahtinen, Renesas Mobile, Finland
- Rainer Leupers, RWTH Aachen, Germany
- Oz Levia, independent consultant, USA
- Dragomir Milojevic, ULB/IMEC, Belgium
- Fernando Moraes, PUCRS, Brazil

- Tobias Noll, RWTH Aachen, Germany
- Juha Plosila, University of Turku, Finland
- Yang Qu, Renesas Mobile, Finland
- Tero Rissa, Nokia, Finland
- Stefan Rusu, Intel, USA
- Marco Santambrogio, Politecnico di Milano, Italy
- Olli Silven, University of Oulu, Finland
- Gerard J. M. Smit, University of Twente, The Netherlands
- Wonyong Sung, Seoul National University, Korea
- Lionel Torres, LIRMM, France
- Seppo Virtanen, University of Turku, Finland
- Steve Wilton, UBC, Canada

External Reviewers

(in addition to Technical Program Committee members and chairmen)

- Tapani Ahonen
- Roberto Airoldi
- Mohammad Abdullah Al Faruque
- Francescantonio Della Rosa
- Vladimir Guzma
- Ismo Hänninen
- Pekka Jääskeläinen
- Sanna Määttä
- Gianluca Palermo
- Erno Salminen
- Carlos Sanchez de la Lama
- Jarno Vanne

Technical Co-Sponsor

Financial Sponsors

Gold Sponsors

NOKIA

Bronze Sponsors

Renesas Mobile

Accepted Papers by Country

(Includes contributed scientific papers and invited papers)

Belgium	1
Denmark	1
Finland	6
France	3
Germany	5
Iran	1
Italy	2
Japan	2
Kuwait	1
Netherlands	1
Romania	1
Russia	1
Spain	1
Sweden	3
United Kingdom	1
United States	1
Total	**31**

Accepted Papers by Type

Scientific - oral	15
Scientific - poster	10
Invited	6
Total	**31**

Invited talk abstracts and biographies

Heterogeneous Concurrent Modeling and Design in Java
Dr. Patricia Derler, UC Berkeley, USA

This presentation describes an open-source framework in Java for modeling and simulating heterogeneous systems, called Ptolemy. Models are defined using the actor-oriented design principle. Actors transform values on inputs, update their state and/or produce values on outputs. The implementation of an actor can be manifold; examples are imperative programs, finite state machines, or physical processes. In Ptolemy, a special actor, the director, coordinates the execution of all other actors in a model. This coordination includes actor communication as well as concurrency mechanisms. The definition of an actor together with the coordination of actor executions describe a model of computation (MoC). Some MoCs that are contained in Ptolemy are discrete event, continuous time, dataflow, synchronous reactive and process network. Different MoCs can be composed hierarchically, however, some restrictions apply. For instance, a timed MoC cannot be embedded in an un-timed MoC. This talk will describe some MoCs and the concurrency mechanisms that govern interactions between actors.

Bio:

Patricia Derler is a postdoctoral researcher at the UC Berkeley. She received her Ph.D. in Computer Science from the University of Salzburg, Austria and she did her undergraduate studies at the University of Hagenberg, Austria. Her research interests are in design and simulation of cyber-physical systems, deterministic models of computation and the use of predictability in software, hardware and the environment towards efficient simulations and executions.

Design challenges in SoCs for mobile devices
Toshihiro Hattori, Renesas Mobile Corporation, Japan

Renesas Mobile Corporation (RMC), established on the first of December 2010, comes to the global chipset market with advanced and innovative products and services for mobile phones, car infotainment solutions, consumer electronics and industrial applications. The modem group in RMC comes with a strong pedigree from Nokia. The group has developed all Nokia?s in-house modems and formed an essential part of the chipset development for Nokia products since the time of NMT and the first generation of GSM. The world-class and leading wireless connectivity expertise is visible today as widely accepted modem technology and IP in billions of handsets. Renesas Mobile continues on this path by combining the modem asset with Renesas?s unique experience in the field of applications processors, microprocessors and controllers to form a base for highly integrated single- or multichip mobile platforms. This presentation introduces design challenges for Baseband LSI, Application LSI, and Onechip LSI for mobile world.

Bio:

Toshihiro Hattori received the B.S. and M.S. degrees in electrical engineering from Kyoto University, Japan, in 1983 and 1985, respectively. He received the Ph.D in informatics from Kyoto University, Japan, in 2006. He joined the Central Research Laboratory, Hitachi, Ltd., Tokyo, Japan, in 1985. He engaged in logic/layout tool development. From 1992 to 1993 he was a Visiting Researcher at the University of California Berkeley, with a particular interest in CAD. He joined the Semiconductor Development Center in the Semiconductor Integrated Circuits Division in Hitachi Ltd. in 1995. He moved to Renesas Technology Corp. in 2003. He was belonging to SuperH (Japan), Ltd. from 2001 to 2004 to conduct SH processor licensing and development. He moved to Renesas Electronics Corp. in 2010. He is currently working with Renesas Mobile Corp. as VP of SoC design. He is a member of IEEE(SSCS), ACM, IEICE, and IPSJ.

Fast and accurate system-level model of a NoC-based MPSoC supporting real-time applications
Leandro Soares Indrusiak, University of York, UK

Time-predictable architectures can be useful for a variety of reasons. Firstly, their time-predictability allow us to forecast its worst-case performance, making them suitable for real-time applications. Secondly, such architectures are more amenable to abstraction: their behaviour does not exhibit excessive variability and can therefore be captured by a simpler model with small loss of accuracy. This talk will present one of such architectures - a network-on-chip (NoC) with priority preemptive virtual channel arbitration - and will show how its predictability can be used to enable the creation of a transaction-level model (TLM) that can significantly reduce its validation time. Experimental results will be discussed, showing that the proposed TLM model can produce accurate communication latency figures more than two orders of magnitude faster than a cycle-accurate model.

Bio:

Leandro Soares Indrusiak received a BEng in Electrical Engineering (UFSM, 1995) and a MSc in Computer Science (UFRGS, 1998) in Brazil, where he has also worked as assistant professor (PUCRS, 1998-2000) and started his doctoral studies. He moved to Germany and finished a binational doctoral degree in Computer Science (jointly awarded by UFRGS and TU Darmstadt, 2003). He worked at TU Darmstadt as a research fellow until 2008, leading a research group on System-on-Chip design. He then moved to the United Kingdom as a permanent faculty member of University of York's Computer Science department. His current research interests include specification, design and analysis of multiprocessor, real-time and distributed embedded systems.

The Promises and Limitations of 3-D Integration
Prof. Axel Jantsch, KTH, Sweden

The intrinsic computational efficiency (ICE) of silicon defines the upper limit of the amount of computation within a given technology and power envelope. The effective computational efficiency (ECE) and the effective computational density (ECD) of silicon, by taking computation, memory and communication into account, offer a more realistic upper bound for computation of a given technology. Among other factors, they consider how distributed the memory is, how much area is occupied by computation, memory and interconnect, and the geometric properties of 3-D stacked technology with through silicon vias (TSV) as vertical links. We use ECE and ECD to study the limits of performance under different memory distribution constraints of various 2-D and 3-D topologies, in current and future technology nodes. Among other results, our model shows that in a 35 nm technology a 16 stack 3-D system can, as a theoretical upper limit, obtain 3.4 times the performance of a 2-D system (8.8 Tera OPS vs 2.6 TOPS) at 70% reduced frequency (2.1 GHz vs 3.7 GHz) on 1/8 the total area (50 mm2 vs 400 mm2).

Bio:

Axel Jantsch (M?97) received a Dipl.Ing. (1988) and a Dr. Tech. (1992) degree from the Technical University Vienna. Between 1993 and 1995 he received the Alfred Schrödinger scholarship from the Austrian Science Foundation as a guest researcher at the Royal Institute of Technology (KTH). From 1995 through 1997 he was with Siemens Austria in Vienna as a system validation engineer. Since 1997 he is Associate Professor at KTH, since 2000 he is Docent, since December 2002 he is full professor in Electronic System Design. A. Jantsch has published over 200 papers in international conferences and journals and one book in the areas of VLSI design and synthesis, system level specification, modeling and validation, HW/SW codesign and cosynthesis, reconfigurable computing and networks on chip. He has served on a large number of technical program committees of international conferences such as FDL, DATE, CODES+ISSS, SOC, and NOCS and others. He has been TPC chair of SSDL/FDL 2000, TPC cochair of CODES+ISSS 2004 and general chair of CODES+ISSS 2005 and TPC co-chair of NOCS 2009. From 2002 to 2007 he was Subject Area Editor for the Journal of System Architecture. At the Royal Institute of Technology A. Jantsch is heading a number of research projects involving a total number of 10 Ph.D. students, in two main areas: System Modeling and Networks on Chip.

Addressing Risk Management during Design Space Exploration
Prof. Jan Madsen, DTU Informatics, Technical University of Denmark

One of the challenges in modern embedded system design is to map the application onto a multi-core platform such that essential requirements are met. This process is called Design Space Exploration (DSE). In order to do DSE at an early stage in the design process, where not all parts have been implemented or even designed, a system-level model of the application executing on the multi-core platform is needed. This model should allow for an accurate modeling of the global performance of the system, including the interrelationships among the diverse processors, software processes and physical interfaces and inter-connections. This talk will focus on risk management in relation to DSE.

Bio:

Jan Madsen is a full Professor in computer based systems at the Department of Informatics and Mathematical Modeling at the Technical University of Denmark, where he is heading the System-on-Chip group. His research interests include high-level synthesis, hardware/software codesign, System-on-Chip design methods, and system level modeling, integration and synthesis for embedded computer systems. Jan Madsen is Program Chair for DATE?07, Vice-Program Chair and Tutorial Chair for DATE?06, and Workshop Chair for CODES+ISSS?05. He was General Chair of CODES 2001 and Program Chair of CODES 2000. He is an editorial board member of the journal ?IEE Proceedings ? Computers and Digital Techniques? and is a member of the steering committee of the IEEE NORCHIP conference. He has served as a technical program committee member on several international (ACM/IEEE) conferences, including DAC, ISSS+CODES, SoC Symposium, ISSS, CODES, DATE, FTRTFT, ED&TC, ACSD and Euromicro DSD.

SoC, MPSoC, RSoC, ... , Design Challenges, The Industrial Point of View
Prof. Yves Leduc, TI chair at University of Nice, France

After a presentation of the new business landscape, we will review the major issues that the semiconductor companies are facing in the nanoelectronic era. System size and complexity are the first barriers. We will identify and understand these barriers and elaborate on ad hoc solutions. A second issue is associated to the ultimate technology scaling. Atoms and molecules in nanoelectronics are no more contributing in a safe average cooperation but tend to act as individuals to perturb the electrical behavior of the basic transistors. Again we will identify and understand the issues and review how the exacerbated process variabilities is influencing the work of our designers. Last but not least, the thermal dissipation in large SOCs will be discussed. It represents, de facto, the actual limits to the realization of our business dream. To conclude, we will detail how 3D-IC will or will not bring new solutions to the designs of our next future.

Bio:
Yves Leduc
Yves got his PhD in electrical engineering from the University of Louvain in 1979. With Texas instruments France for the last 30 years, he created in 1994 the mixed signal development team of TI France and was elected TI Fellow in 1998. Yves then led the advanced system technology team paving the way to the future development of the new SOCs. Yves is currently holding the TI Chair at the University of Nice and participate to several organizations to promote the creation of start-ups in microelectronics.

SOC 2011: Advanced Program

Tuesday November 1

Opening

9:00 WELCOME TO SOC 2011

- Studio

Chair: Jari Nurmi

Tuesday November 1

Keynote

9:15 3D Integration of NoC-based Systems, Axel Jantsch, KTH, Sweden

- Studio

Tuesday November 1

Coffee

10:00

- Rondo

Tuesday November 1

System-level design flow and methodology

10:30

- Studio

Analyzing Transport and MAC Layer in System-Level Performance Simulation

Subayal Khan[1], Jukka Saastamoinen[1], Mikko Majanen[1], Jari Nurmi[2]
[1]VTT Technical Research Centre of Finland, [2]Tampere University of Technology. Department of Computer Systems

Moonrake Chip - GALS Demonstrator in 40 nm CMOS Technology

Miloš Krstić[1], Xin Fan[1], Eckhard Grass[1], Luca Benini[2], M. R. Kakoee[2], Christoph Heer[3], Birgit Sanders[3], Alessandro Strano[4], Davide Bertozzi[4]
[1]IHP, [2]University of Bologna, [3]Intel Mobile Communications, [4]University of Ferrara

Analyzing Synchronous Dataflow Scenarios for Dynamic Software-defined Radio Applications

Firew Siyoum[1], Marc Geilen[1], Orlando Moreira[2], Rick Nas[2], Henk Corporaal[1]
[1]Eindhoven University of Technology, [2]ST-Ericsson Eindhoven

Tuesday November 1

Processor architectures

11:30

- Studio

A Hybrid Model of Speculative Execution and Scout Threading for Auto-Memoization Processor

Tomoki IKEGAYA[1], Ryosuke ODA[1], Tatsuhiro YAMADA[1], Tomoaki TSUMURA[1], Hiroshi MATSUO[1], Yasuhiko NAKASHIMA[2]
[1]Nagoya Inst. of Tech., [2]Nara Inst. of Sci. and Tech.

Customizable Datapath Integrated Lock Unit

Pekka Jääskeläinen, Erno Salminen, Otto Esko, Jarmo Takala
Tampere Univ. of Tech.

Exploring Instruction caching strategies for tightly-coupled shared-memory clusters
Daniele Bortolotti, Francesco Paterna, Christian Pinto, Andrea Marongiu, Martino Ruggiero, Luca Benini
University of Bologna - DEIS

Tuesday November 1

Lunch

12:30

- Restaurant Fuuga

Tuesday November 1

Invited 2

13:30 Fast and accurate system-level model of a NoC-based MPSoC supporting real-time applications, Leandro Soares Indrusiak, University of York, UK

- Studio

Tuesday November 1

SoC design and analysis methods

14:15

- Studio

Static Analysis Method for Deadlock Detection in SystemC Designs
Mikhail Moiseev[1], Alexey Zakharov[1], Ilya Klotchkov[2], Sergey Salishev[2]
[1]SPbSPU, [2]Intel Labs

SAMOSA: Scratchpad Aware Mapping Of Streaming Applications
Zubair Wadood Bhatti[1], Davy Preuveneers[1], Narasinga Rao Miniskar[2], Roel Wuyts[2], Yolande Berbers[1]
[1]K.U.Leuven, [2]IMEC

A System Level Power Consumption Estimation for MPSoC
Santhosh Kumar Rethinagir[1], Rabie ben Atitallah[2], Jean-Luc Dekeyser[3]
[1]Affiliation, [2]Univ.Valenciennes, [3]Univ. Lille

Tuesday November 1

Posters and coffee

15:15

- Rondo

OpenCL Implementation of Cholesky Matrix Decomposition
Claudio Brunelli, Eero Aho, Heikki Berg
Nokia

Low-power Arithmetic Unit for DSP Applications
Mehdi Modaressi[1], Hossein Nikonia[2], Amir-Hossein Jahangir[1]
[1]Sharif University of Technology, [2]Tehran University

Co-designs of Parallel Rijndael
Issam Damaj
American University of Kuwait

A Set of Traffic Models for Network-on-Chip Benchmarking
Esko Pekkarinen, Lasse Lehtonen, Erno Salminen, Timo D. Hämäläinen
TUT

Effects of Loop Unrolling and Use of Instruction Buffer on Processor Energy Consumption
Vladimír Guzma, Teemu Pitkänen, Jarmo Takala
Tampere University of Technology

Applying IP-XACT in Product Data Management
Erno Salminen, Timo D. Hämäläinen, Marko Hännikäinen
TUT

Increasing Energy Efficiency of Automotive E/E-Architectures with Intelligent Communication Controllers for FlexRay
Christoph Schmutzler[1], Abdallah Lakhtel[1], Martin Simons[1], Jürgen Becker[2]
[1]Daimler AG, [2]Karlsruhe Institute of Technology, ITIV

AN AUTOMATIC EXPERIMENTAL SET-UP FOR ROBUSTNESS ANALYSIS OF DESIGNS IMPLEMENTED ON SRAM FPGAS
Uli Kretzschmar, Armando Astarloa, Jesús Lázaro, Jaime Jimenez, Aitzol Zuloaga
Universidad del Pais Vasco UPV/EHU

A Coarse-Grained Reconfigurable Protocol Processor
Mohammad Badawi and Ahmed Hemani
Royal Institute of Technology, Stockholm, Sweden

moviTest: A Test Environment Dedicated to Multi-Core Embedded Architectures
Teodor Tite[1], Adelina Vig[1], Nicolae Olteanu[2], Cristian Cuna[2]
[1]University "Politehnica" of Timisoara, Department of Computer and Software Engineering, Timisoara, Romania, [2]Movidius SRL, Tools Department, Timisoara, Romania

Tuesday November 1

Invited 3

16:00 Jan Madsen DTU, Denmark

- Studio

Tuesday November 1

Panel Discussion

16:45

- Studio

Tuesday November 1

Reception

18:00

- Rondo

Wednesday November 2

DASIP 2011 Opening

9:00 WELCOME TO DASIP 2011

- Studio

Chair: Tapani Ahonen and Jari Nurmi

Wednesday November 2

Invited 4

9:20 Yves Leduc, University of Nice, France

- Studio

Wednesday November 2

Posters and coffee

10:05

- Rondo

Wednesday November 2

Network-on-chip and inter-chip communications

11:00

- Studio

Mismatch Characterization of High-Speed NoC Links using Asynchronous Sub-sampling
Sebastian Höppner, Dennis Walter, Georg Ellguth, René Schüffny
TU Dresden

Impact of Proactive Temperature Management on Performance of Networks-on-Chip
Tim Wegner, Martin Gag, Dirk Timmermann
University of Rostock

Bringing Network-on-Chip Links to 45nm
Marco Ferraresi, Giuseppina Gobbo, Daniele Ludovici, Davide Bertozzi
University of Ferrara

Synchronizing Distributed State Machines In A Coarse Grain Reconfigurable Architecture
Omer Malik and Ahmed Hemani
Royal Institute of Technology - KTH

Wednesday November 2

Lunch

12:20

- Restaurant Fuuga

Wednesday November 2

Invited 5

14:00 Design challenges in SoCs for mobile devices, Toshihiro Hattori, Renesas Mobile Corporation, Japan

- Studio

Wednesday November 2

Software tools

14:45

- Studio

Automatic Calibration of Streaming Applications for Software Mapping Exploration
Weihua Sheng, Stefan Schürmans, Maximilian Odendahl, Rainer Leupers, Gerd Ascheid
RWTH Aachen University

Building a RTOS for MPSoC Dataflow Programming

Yaset Oliva[1], Maxime Pelcat[1], Jean-Francois Nezan[1], Jean-Christophe Prevotet[1], Aridhi Slaheddine[2]
[1]IETR/INSA Rennes, [2]Texas Instruments

Wednesday November 2

DASIP posters and coffee

15:25

- Rondo

Wednesday November 2

Invited 6

16:15 Heterogeneous Concurrent Modeling and Design in Java, Patricia Derler, UC Berkeley, USA

- Studio

Wednesday November 2

Closing

17:00

- Studio

Wednesday November 2

Banquet

19:00

Analyzing Transport and MAC Layer in System-Level Performance Simulation

Subayal Khan, Jukka Saastamoinen,
Mikko Majanen, Jyrki Huusko
VTT Technical research Center of Finland,
FI-90570, Oulu, Finland
email:{subayal.khan,jukka.saastamoinen,
mikko.majanen,jyrki.huusko}@vtt.fi

Jari Nurmi
Tampere University of Technology,
Department of Computer Systems
P.O.Box 553 (Korkeakoulunkatu 1),
FIN-33101 Tampere, FINLAND
jari.nurmi@tut.fi

Abstract:

The modern mobile embedded devices support complex distributed applications via heterogeneous multi-core platforms. For the successful deployment of these applications, the scalability and performance analysis must be performed at all the layers of OSI model. This helps to identify the potential bottlenecks at different layers to perform the necessary optimizations. To achieve this goal, a framework is needed which accurately models the functionalities at different layers. The technical contributions described in this article include the extensions of ABstract inStruction wOrkLoad & execUtion plaTform based performance simulation (ABSOLUT) for the performance and scalability analysis of Transport and Medium Access Control (MAC) layers in the system level performance simulation. The article elaborates the design accuracy of the modeled components and their application in the context of M3 (multi-device, multi-vendor, multi-domain), which is a tri-layered conceptual interoperability architecture for embedded devices. These extensions pave the way towards the full coverage of the OSI model in the system-level performance simulation of distributed embedded systems. The network simulators for example ns-2, OMNeT++ and OPNET though provide detailed models of transport and MAC protocols but do not provide any framework such that these models can be used by the application workload models to mimic the real world use-cases. Also these models do not model the execution workload of these protocols on a particular execution platform and hence cannot be used in the architectural exploration of distributed embedded systems.

I. INTRODUCTION

Modern nomadic devices support computationally intense distributed applications by using heterogeneous multiprocessor architecture platforms. Deployment of a new distributed application is challenging not only due to the heterogeneous parallelism in the platforms [1], but also due to performance and energy constraints. Furthermore, these devices employ diverse communication, transport technologies and application-level protocols to enable information sharing and synchronization among the processes of distributed applications. These technologies enable complex use-cases spanning multiple devices. To evaluate the feasibility of these use-cases, the system-level performance simulation methodology must identify the potential bottlenecks at each layer of the OSI model so that the appropriate optimizations can be performed. The performance analysis of protocols spanning different layers of OSI model require a framework which models these protocols with reasonable accuracy while maintaining a good simulation speed.

In case of non-distributed applications, the processes use inter-process communication (IPC) for synchronization of tasks and the transport, Datalink and Physical layers of the OSI-Model do not play any role. The system-level performance simulation of non-distributed application requires application workload models, platform capacity models and workload models for external libraries. ABSOLUT performance simulation methodology has been already used to evaluate such use-cases [2] and [3].

In the case of distributed applications, the use-cases span multiple devices and the processes communicate with each other via a transport API. The transport layer APIs are implemented on top of Layer 2 of the OSI model which in turn makes use of physical layer. Therefore apart from the performance evaluation of the platform components, the performance evaluation of the applied transport technology, MAC protocol and transmission techniques must be performed. Afterwards, the performance of same application level workload models can be evaluated with other available alternatives of transport, MAC and transmission technologies. The design space is thus much bigger and spans all layers of the OSI model.

The main contribution of this article is to describe the extensions made to ABSOLUT for the performance evaluation of distributed applications. A detailed survey of performance simulation techniques has been presented in [4], therefore the landmark performance simulation methodologies are not discussed in this article.

Rest of paper is organized as follows: Section 2 briefly explains ABSOLUT methodology. Section 4 describes the extensions made to ABSOLUT for enabling the analysis of Layer-2 in System-level performance evaluation. We first elaborate a general method for developing operating system (OS) services for Transport, MAC and physical layers of OSI models. These extensions pave the way towards the full coverage of the OSI model in the system-level performance simulation. These services are implemented by using freely available tools and libraries such as SystemC [5] and itpp library [6]. Afterwards we focus on the modeling and integration of IEEE 802.11 MAC and transport layer of OSI model to ABSOLUT. In section 5, the modeled components are used for the MAC and transport layer scalability and performance analysis of M3 [7]. The obtained simulation results are compared with the ns-2 [8] and OMNeT++ [9] network simulators under different simulation scenarios. Conclusions and Future work are mentioned in Section 6.

978-1-4577-0671-4/11 $26.00 © 2011 IEEE

II. ABSOLUT

ABSOLUT follows the Y-chart model consisting of application workload and platform model [10]. The workload models are mapped to the platform for transaction-level performance simulation in SystemC [2].

3.1 Application Workload Model

The workloads consist of four layers .i.e., main workload, application workload, process workload and function workload as shown in Figure 1.

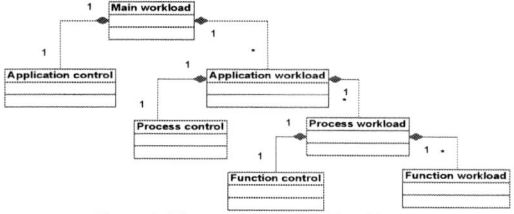

Figure 1: The application workload layers

3.2 Execution Platform Model

The platform model is an abstract hierarchical representation of actual platform architecture. It is composed of three layers: component layer, subsystem layer, and platform architecture layer as shown in Figure 2. Each layer has its own services, which are abstract views of the architecture models. Services in subsystem and platform architecture layers are invoked by application workload models.

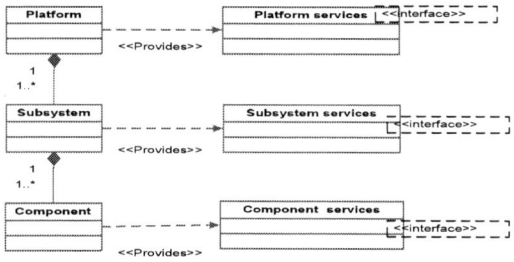

Figure 2: The platform architecture model layers.

III. EXTENDING ABSOLUT FOR ANALYSIS OF LAYER-2

The performance modelling of distributed applications via ABSOLUT demands the modelling and integration of Transport and MAC protocols, modulation techniques, coding schemes and channel models. We now elaborate the modelling and integration of these components to ABSOLUT.

3.1 Design and integration of OS Services

Extension of ABSOLUT for analysing the performance of protocols operating at different layers of OSI model during system-level performance simulation for architectural exploration demands a mechanism for instantiating new H.W and S.W services. These services are registered to the operating system and are used by the application workload models. Furthermore the services operating at a higher layer for example transport-level services (such as TCP) use Data-link level services such as IEEE 802.11 MAC protocols for the transmissions of frames of a packet. These services are created by deriving them from the *OS_Service* base class as shown in Figure 3 which implements the *Generic_Serv_IF*.

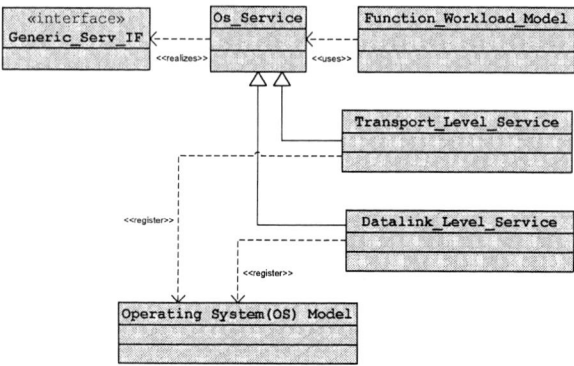

Figure 3: OS services implemented to model use cases spanning multiple devices and for modeling BSD API as OS services.

Implementation of *Generic_Serv_IF* by the *OS_Service* base enables the process-level application workload models or higher level services request a certain service by its name and invoke the functionality implemented by the derived service. This is shown in Figure 4.

```
SID=Use_Service("Serv_Name");
Wait_Service(SID);
```

Figure 4: Accessing an OS_Service via Generic_Serv_IF

3.2 Accessibility and Hierarchy of *OS_Services*

The *OS_Service* base class implements the functionality related to scheduling of service requests of processes via priority queues and informs the requesting process on service completion after taking it to running state again. This is shown in Figure 5.

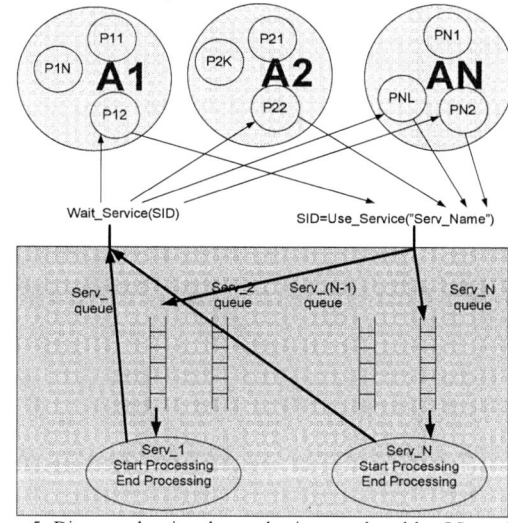

Figure 5: Diagram showing the mechanism employed by OS services to execute requests of processes.

The derived services implement the service-specific functionality making service modelling straight forward. The services at upper layers make use of services at lower layer .e.g., Transport layer use Data-Link layer services.

The services are accessed from the platform using the service name assigned while registration to the OS model. For example BSD socket API function "send()" can be modelled as an *OS_Service* and registered by a unique service name for example "PktTx" to the OS. It can then accessed by the process workload models by using its unique service name via *Generic_Serv_IF*. This is shown

978-1-4577-0671-4/11 $26.00 © 2011 IEEE

in Figure 4 and Figure 6.

```
//The processor model
ARM_Crtx_Proc_ptr=new Scalable_MultiCore_CPU
"m_ARMCortex_A9_MP_Processor"};

//Creating the operating system (OS) model
m_os = new Generic_serv_op_sys
("os",ARM_Crtx_Proc_ptr->GetProcessorCores(),
m_os_addr);

//Send() API function registered as "PacketTx" to OS
Pkt_Tx_Service= new Packet_Tx_Serv("Transmit_Packet",m_os);
Serv_type Msg_serv_type = { SERV_TYPE_LOCAL };
m_os->register_service(Pkt_Tx_Service,"PacketTx",
Msg_serv_type);

//This service handles transmission of single frame via
//IEEE 802.11 DCF. Registered by name "FrameTx" to OS
Frame_Tx_Service= new Frame_Tx_Serv("Transmit_Frame",m_os);
Serv_type Frame_serv_type = { SERV_TYPE_LOCAL };
m_os->register_service(Frame_Tx_Service,"FrameTx",
Frame_serv_type);
```

Figure 6: Registration of services to the operating system (OS) model

3.3 Implementation and integration of MAC and Transport Level OS_Services

IEEE 802.11 distributed coordination function (DCF) requires a station wanting to transmit, to first listen to the channel to check its status (occupied or not) for a DCF Interframe Space (DIFS) interval. If the channel is found busy during the DIFS interval, the station defers its transmission. In a network where a number of stations contend for the wireless medium, if multiple stations sense the channel busy and defer their access, they will also virtually simultaneously find that the channel is released and then try to seize the channel. As a result, collisions may occur. In order to avoid such collisions, DCF also specifies random back off, which forces a station to defer its access to the channel for an extra period. DCF also has an optional virtual carrier sense mechanism that exchanges short Request-to-send (RTS) and Clear-to-send (CTS) frames between source and destination stations during the intervals between the data frame transmissions. The IEEE 802.11 DCF can be shown in the form of a flow chart as in Figure 7.

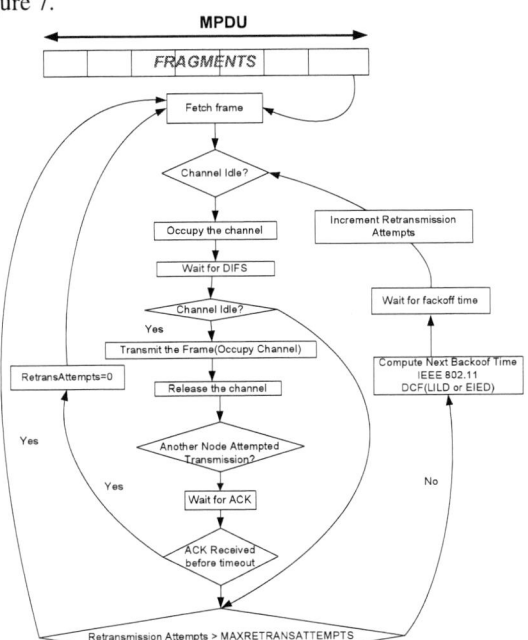

Figure 7: Flow chart of IEEE 802.11 DCF

The IEEE 802.11 DCF is implemented by the following M3_Frame_Tx class derived from the *OS_Service* base class. Only the constructor of the class is shown. The DIFS, SLOT_TIME [11] and the initial value of the contention window are assigned in the constructor as shown in Figure8.

```
M3_Frame_Tx::M3_Frame_Tx
(sc_module_name _name,
Proc_ctl_IF * host)
:OS_Service(_name){

//Initializing values of

//Contention window
CwCurrent=CWMIN;

//DIFS
DIFS=sc_time(128,SC_US);

//Slot Time
SLOT_TIME=sc_time(50,SC_US);

//Backoff time
BackOffTime=SLOT_TIME;

//Retransmission attempts
RetransmissionAttempts=0;
}
```

Figure8: The Frame Transmit Service initializing MAC parameters

In ABSOLUT the MAC protocols as well as transport-level protocols such as TCP are also modeled as servies by deriving them from the same base class *OS_Service* which provides the scheduling and synchronization mechanism. The transport-level services then request the Layer-2 services such as IEEE 802.11 for frame transmission. The do_service() method of the *OS_Service* base class is implemented by the derived class to provide the functionality of IEEE 802.11 DCF as shown in Figure 10. The do_service() method spams a separate frame transmission function for handling each request of frame transmission from the transport as shown in Figure 9.

```
void M3_Frame_Tx::do_service(){
while(true)
  {

  //Get Service attributes from OS
  m_crnt_attr=
  const_cast< Serv_attributes *>
  Current_Service_Attributes);

  //Initial retransmission Attempts
  RetransmissionAttempts=0;

  //Spawn Frame Transmission Function
  SC_FORK
  sc_spawn
  sc_bind(&M3_Frame_Tx::Tx_Crnt_Frame,
  this,static_cast<Smart_M3_Attr *>
  m_crnt_attr))
  SC_JOIN

  //Inform OS about service completion
  m_service_complete_ev.notify();

  //Wait for next Frame Tx request
  wait();
  }
}
```

Figure 9: Implementation of Frame Transmit service. It invokes a spammed function for the transmission of a single frame.

The spammed function implements the flow chart shown in Figure 7. It attempts the retransmission transmission for up to a maximum number of retransmissions, the frame is considered lost and the transport-level service (transport protocol) is informed. For connection oriented protocols, the remaining frames are not transmitted. For connection-

978-1-4577-0671-4/11 $26.00 © 2011 IEEE

less protocols this information is neglected.

The transport-Level services .i.e., TCP and UDP are also modeled as *OS_Services*. Both the services make use of MAC level services for frame transmission. TCP calls the Frame Transmission service of MAC (which receives a single frame at a time for transmission) as many times as the number of frames in the transport packet. If one frame is lost (due to errors or collisions) the MAC informs transport and the rest of the frames are dropped and a packet loss is recorded. In the UDP transport-level service, the MAC does not inform the transport layer about the frame loss and all the frames are transmitted even if one or more frames of a packet are lost (due to collisions or errors). The accuracy of the modelled components is elaborated in the next section.

IV. ACCURACY OF SIMULATION RESULTS

To study the MAC protocols in isolation under a particular scenario, we abstract out the Application workloads with delays obtained after profiling or use traffic generators. Three types of traffic generators .i.e., pareto on off, exponential and constant-bit rate available in ns-2 have been integrated to ABSOLUT. The different modulation techniques like QPSK and BPSK have been modeled along-with MC-CDMA. Two channel coding techniques .i.e., convolutional and Reed Solomon codes and two channel models .i.e., binary symmetric channel and additive white Gaussian noise (AWGN) channels have been integrated by using models available in itpp library. The performance model is configured with a certain type of modulation scheme, coding scheme and channel model. Bit errors are computed using the functions available in itpp library. Frame lengths can be chosen randomly or fixed to a value before simulation to analyze MAC and transport protocols in a particular scenario.

4.1 Analysing accuracy of bit error rate calculation

Different modulation schemes available in itpp library have been used without modification. We present the results for Multi-Code CDMA with QPSK modulation. For $1e^6$ bits the results are over 99.8% accurate (when compared to theoretical results) as shown in Figure 10.

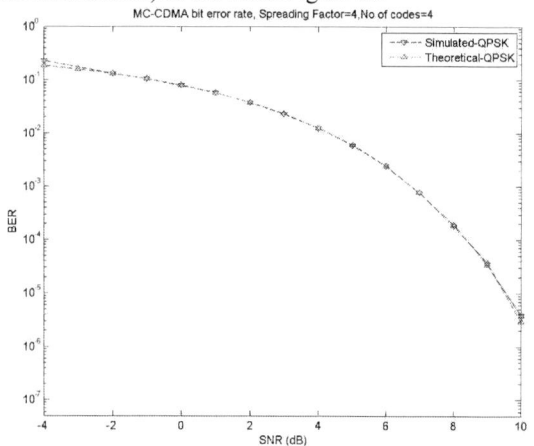

Figure 10: Theoretical versus simulation bit error rate for MC-CDMA with QPSK. Number of codes (M) = 4 .Spreading Factor (k=4). Number of bits =100,000.

4.2 Analysing accuracy of frame error rate calculation

In the absence of any encoding in IEEE 802.11, the fragment and the bit error rate are related by Equation 1.

$$P_e = 1 - (1 - BER)^s \quad (1)$$

Where s is the fragment size and *BER* is the Bit Error Rate and *Pe* is the probability of frame error. The bit error rates are plotted against frame error rates for different values of frame lengths as is shown in Figure 11.

The frame and bit-error rates can be recorded directly from simulation and plotted for different values of bit error rates as shown in Figure 11. The recorded simulation results are over 92% accurate when averaged after 20 simulation runs. The simulation results are compared to analytical results for Packet Lengths of 228 and 2228 as shown in Figure 11.

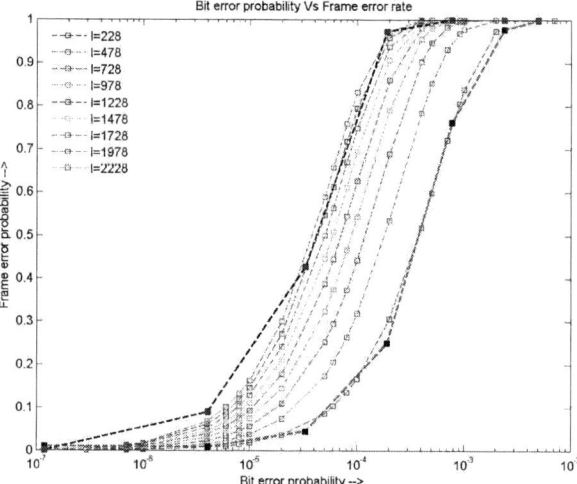

Figure 11: Frame error probability versus bit error rate. Theoretical results compared to simulation results for frame lengths 228 and 2228.

4.3 Analysing accuracy of packet error rate

In case of IEEE 802.11, one MAC service data unit (MSDU) can be partitioned into a sequence of smaller MAC protocol data unit (MPDUs) in order to increase reliability. Fragmentation is performed at each immediate transmitter. The process of recombining MPDUs into a single MSDU is called defragmentation. Defragmentation is also done at each immediate recipient. When a directed MSDU is received from the LLC with a length greater than a Fragmentation-Threshold, the MSDU is divided into MPDUs. Each fragment's length is smaller or equal to a Fragmentation-Threshold [11]. The MPDUs are sent as independent transmissions, each of which is separately acknowledged. The loss probability of transmitting a transport packet fragmented at the MAC layer into N fragments is given by the Equation 2 [12].

$$P_{wl} = 1 - \left(\sum_{i=1}^{i=M} P_l^{i-1} (1 - P_l) \right)^N = 1 - (1 - P_l^{M-1})^N \quad (2)$$

Where P_l denotes the successful transmission probability of one attempt, i denotes the retransmission attempts and M is the maximum number of retransmission attempts. Figure 12 shows the transport packet loss rate as a function of the MAC frame loss probability during each transmission retry for a fixed number of fragments *(N=10)* and for different values of maximum retransmission attempts[12]

(M=1→10). The simulation results are compared to the analytical results as shown in Figure 12. The values of *M* and *N* were fixed, the value of signal to noise ratio (SNR) was varied and the simulation was repeated several times. The results for each value of SNR were averaged to obtain each point on the two curves. The simulation was run 20 times and the averaged results achieve an accuracy of over 85% when compared with analytical results as shown in Figure 12.

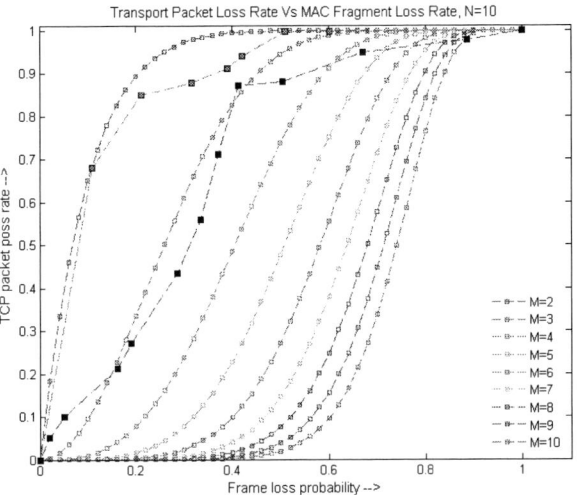

Figure 12: Theoretical versus simulation results. MAC Frame loss probability versus transport packet loss rate, for Maximum Retransmission attempts (M=2 and 3) and number of fragments (N=10).

4.4 Comparing MAC and Transport layer models with ns-2 and OMNeT++ network simulators

We now compare the throughput, Frame delays and Packet delays of ABSOLUT MAC and Transport models with ns-2 and OMNeT++ network simulators. Different Packet lengths, transport protocols and Packet transmission rates are used in both the case studies. The simulations are carried out under saturated conditions. The simulation parameters are mentioned in Table 1.

Table 1: Experiment parameters

Parameters	Values
SIFS	10 micro seconds
DIFS	50 micro seconds
Slot Interval	20 micro seconds
Preamble Length	144 bits
PLCP header Length	48 bits
Channel bit rate	2 Mbps
CWmin	32
CWmax	2048
CWo	32
EW	16

The simulations are carried out in WLAN environment in the context of M3. The abbreviation M3 means multi-device, multi-vendor, multi-domain to highlight the flexibility and portability of the technology [7]. It means that all the network client nodes called Knowledge Processors (KPs) at information level in M3 are within the transmission range of a single server called Semantic Information Broker (SIB) which acts as the only destination for the KPs.

1) Case Study 1: Comparison of simulation results at MAC and Transport Layer with ns-2

In the first case study we compare the results of our MAC and transport models with ns-2. The data traffic is

generated using the Constant bit rate (CBR) traffic generator available in ns-2 simulator and the transport protocol is TCP. [12]. The traffic generators can be configured by using the simple interface as shown in Figure 13.

```
//CBR traffic generator parameters

// 2.5 milliseconds
double AverageBurstTime=.0025;
// 1 second
double InterFrameTime  = 1   ;
// 2048 bytes
double SinglePktSize=2048;
// 2 Mega bits per second
double Bitrate=2000000;

//Instantiate CBRtraffic generator
SimConfig::Instance()->SetTrafficGenerator(
CBR_TRAFFIC_GENERATOR,
AverageBurstTime,
InterFrameTime,
Bitrate,SinglePktSize
);
```

Figure 13: An example configuration of the CBR traffic generator. Packet Length=2048 Bytes. Data rate= 2 Mbps. Average Burst Time=.0025 seconds. Inter Frame Space=1 second.

With the same experimental parameters mentioned in Table 1, we perform our simulations in two different frame lengths .i.e., 512 bytes and 1024 bytes. In both the cases, the transport-level packets are not fragmented into multiple MAC frames; therefore we only use the MAC Frame delays and throughput for comparison. The average Delays for both the frame lengths are shown in Figure 14 for different number of active nodes (20→100 KPs and one SIB in smart spaces) in the network (Smart Space).

The ns-2 and ABSOLUT simulations were run 50 times and the average values were computed. The results indicate that if ns-2 is used as a reference bench mark, the results of ABOLUT Mac and transport are 70-80% accurate. The inaccuracy is due to absence of the RTS/CTS mechanism in ABSOLUT models. The results show that ABSOLUT models always produce pessimistic results, .i.e., less throughput and more delays for the same simulation scenario.

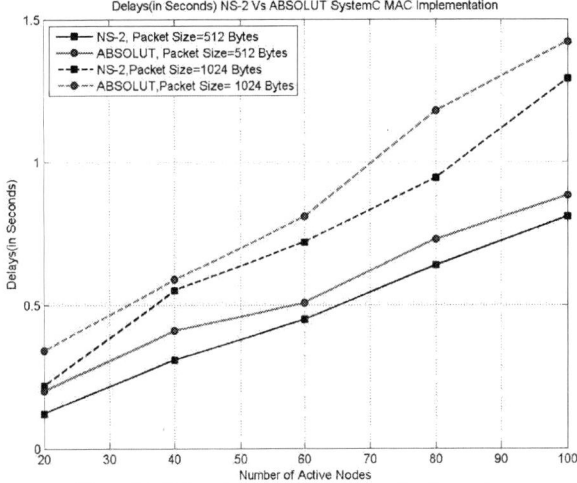

Figure 14: Delays (seconds) Vs number of active nodes (Ns-2 versus ABSOLUT)

The normalized throughput for both the frame lengths is

978-1-4577-0671-4/11 $26.00 © 2011 IEEE

shown in and Figure 15.

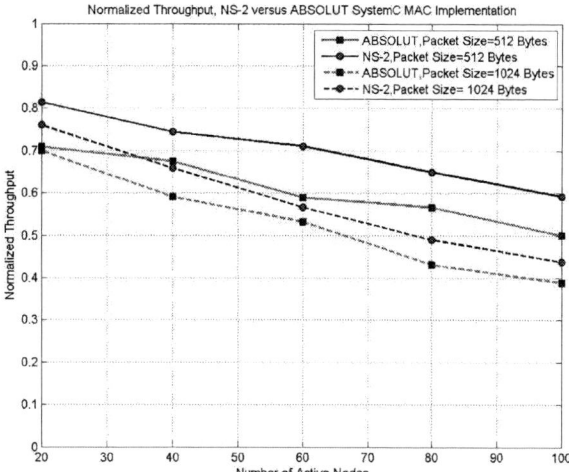

Figure 15: Normalized Throughput versus number of active nodes (Ns-2 Vs ABSOLUT)

The average collisions times (Number of collisions/100 seconds) are shown in Figure 16.

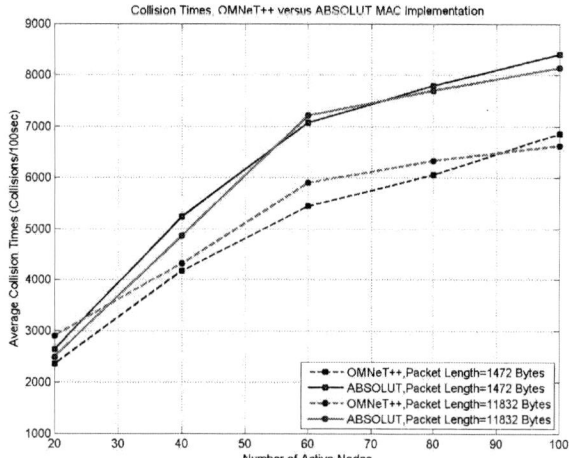

Figure 16: Average collision times Vs number of active nodes (Ns-2 Vs ABSOLUT).

2) Case Study 2: Comparison of simulation results at MAC and Transport Layer with OMNeT++

In the second case study, we compare the results of ABSOLUT MAC and transport models with OMNeT++. No traffic generators were used. The application sends packets at 2 milli-second interval. The simulations are performed under two scenarios. In the first scenario, the application sends 11832 Bytes long packets. The packet is fragmented into 8 fragments. As a consequence MAC sends 8 fragments, each of length 1500 Bytes. In the second scenario, the application sends a 1472 Byte packet at 2 milli-second interval. There is no fragmentation on any layer and as a result, MAC sends a single frame of length 1534 Bytes for each Application packet. Since the packet transmission rate is too fast, the collision rate is quite high which significantly increases the delays and reduces the throughput.

The maximum and average Delays for the packet length of 11832 Bytes(8 Frames/Packet) is shown in Figure

17 as the number of nodes (KPs) is varied (20→100) in the network (Smart Space).The goal is to investigate the case where multiple frames are transmitted for a single transport packet.

The OMNeT++ and ABSOLUT simulations were run 20 times and the average values were computed. The results indicate that if OMNeT++ is used as a reference bench mark, the results of ABOLUT MAC and transport are 75-90% accurate. The inaccuracy is due to the absence of the RTS/CTS mechanism in ABSOLUT models. The results show that ABSOLUT models always produce pessimistic results, .i.e., less throughput and more delays for the same simulation scenario.

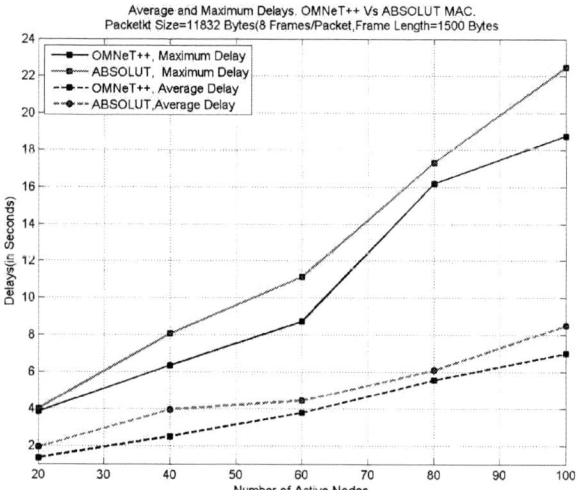

Figure 17: Maximum and Average Delays (seconds) Vs number of active nodes (OMNeT++ Vs ABSOLUT).

The normalized throughput for both the packet lengths is shown in Figure 18.

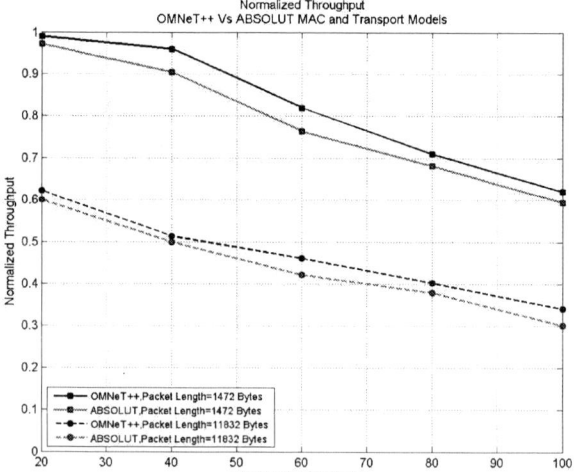

Figure 18: Normalized Throughput Vs number of active nodes (OMNeT++ Vs ABSOLUT).

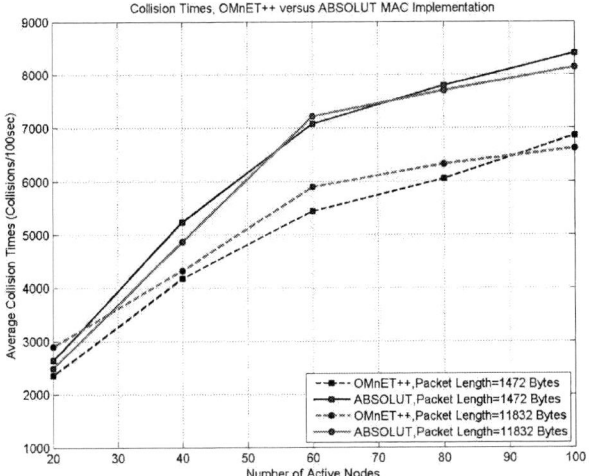

Figure 19: Average collision times Vs number of active nodes (OMneT++ Vs ABSOLUT).

4.5 Platform utilization of Transport and MAC

Each network node (KPs or SIB) were mapped to a separate ABSOLUT platform model. Each platform model used in the both case studies is a modified OMAP-44x platform model. The MAC and Transport services were registered to the OS model of the platform. The platform model consists of two ARM Cortex-A9 processors consisting of four and three processing cores respectively instead of two (as in case of original TI OMAP44-x platform [13]), SDRAM, a POWERVR SGX40 graphics accelerator and an Image signal processor. This is shown in Figure 20. The Network-on-Chip (NoC) infrastructure was abstracted out and replaced with on-chip bus as shown in Figure 20.

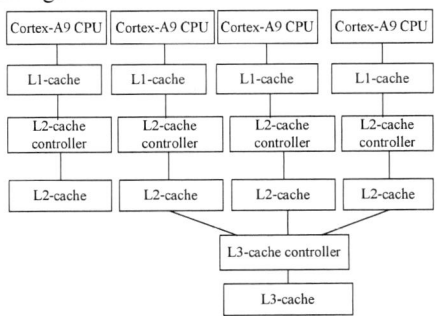

Figure 20: OMAP 44x Platform ABSOLUT model.

Each processor core (Cortex-A9 CPU model) has an L1 and L2 cache and can possibly share an L3 cache with one or more cores in the Multi-Core Processor model. This is shown in Figure 21.

Figure 21: Diagram showing the quad-core processor model used in the performance platform model.

For performance utilization of Transport and MAC layer we only model the workload of the TCP since it is a connection-oriented, complex and more computationally intensive transport protocol then UDP. UDP is a light

weight non-connection oriented transport protocol used primarily for real-time streaming applications; therefore processing costs of UDP are lower than TCP.

The workload for TCP *OS_Service* was extracted from [14]. The processing times of TCP functions were scaled down since ARM-Cortex-A9 processors operate at a much faster clock then DECstation 500/200. The approximate number of abstract instructions [2] for the TCP OS_Service processing workload were extracted and modelled as a function workload model [2]. This function workload model mimics the execution workload of the TCP transport protocol. It executes inside a Process workload model triggered by the TCP *OS_Service* during the processing of the service request. The Application models were abstracted out by using constant bit rate traffic generator and constant delays to measure the performance of TCP in isolation.

The average busy time of any processor core of any network node (KP or SIB in the case of M3) involved in the experiment was less than .00001% even for the case of 100 KPs. This is the processing time for TCP. The processing time of IEEE 802.11 is merely a subset of this processing time. The performance costs even after using NoTA DIP over TCP were found to be less than .0001% [15], thus confirming the results.

V. CONCLUSIONS AND FUTURE WORK

The analysis performed in [16] concludes that the network simulators .i.e., ns-2 and OPNET though give different absolute values under the same simulation scenarios but the trend of the results obtained from both the simulators are the same. The ABSOLUT models also validate these results. As the number of KPs is increased, OMNeT++ and ns-2 as well as ABSOLUT show a similar trend in the change in values of delays, throughput and collisions. The following conclusions can be drawn on the basis of the case studies in the previous section.

1. ABSOLUT MAC and Transport models can be used for the system-level performance simulation of distributed embedded systems. In case of distributed applications, the end-end delays and throughput play important role and must be met for providing a pleasant and acceptable end-user experience. The results are reliable since they are pessimistic when compared to state of the art network simulators for example ns-2 and OMNeT++. In other words the values of delays are always higher than OMNeT++ and ns-2. This guarantees that if the ABSOLUT MAC and transport models validate the values of delays and throughput, they are definitely validated by OMNeT++ and ns-2. Thus the modelled MAC and Transport layer services can be confidently used for system-level performance simulation of distributed applications and architectural exploration.

2. The application models can be replaced by traffic generators or constant delays to study the behaviour of Transport and MAC protocols in isolation. This helps to identify the potential bottlenecks at different layers in the OSI model giving full coverage of the OSI protocol stack in

architectural exploration of distributed embedded systems.

3.   The performance costs of the MAC and transport layers are negligible and can be abstracted out for system-level performance simulation.

4.   In the case of M3 architecture, according to ABSOLUT MAC and transport models, the following conclusions can be drawn in the traffic conditions considered in the two case studies. In first case study, the average delay per packet increases from .24 and .33 seconds to .9 and 1.39 seconds and the normalized throughput decrease from .71 and .7 to .5 and .41 for packet lengths 512 and 1024 bytes. In second case study, the average delay per packet increases from 2 and 4 seconds to 8.1 and 23 seconds and the normalized throughput decrease from .98 and .6 to .6 and .4 for packet lengths 111832 and 1472 bytes respectively. The platform utilization is negligible in both cases. Therefore IEEE 802.11 MAC and TCP do not show any significant performance bottlenecks as far as platform utilization is concerned but the delays increase significantly as more and more KPs join the smart space under the considered traffic conditions. The throughput also decrease significantly as the number of KPs increases in a smart space under these traffic conditions. Hence we conclude from our analysis that for small scale smart spaces such as smart cars and class rooms M3 will operate pretty well with IEEE 802.11 standard operating at Layer-2 and TCP. On the other hand in large scale smart spaces such as smart hockey stadiums operating 1000s of KPs to share information, IEEE 802.11 has to be either optimized or replaced by another potential solution at MAC-layer.

In the future, it is planned to further extend the ABSOLUT methodology to the incorporate multithreading application workload modelling support and C++ workload extraction methodology. These extensions will enable the seamless integration of design and performance simulation of distributed applications. The extended ABSOLUT framework will then employed for the system-level performance simulation of distributed GENESYS, NoTA and M3 applications. In case of non-distributed GENESYS applications that milestone has already been achieved [3].

VI. ACKNOWLEDGEMENTS

This work was performed in the Artemis SOFIA project partially funded by Tekes – The Finnish Funding Agency for Technology and Innovation and the European Union. The work was performed in cooperation with Finnish ICT SHOK research project Future Internet. Authors would like to thank all their colleagues for valuable discussions about the topic.

REFERENCES

[1] K. Keutzer, A. Newton, J. Rabaey and A. Sangiovanni-Vincentelli,"System-level design: orthogonalization of concerns and platformbased design", IEEE Transactions on Computer-Aided Design of Integrated Circuits and Systems, 19 (12), 2000, pp. 1523 - 1543.

[2] J. Kreku, M. Hoppari, T. Kestilä, Y. Qu, J.-P. Soininen, P. Andersson and K. Tiensyrjä. Combining UML2 Application and SystemC Platform Modelling for Performance Evaluation of Real-Time Embedded Systems. Hindawi Publishing Corporation. EURASIP Journal on Embedded Systems. Volume 2008, Article ID 712329, 18 pages, doi:10.1155/2008/712329.

[3] Linking GENESYS Application Architecture Modelling with Platform Performance Simulation. Subayal Khan, Susanna Pantsar-Syväniemi, Jari Kreku, Kari Tiensyrjä and Juha-Pekka Soininen. 12th Forum on specification and Design Languages, FDL2009. Sep 22- 24, 2009. Sophia Antipolis, France.

[4] S. Khan et al., "From y-chart to seamless integration of applicationdesign and performance simulation," in System on Chip (SoC), 2010International Symposium on, 2010, pp. 18 –25.

[5] http://www.systemc.org/home/

[6] http://itpp.sourceforge.net

[7] Lappeteläinen, Antti et. al., 'Networked Systems, Services, and Information - The Ultimate Digital Convergence'. 1st International Conference on Network on Terminal Architecture, June 11, 2008, Helsinki.

[8] http://www.isi.edu/nsnam/ns/

[9] http://www.omnetpp.org/

[10] B. Kienhuis, E. Deprettere, K. Vissers and P. van der Wolf. Approach for quantitative analysis of application-specific dataflow architectures," in Proceedings of the IEEE International Conference on Application- Specific Systems, Architectures and Processors (ASAP '97), pp. 338– 349, Zurich, Switzerland. July 1997. USC Center for Software Engineering, Guidelines for Model-Based (System) Architecting and Software Engineering, http://sunset.usc.edu/research/MBASE, 2003.

[11] J.M. Paul, D.E. Thomas, and A.S. Cassidy. High-Level Modeling and Simulation of Single-Chip Programmable Heterogeneous Multiprocessors. ACM Transactions on Design Automation of Electronic Systems, Vol. 10, No. 3, 2005, pp. 431-461.

[12] Fethi Filali, Impact of Link-Layer Fragmentation and Retransmissions on TCP performance in 802.11-based Networks. in Proc. of MWCN 2005, 7th IFIP/IEEE International Conference on Mobile and Wireless Communications Networks, September 19th-21st 2005,Marrakech, Marrocco.

[13] www.ti.com

[14] Jonathan Kay, Joseph Pasquale: The Importance of Non-Data Touching Processing Overheads in TCP/IP. SIGCOMM 1993: 259-268

[15] Instantiation and feasibility evaluation of NoTA SOAD via MARTE profile and binary instrumentation. Khan, Subayal; Tiensyrjä, Kari; Nurmi,Jari. The 7th International Conference and Expo on Emerging Technologies for a Smarter World. CEWIT 2010. Incheon, 27 - 29 Sep. 2010

[16] ns-2 vs. OPNET: a comparative study of the IEEE 802.11e technology on MANET environments, P. Pablo Garrido, Manuel P.Malumbres and Carlos T.Calafate. SIMUTOOLS 2008 - 1st International ICST Conference on Simulation Tools and Techniques for Communications, Networks and Systems.

Moonrake Chip - GALS Demonstrator in 40 nm CMOS Technology

Miloš Krstić, Xin Fan, Eckhard Grass
IHP
Frankfurt (Oder), Germany
{krstic, fan, grass} @ihp-microelectronics.com

Luca Benini, M.R. Kakoee
University of Bologna
Bologna, Italy
{luca.benini, m.kakoee }@unibo.it

Christoph Heer, Birgit Sanders
Intel Mobile Communications
Neubiberg, Germany
{christoph.heer, birgit.sanders}@intel.com

Alessandro Strano, Davide Bertozzi
University of Ferrara
Ferrara, Italy
{strlsn1, brtdvd}@unife.it

Abstract— **In this paper we present for the first time a complex GALS ASIC demonstrator in 40 nm CMOS process. This chip, named Moonrake, compares synchronous and GALS synchronization technology in a homogeneous experimental setting: same baseline designs, same manufacturing process, same die. The chip validates GALS technology for both point-to-point and network-centric on-chip communications, demonstrating its potentials for different applications.**

Keywords- GALS, NoC, asynchronous, pausible clocking

I. INTRODUCTION

Globally Asynchronous Locally Synchronous (GALS) technology has been proposed many years ago as an alternative to the traditional synchronous paradigm for chip synchronization [1]. Although significant potential was reported by the academia, the GALS methodology has never taken off in the industry. However, the growing challenges, imposed by the unrelenting pace of technology scaling to the nanoscale regime, urge for an efficient and safe system-level integration methodology. Consequently, we have targeted the implementation of a chip in the advanced 40 nm CMOS process, aiming at the assessment of GALS technology for nanoscale designs. The chip was named *Moonrake*.

Our intention was to evaluate GALS vs. standard synchronous technology on the same die, by implementing synchronous and GALS counterparts of the same baseline designs, both in the point-to-point as well as in the network-on-chip (NoC) scenarios for on-chip communication. The two scenarios are very different, hence motivating the different choice of baseline designs for their analysis. *In point-to-point communication,* once an optimized GALS interface is selected, *the focus is on the implications of redesigning an entire system around these links.* In this direction, we took a state-of-the-art multi-million gate synchronous system, an OFDM baseband transmitter developed for a 60 GHz transceiver with a gigabit throughput [2], and re-implemented it with GALS methodology, using the optimized interfaces for pausible (stoppable) clocking in [3]. One major goal was to explore EMI properties of GALS designs, initially analyzed in [4]. *For on-chip networking applications,* the communication landscape

is more heterogeneous since it results from the interconnection of domains with different synchronization assumptions. Therefore, our focus was on the *provision of flexible and cost-effective interfaces for arbitrary composability.* In this direction, the novel synchronization interfaces in [10, 11], aiming at low-area/power/latency overhead while preserving timing robustness, were integrated into NoC test structures exposing (and comparing) a range of flexible GALS solutions.

The contributions of this paper are as follows:

- The GALS partitioning criteria for a state-of-the-art OFDM transmitter is presented, highlighting the optimized asynchronous link crossing scheme and the partitioning granularity and strategy at the system level.

- The design flow followed for different GALS systems is illustrated: from pausible clocking to mesochronous synchronization to mixed-timing systems. Compatibility with mainstream standard cell libraries and design toolflows is discussed.

- The feasibility of GALS NoCs linking sub-systems with heterogeneous timing assumptions by means of area/power/latency optimized interfaces while preserving timing margins has been demonstrated.

- Synchronous and GALS counterparts of the same baseline designs (the OFDM transmitter and a NoC sub-set), implemented in the same demonstrator chip, have been compared in terms of area, pointing out counterintuitive benefits of the GALS design style.

II. GALS SYSTEMS AND DEMONSTRATORS

In past years several GALS chip implementations were reported. Many of them were focused on point to point GALS architectures (several designs from ETHZ [1], WLAN baseband processor from IHP [2]), while more recent implementations have also explored the GALS NoC concept (NEXUS chip [5], recent SpiNNaker [6], and three implementations from LETI, namely FAUST [12], ALPIN [7], and MAGALI [8] chip).

978-1-4577-0671-4/11 $26.00 © 2011 IEEE

TABLE I.	RECENT GALS DEMONSTRATORS			

GALS Chip Demonstrators		*Technology node*	*Year of fabrication*	*Size (mm²)*	*Power (mW)*
Moonrake chip		40 nm	2010	9	220-250
	Main features	GALS demonstrator, based on pausible clocking that serves the function of an OFDM transmitter for 60 GHz communication with the processing throughput of 1.6 -2.5 Gbps. It contains parallel synchronous and GALS implementation and NoC test structures.			
Magali chip [8]		65 nm	2010	32	500
	Main features	Digital baseband circuit for software-defined-radio and cognitive-radio applications that features less than 50 µs for full reconfiguration. Developed for 4G mobile phones, is based on mesh asynchronous network-on-chip infrastructure delivering 2.2GB/s/link. The chip includes 23 integrated signal processors and processing with an ARM1176 processor. The circuit exhibits less than for up to 40 GOPS performance. MAGALI has been tested on a 3GPP-LTE application, delivering 50Mbits/s on a 4×2 MIMO scheme.			
Alpin chip [7]		65 nm	2008	11.5	20-81
	Main features	Aiming at reducing both dynamic and static power consumption. Based on GALS methodology, each synchronous island is an independent frequency and power domain. The proposed DVFS architecture is based on the association of pausible clocking and supply voltage management unit.			
SpiNNaker chip [6]		130 nm	2011	~100	500-1000 (estimate)
	Main features	The chip, emulating neural system, contains 18 ARM9 processor cores, each running at around 200 MHz. These cores must all communicate 166MHz mobile DDR SDRAMs. The chip uses two distinct NoCs based on Chain, a delay-insensitive (DI) communication technology. The information regarding Spinnaker chip are just provisional since the design is just produced and the results are not yet published.			

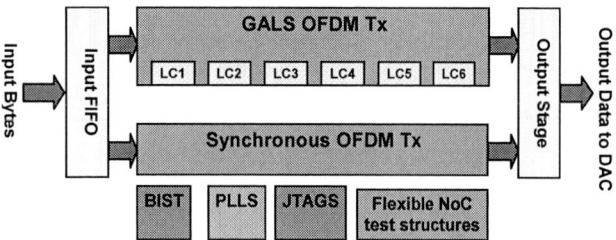

Figure 1: Block diagram of the Moonrake Chip Architecture

For the sake of comparison, Table 1 illustrates an overview of basic features of the GALS demonstrators in the last 3 years. The Magali chip is the most complex one, followed by the Moonrake chip. Improvement of this latter upon state-of-the-art concerns the most aggressive manufacturing process, the higher industrial relevance of its "galsified" designs, the maturity of implemented synchronization interfaces. Above all, both the baseline synchronous and the GALS counterparts of the same designs are now available on the same chip, thus paving the way for benchmarking in a truly homogeneous experimental setting (both at the architecture and at the technology level).

III. MOONRAKE ARCHITECTURE

As anticipated, the Moonrake architecture is based on the parallel implementation of the synchronous and the GALS variants of the same baseline designs. In order to reduce chip area, the same pad frame for both was re-used, and therefore data input and output pins were in most cases multiplexed. The chip architecture is pictorially illustrated in Fig. 1. Depending on the functional mode, input data will be processed by the synchronous or the GALS OFDM transmitter. In addition to the global dataflow, there is also a BIST logic used for testing of both synchronous and GALS transmitters, PLLs that trigger the synchronous Transmitter and the NoC parts, and JTAG interfaces used for programming the PLLs, setting the

parameters of local clock generators, programming NoC traffic generators and controlling the different modes of operations. It should be observed that the OFDM sub-system and the NoC test structures are independent from each other, except for the sharing of some input pins where possible: 37% of NoC pins are shared with the OFDM designs.

IV. IMPLEMENTATION OF OFDM GALS TRANSMITTERS

When galsifying functional blocks in the signal processing domain, it is very frequent to end up exposing point-to-point asynchronous links between partitions, thus conditioning area, power and performance metrics of the entire design. For this reason, the Moonrake framework presents a pilot partitioning experience of an industry-relevant design, where the system-level GALS partitioning methodology is as important as the design of efficient signaling schemes at the link level.

A. Starting point – the Synchronous Transmitter

The foundation of this implementation was the OFDM transceiver system in [2], developed for 60 GHz communications and offering gigabit throughput. The OFDM transmitter is a fairly complex structure that includes, among the other blocks, 256-point FFT and six different interleaver units. This circuit was originally implemented in Xilinx FPGA, with the estimated complexity of 7.8 M gates. The design is dominated by the various memory structures (64 RAM blocks, and 43 ROMs), occupying around 70% of the total area.

B. GALS Partitioning Methodology

The main methodology for partitioning was the area/power equalization between the GALS blocks, thus aiming at easier timing closure and good supply noise profile. It is conceivable to expect that such balanced partitioning would result in the set of the compact blocks that can be much easily optimized by the CAD tools. As a result it is expected to achieve relaxed timing closure, the clock trees of significantly reduced complexity, and reduced gate insertion penalty introduced during the timing

978-1-4577-0671-4/11 $26.00 © 2011 IEEE

optimization by the back-end tools. The GALS interfacing that is used here is based on pausible clocking. We have used our optimized version of interface circuits that can offer improved performance under increased communication load and improved safety of the generated clocks (see our past work in [3]). Although our optimized GALS interfaces could reduce the performance losses introduced by the concurrent arbitration of the parallel asynchronous links up to 33% [3], extensive arbitration might still result in reduced throughput. Therefore, an additional goal of the partitioning process was to reduce the number of communication links between the mutually-asynchronous blocks in order to preserve performance. As a result, the complete transmitter was partitioned into 6 independent GALS blocks (Fig. 2), mutually connected over asynchronous links. It can be observed that the complete set of front-end processing (symbol mapping, scrambler, encoder, mapper, pilot insertion etc.) is integrated into a single block; the three pairs of interleavers build up GALS blocks 2 – 4, and finally the 256-point FFT was partitioned into the last two GALS blocks (5-6). We ended up implementing 16 different asynchronous links between the locally synchronous blocks (and consequently 32 asynchronous ports), without any significant performance degradation.

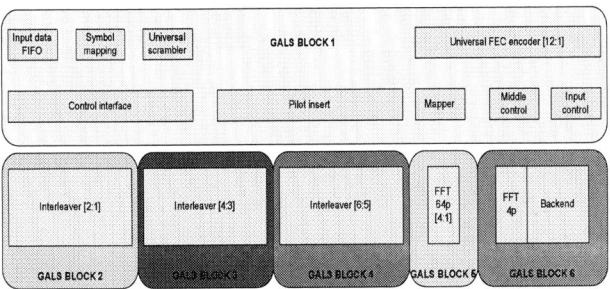

Figure 2: GALS partitioning of the transmitter

C. Design Flow Concept

Being a GALS design, the Moonrake chip was implemented using the mixed synchronous-asynchronous design flow. The asynchronous components, such as port controllers, have been modelled as signal transition graphs and correspondingly synthesized in hazard-free combinational logic using the tool Petrify [9]. The port controllers and pausible clock generators have been separately laid out as hard-macros, in order to fully control the timing. For the implementation, we have used an extended standard cell library including additionally developed asynchronous components - various C-element types and mutual exclusion elements. The top level system implementation was relatively straightforward, starting with the floorplaning of the memories and hard macros. We have generated top level constraints in such a way that also global asynchronous paths have been covered. This has been achieved by the careful definition of minimum and maximum delays of the asynchronous paths and generation of the virtual clocks between the GALS controllers. This approach is usually difficult for pure asynchronous designs due to the number of paths that have to be constrained. However, since in many GALS systems (like the one considered here) the number of asynchronous paths is limited, the constraining could be successfully applied. In this way, we have been able to use the

standard tools for timing closure and STA. This was a very important achievement that enabled effective usage of the timing optimization, placement and routing capabilities offered by the the standard front-end and back-end tools, thus exploiting the relaxation of the global timing introduced by the balanced partitioning.

V. TEST STRUCTURES FOR GALS NOCS

Other application domains currently call for a larger communication parallelism, which can be efficiently provided by on-chip interconnection networks in a structured and scalable way. In these cases, applying pausible clocking to multi-port components like a NoC switch brings significant performance penalty. In contrast, it is important to engineer synchronization interfaces able to marginally impact NoC quality metrics while at the same time enabling arbitrary composition of timing-heterogeneous sub-systems. Moreover, timing the NoC itself is an additional problem that point-to-point links do not have to address. Therefore, providing efficient interfaces for truly heterogeneous and cost-effective GALS partitioning is currently more critical than the assessment of system-level partitioning schemes. For this reason, the Moonrake chip aims at validating the new synchronization interfaces for *synchronizer-based GALS NoCs* presented in [10,11]. Their distinctive features are as follows:
- Besides enabling the traditional loose coupling between NoC building blocks and the synchronizers (i.e., synchronizers are placed in front of the switch/network interface they serve), they enable a tight coupling design style (i.e., synchronizer and switch input buffer are merged together) and other variants for increased timing robustness (e.g., the hybrid coupling, implementing partial merging and link retiming).
- They target an heterogeneous timing landscape, made possible by dual-clock FIFOs, mesochronous synchronizers or a mix thereof. The focus is on synchronizer-based GALS NoCs and not on fully asynchronous interconnect fabrics, since these latter feature a wider gap with respect to traditional synchronous architectures and design tools. In fact, they are typically available only through hard macros [13].
For the sake of validating these interfaces and of their comparison with traditional synchronous ones, the Moonrake chip implements simple test structures. The baseline structure consists of a 2-ary 1-mesh topology with 4 cores (two traffic generators and two memory cores) ensuring bi-directional communication across the simple topology. Each traffic generator is programmable and connected to a JTAG interface for I/O. See Fig.3 for a pictorial overview.

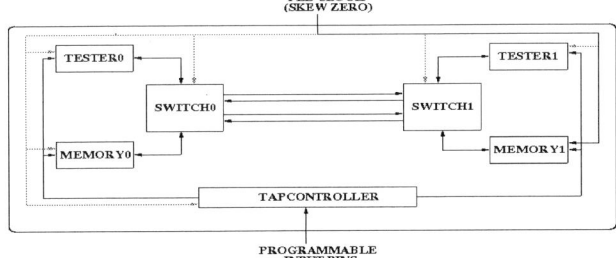

Figure 3: Baseline synchronous test structure for NoCs

978-1-4577-0671-4/11 $26.00 © 2011 IEEE

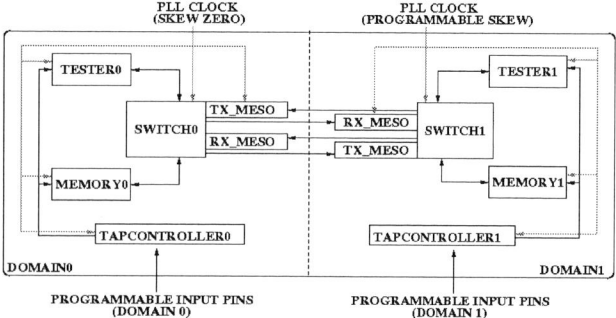

Figure 4:   Mesochronous synchronization with loose coupling with the NoC

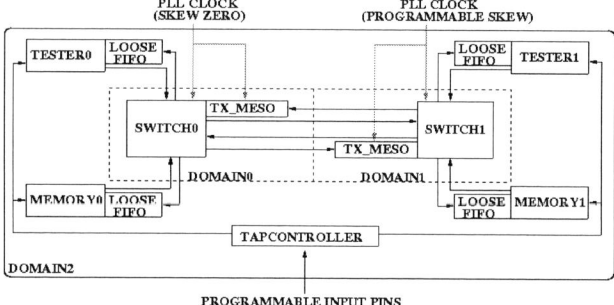

Figure 5: NoC test structure where the NoC is an independent clock domain featuring hybrid coupling.

To demonstrate the feasibility of different partitioning options, we re-implemented the baseline sub-system three times:

- **Loosely coupled mesochronous synchronization.** This sub-system has two clock domains having the same clock frequency but unknown phase offset (mesochronous synchronization). This platform uses loosely coupled (i.e., not embedded into the switches) mesochronous synchronizers (RX and TX) placed in the bidirectional link next to the switches. The RX module synchronizes the data while the 1-bit TX module synchronizes the flow control signal. This is the scheme of Fig.4. Mesochronous synchronization could be a promising option to reduce the complexity and enable feasibility of the top-level clock tree through relaxation of the global skew constraint [14].

- **Mesochronous synchronization with hybrid coupling.** Same as before, but only the 1-bit TX synchronizer is not embedded into the switch, while the most complex RX one is. As proven in [11], this approach can be used to push a higher operating speed while still preserving the low area/power/latency footprint of tight coupling.

- **Mixed timing.** In this sub-system, the network is an independent clock domain with respect to traffic injectors and memory cores, hence separated by means of dual-clock FIFOs. In turn, the network is split into two mesochronous sub-domains, implemented with hybrid coupling. See Fig.5. Only one pair of dual-clock FIFOs was tightly coupled (with the NoC switch), while the remaining pair was not, since this would have implied their integration into the network interfaces, which could not be the general case.

The PLL outputs a 250 MHz clock together with different phase shifted replicas, hence enabling to test mesochronous synchronization under 0%, 25%, 50% and 75% clock phase offset. For the mixed timing sub-system, the PLL clocks the network mesochronous domains, while cores and memories are clocked externally with a slow 25 MHz clock.

The entire design flow was based on standard cells (i.e., no full-custom designed components). Only during place-and-route we forced rigid fences for placement of external synchronizers, and used native CAD tool support for bundled routing of the source-synchronous links (for delay matching). Link length/switch separation was set to be 0.5 mm. As a result, the entire design flow was within reach of current industrial design tools.

VI.   RESULTS

The area breakdown after the chip back-end is provided in Table 2. It is quite noticeable that the area utilized by the synchronous transmitter is larger than the GALS transmitter by around 5% including PLL, although the GALS version contains additional asynchronous controllers and six pausible clock generators, and the design is largely dominated by fixed size memory blocks. Indeed after the logic synthesis we have observed that those asynchronous blocks contributed to the slight GALS overhead (1% w/o PLL), even when the synchronous transmitter was extended with scan DFT logic. However, during the back-end and IPO (in place optimization), the effort for timing closure on the synchronous side was much more challenging due to the more complex global synchronous domain and chip-wide floorplan. This led to the insertion of additional buffers and to the resizing of existing cells, and resulted, together with the PLL overhead, in increased complexity of the synchronous transmitter (0.7% w/o PLL). Additionally, with the removal of the global clock constraint and with the balanced GALS block complexity, the clock network complexity was reduced by a factor 3.6, comparing the number of clock tree levels in the fully synchronous tree and average number of clock tree levels for the GALS blocks. This also contributed to the better area results of the GALS blocks. The observed area and clock tree complexity reduction on the GALS side was a clear effect of our GALS partitioning methodology, where no block has significantly higher area (the most complex block is only 15% larger than the average) or power (the most power hungry block consumes only 11% more than the average value) than others. Also, it was an effect of constraining the design using standard back-end design tools. Combining these two features together, the engineered GALS partitions proved a very good starting point for the back-end design steps. When it comes to the NoC test structures, the area breakdown is illustrated in Fig.6. Evolving the synchronous system to the mesochronous one with the loosely coupled design style clearly comes at a significant 17% area penalty. However, the newer hybrid coupled design style enables to contain this penalty within 5% due to the intensive buffer reuse for different purposes (synchronization, performance buffering, flow control). When considering switches in isolation (i.e., excluding testers, memories, etc.), the savings of hybrid over loose coupling amount to 40%. Also, dual-clock FIFOs have been demonstrated to be clearly more expensive than meso-

978-1-4577-0671-4/11 $26.00 © 2011 IEEE

TABLE II. AREA DISTRIBUTION IN MOONRAKE CHIP AFTER BACK-END

| Total | OFDM GALS and Synchronous Transmitter | | | | NOC Test Structures | Pads |
	GALS Tx	SYNC Tx	Others (IO stage, BIST)	Total		
5406853 (100%)	2220080 (41%)	2334712 (43.2%)	91916 (1.7%)	4643900 (85.9%)	227374 (4.2%)	537075 (9.9%)

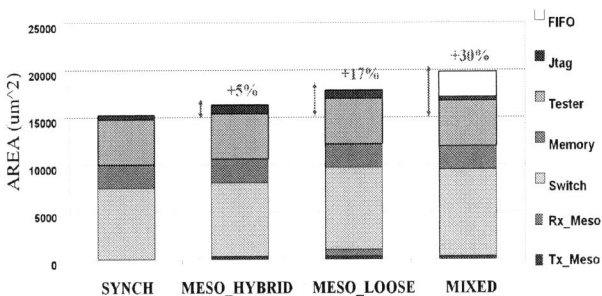

Figure 6: Area breakdown for the NoC test structures

-chronous synchronizers, thus suggesting their mild use and fostering their combination with other synchronizers.

The Moonrake chip was taped-out in April 2010, and afterwards fabricated in 40 nm technology run. The chip was recently successfully tested. Main part of testing was devoted to the functional verification of the system including BIST transmitter tests in the synchronous and GALS mode, and test of various NoC modes.

VII. CONCLUSIONS

The Moonrake chip is the first GALS demonstrator in a state-of-the-art 40 nm CMOS technology. The design includes various components for evaluating point-to-point and NoC GALS architectures in an architecture- and technology-homogeneous experimental setting. The chip is functionally tested and verified with success. For point-to-point links, the Moonrake chip has provided a pilot partitioning experience with the GALS methodology, relying on standard CAD tools, highlighting an industry-relevant co-design example between the system layer and the link layer. Also, counterintuitive area benefits of the GALS methodology have been, to our knowledge, for the first time demonstrated, due to the lower burden for timing closure.

On-chip networking calls for different considerations, hence the Moonrake chip devoted several test structures to it. The outcome is the validation of the timing robustness of a newer synchronization technology merging synchronizers with NoC building blocks. Above all, the capability of the newer technology to contain area penalty for the upgrade of synchronous networks into GALS ones has been quantified. Also, it has been proved that this technology is affordable to industry-standard technology libraries without any custom extensions and that the design flow is within reach of

mainstream industrial tools. We have planned the further measurements of the Moonrake chip, including power, performance and EMI evaluation, that will be described in detail in our future publications.

ACKNOWLEDGMENT

This work has been supported by the European Project GALAXY under grant reference number FP7-ICT-214364. The authors would like to thank Dr. Luis Plana for preliminary information about the SpiNNaker chip.

REFERENCES

[1] M. Krstic, et al., "Globally Asynchronous, Locally Synchronous Circuits: Overview and Outlook", IEEE Design & Test of Computers, pp: 430 - 441 , Volume: 24 Issue: 5, Sept.-Oct. 2007

[2] M. Krstic, et al., "OFDM Datapath Baseband Processor for 1 Gbps Datarate", Proc. IFIP/IEEE VLSI-SoC Conference 2008, pp. 156-159.

[3] X. Fan, M. Krstić, E. Grass, "Analysis and Optimization of Pausible Clocking based GALS Design", In Proc. of IEEE ICCD 2009, pp 358-365.

[4] X. Fan, et al., "A GALS FFT Processor with Clock Modulation for Low-EMI Applications", In Proc. IEEE ASAP 2010.

[5] Lines, A., "Asynchronous interconnect for synchronous SoC design", IEEE Micro, 2004, 24, (1), pp. 32–41.

[6] L. Plana, et al., "A GALS Infrastructure for a Massively Parallel Multiprocessor", IEEE Design & Test of Computers, Sept.-Oct. 2007, Volume: 24 Issue:5, pp.: 454 - 463

[7] E. Beigne, et al, "An asynchronous power aware and adaptive NoC based circuit", 2008 IEEE Symposium on VLSI Circuits, pp. 190 – 191.

[8] R. Lemaire, Y. Thonnart, Magali, "A Reconfigurable Digital Baseband for 4G Telecom Applications based on an Asynchronous NoC", In Proc ACM/IEEE International Symposium on Networks-on-Chip, 2010.

[9] J. Cortadella, et al, Petrify: "A tool for manipulating concurrent specifications and synthesis of asynchronous controllers", IEICE Transactions on Information and Systems, March 1997, pp. 315-325.

[10] A.Strano, D.Ludovici and D. Bertozzi, "A Library of Dual-Clock FIFOs for Cost-Effective and Flexible MPSoCs Design", SAMOS, 2010.

[11] D.Ludovici, et al, "Design Space Exploration of a Mesochronous Link for Cost-Effective and Flexible GALS NOCs", DATE'10, pp. 679–684, Dresden, Germany, 2010.

[12] P.Vivet, et al, "FAUST, an Asynchronous Network-on-Chip based Architecture for Telecom Applications", DATE, 2006.

[13] Y.Thonnart, P.Vivet, F.Clermidy, "A Fully Asynchronous Low-Power Framework for GALS NoC Integration", pp.33-38, 2010.

[14] I.M.Panades, F.Clermidy, P.Vivet, A.Greiner, "Physical Implementation of the DSPIN Network-on-Chip in the FAUST Architecture", Int. Symp. on Networks-on-Chip, pp.139-148, 2008.

978-1-4577-0671-4/11 $26.00 © 2011 IEEE

Analyzing Synchronous Dataflow Scenarios for Dynamic Software-defined Radio Applications

Firew Siyoum[1], Marc Geilen[1], Orlando Moreira[2], Rick Nas[2], Henk Corporaal[1]

[1]Eindhoven University of Technology
[2]ST-Ericsson Eindhoven

Abstract—Contemporary embedded systems for wireless communications support various radios. A software-defined radio (SDR) is a radio implemented as concurrent software processes that typically run on a multiprocessor system-on-chip (MPSoC). SDRs are real-time streaming applications with throughput requirements. One efficient approach for timing analysis of concurrent real-time applications is the dataflow model of computation (MoC). Nonetheless, the dataflow modeling of SDRs is challenging due to their dynamically changing data processing workload. A dataflow MoC that is not expressive enough to capture this dynamism gives pessimistic throughput results. On the other hand, if it is too expressive and detailed, it may not be analyzable at all. In this paper, we address the challenge of dataflow modeling of SDRs such that their timing behavior can be accurately analyzed to guarantee real-time requirements without unnecessarily over-allocating MPSoC resources.

The basis of our modeling approach is splitting the dynamic data processing behavior of a SDR into a group of static modes of operation. Each static mode of operation is then modeled by a Synchronous Dataflow (SDF), which we refer to as *scenario*. This paper has two main contributions: 1) a scenario-based dataflow model of Long Term Evolution (LTE), which is the latest standard in cellular communication, and 2) investigation of existing throughput analysis techniques of SDF scenarios for our LTE model. Our results show that scenario-based worst-case throughput computation is 2 to 3.4 times more accurate than a state-of-art SDF analysis technique. Our investigation also shows that existing timing analysis techniques of SDF scenarios have very low run-time that scales very well with increase in graph size. This makes SDF scenarios suitable in practice for modeling and analyzing SDRs as well as similar dynamic applications.

Index Terms—Synchronous Dataflow, Long Term Evolution, Software-defined Radio, Throughput, Scenario-aware Dataflow

I. INTRODUCTION

Present-day embedded systems for cellular, home and automotive communications support various wireless communication standards. We refer to these communication standards as *radios*. Smartphones, for instance, include different radios such as WCDMA, LTE and IEEE 802.11x. Modern radios are naturally dynamic in the sense that their transmission resource allocation change with channel conditions. As a result, radios have variable workload that is driven by control channels and signals. In addition, radios are real-time streaming applications with latency and throughput requirements. Therefore, radio designs must be *predictable* to ensure that temporal requirements are satisfied in all operating conditions.

The current trend in radio design shows that the implementation of physical layer functionalities of radios is shifting from dedicated hardware architectures to software processes for better flexibility and efficiency [1]. In *software-defined radio (SDR)* [2], some or all of the physical layer functionalities of a radio are implemented as concurrent software processes. These software processes typically run on a multiprocessor system-on-chip (MPSoC) for power and performance reasons. MPSoC architectures of SDR combine homogeneous and heterogeneous multiprocessing, including general purpose processors, vector processors and weakly programmable accelerators.

The mapping of the software processes onto the MPSoC and allocation of resources, such as memories and interconnects, is decided at the early stages of the SDR design. After such design decisions, the system must be analyzed to check if temporal requirements are met. This is not a trivial task due to the vastness of the design space and the complexity of SDRs. As a result, system-level modeling techniques are needed to efficiently analyze SDRs and ensure predictability. One established approach in the embedded domain for modeling and analyzing real-time streaming applications, such as SDRs, is the *dataflow* model of computation (MoC) [3].

Dataflow MoCs allow modeling of concurrent streaming applications as directed task graphs. Modeling and analyzing SDRs using dataflow MoCs has mainly two complementary challenges: the *dynamism of radios* and the *scarcity of MPSoC resources* [4]. On one hand, if the selected dataflow MoC is highly expressive in order to capture the dynamic behavior in detail, temporal analysis may not be possible at all. On the other hand, if it is made simplistic for the sake of analyzability, it may give pessimistic, if not invalid, results. Pessimistic results lead to unnecessary over-allocation of scarce MPSoC resources to satisfy temporal requirements. In this paper, we address the challenge of dataflow modeling of SDRs such that their timing behavior can be accurately analyzed to guarantee real-time requirements without unnecessarily over-allocating MPSoC resources.

The basis of our approach is splitting the dynamic data processing behavior of a SDR into a group of static modes of operation. Each static mode of operation is then modeled by a Synchronous Dataflow (SDF) [5] graph, which we refer to as *scenario* [6]. The possible orders of executions of these scenarios are specified by a finite-state machine (FSM) [7] [8]. In this work, we show the applicability of this MoC for SDRs by modeling and analyzing the baseband (physical layer) processing of Long Term Evolution (LTE) [9], which is the latest standard in cellular communication. This paper has two main contributions: (1) a scenario-based dataflow model of the baseband processing of LTE that accurately models dynamic behavior, and (2) investigation of existing worst-case throughput analysis techniques of SDF scenarios for our LTE model.

This scenario-based dataflow modeling is not new [6] [7]. This work extends previous works by presenting a technique for modeling control information exchanged between scenarios. In SDRs, the transfer of configurations and data from one mode of operation to the next is quite common. In LTE, for instance, a mode detection scenario broadcasts the type of a received frame to subsequent scenarios for configuration. We call such type data dependency between scenarios *scenario dependency*. Scenario dependencies that are not properly modeled result in an early start of execution of processes in subsequent scenarios. This ultimately leads to invalid worst-case temporal analysis. In this work, we present a technique to model scenario dependencies between different SDF graphs, and then show its applicability in our LTE model.

978-1-4577-0671-4/11 $26.00 © 2011 IEEE

Our investigation on LTE's baseband processing shows that the scenario-based worst-case throughput computation is at least two times more accurate than a state-of-the-art SDF analysis technique. Our results also show that existing timing analysis techniques of SDF scenarios have very low run-time that scales very well with increase in graph size. This makes SDF scenarios suitable in practice for modeling and analyzing SDRs as well as similar dynamic applications.

The remaining part of this paper is organized in six sections. First, Section II recaps dataflow modeling. Then, Section III presents a scenario-based dataflow model for LTE. Section IV introduces a new technique for modeling scenario dependencies. Accuracy of throughput analysis techniques for our LTE model is investigated in Section V. We postpone the review of related works to Section VI, as preceding sections help to better present the literature study. Finally, the paper concludes in Section VII, summarizing this work.

II. PRELIMINARIES

This section recaps the basics of dataflow modeling. It discusses basic SDF concepts that are important for a complete understanding of this paper. In this paper, we use $\mathbb{N}$ to denote natural numbers and $\mathbb{N}^0$ natural numbers including zero.

A. Synchronous Dataflow Graph

A Synchronous Dataflow Graph (SDFG) [5] is a directed graph that can model concurrent tasks. It can capture cyclic data dependencies between tasks. It also has efficient analysis techniques to compute throughput and buffer sizes [10] [11]. Due to its analyzability, it is widely used in system-level design flows for modeling and analyzing real-time streaming applications, running on MPSoCs [12] [13].

A SDFG consists of *actors* that are connected through *channels*. A channel represents a FIFO buffer through which actors communicate by sending *tokens*. A channel may have some *initial tokens* at the start. Every channel is exactly connected to one source actor and one destination actor. The connection point between an actor and a channel is referred to as a *port*. Each port of an actor is annotated with a fixed number, called the *port rate*.

An actor models a given system task, for example a software process. The duration of one complete execution of an actor is termed as its *execution time*, measured in any preferred *time-unit*. When an actor *fires*, i.e. starts execution, it reads tokens from all of its input ports. At the end of the execution, it produces tokens in all of its output ports. For each port of an actor, the number of tokens consumed or produced in every single execution of an actor is fixed, and it is equal to the port rate. A formal definition of SDFG is given in Definition 1.

Definition 1 (SDFG). *A SDFG is a tuple* $\mathbf{G} = (A, C, \mathcal{X}, \mathcal{I})$, *comprising a set of actors* A, *a set of channels* C, *execution times of actors* $\mathcal{X} : A \rightarrow \mathbb{N}^0$ *and initial tokens of channels* $\mathcal{I} : C \rightarrow \mathbb{N}^0$. *Given the set of all ports* P, *port rate of* $\mathbf{G}$ *is denoted as* $\mathcal{R} : P \rightarrow \mathbb{N}$.

An example SDFG consisting of three actors (x, y, z) and four channels is shown in Figure 1. In SDFG schematics, a black dot represents the number of initial tokens in a channel.

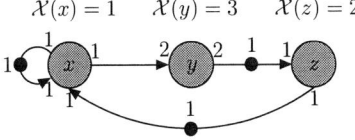

$$\mathcal{X}(x) = 1 \quad \mathcal{X}(y) = 3 \quad \mathcal{X}(z) = 2$$

Fig. 1. Example SDFG: Graph **A**

The execution of a SDFG is a timed simulation of the executions of its actors. A property that ensures a deadlock-free execution of a SDFG is *consistency*. A SDFG is called consistent if the initial tokens configuration can be restored after a finite numbers of firings of its actors. For example, for Figure 1, the numbers of firings of actors x, y and z that bring the graph back to its original state are 2, 1 and 2, respectively. This can be conveniently written in vector form as $[2, 1, 2]$. This vector is termed as the *repetition vector* of the SDFG. The repetition vector determines one complete execution of the graph, referred to as *iteration*, as defined in Definition 2.

Definition 2 (Iteration). *An iteration is defined as an execution of a SDFG where each actor fires exactly as many times as its entry in the repetition vector.*

B. The Time-stamp Vector

The purpose of this section is to show how the total number of initial tokens in a SDFG affects the performance of throughput analysis techniques, discussed later in Section V.

In temporal analysis of SDFGs, we are interested in completion times of iterations. This is because the number of iterations that can be completed per given time interval is a measure of the throughput of the graph. The collection of tokens that exist after each iteration of the graph is the same as the initial tokens configuration. Define a vector γ that has exactly one entry for every initial token in the graph, to record the production times of initial tokens after each iteration. Vector γ is referred to as a *time-stamp vector*. The length of the time stamp vector $|\gamma|$ is equal to the total number of initial tokens of the graph. However, the entries in this vector change between iterations.

A useful mathematical tool to analyze the timing evolution of the time-stamp vector of a SDFG is the $(max, +)$ algebra [14]. In *self-timed execution* of a given SDFG, an actor fires as soon as it has sufficient input tokens. Hence, the start time of an actor's firing is determined by the tardiest input token, i.e. the **maximum** of the production times of all input tokens. The end time of an actor's firing can be obtained by **adding** its execution time on its start time. As a result, the entire timing behavior of a self-timed execution of a SDFG can be analyzed using $(max, +)$ expressions.

The time-stamp vector of the kth iteration of the graph is denoted γ_k. The relationship between any two consecutive iterations is expressed by a matrix multiplication in $(max, +)$ algebra, i.e. $\gamma_{k+1} = \mathbf{M} \cdot \gamma_k$ where $k \in \mathbb{N}^0$. $\mathbf{M}$ is referred to as the *matrix* of the graph and has a size of $|\gamma| \times |\gamma|$. An algorithm to compute the matrix of a SDFG is provided in [15]. The matrix of SDFG $\mathbf{A}$, shown in Figure 1, and its time-stamp vectors for the first three iterations are as follows.

$$\mathbf{M} = \begin{bmatrix} 2 & 3 & 2 \\ 5 & 6 & 5 \\ 7 & 8 & 7 \end{bmatrix} \text{ and } \gamma_0, \gamma_1, \gamma_2, \gamma_3 = \begin{bmatrix} 0 \\ 0 \\ 0 \end{bmatrix}, \begin{bmatrix} 3 \\ 6 \\ 8 \end{bmatrix}, \begin{bmatrix} 10 \\ 13 \\ 15 \end{bmatrix}, \begin{bmatrix} 17 \\ 20 \\ 22 \end{bmatrix}$$

The self-timed execution of a SDFG reaches a periodic phase after a finite number of transient iterations [10]. For the SDFG $\mathbf{A}$ of Figure 1, the periodic phase starts after the first iteration (after γ_1). The *period* of this periodic phase is 7 time-units (the difference between consecutive time-stamp vectors), and hence, the graph has a maximum throughput of $\frac{1}{7}$ iterations/time-unit.

For exact computation of time-stamp vectors, we start from a zero vector γ_0. However, approximations (upper-bounds to γ) are also possible. This is done by starting from a selected γ_0, called a *reference schedule* or a *delay* [15]. This technique is referred to as *delay-period* approximation. As shown later in Section V, the technique allows faster throughput computation, since it makes the entire timing behavior periodic.

The time-stamp vector is a timing interface that separates iterations of a SDFG. It can also analyze a sequence of iterations of different SDFGs. To achieve this, γ has to be extended first to cover all initial tokens that are common between these graphs. This approach is used to compute the throughput of dataflow MoCs that model dynamic applications through a set of SDFGs. One such dataflow model is discussed next in Section II-C.

C. Synchronous Dataflow Scenarios

Real-life streaming applications, such as multimedia codecs and SDRs, go through different operating modes, depending on the processed data. A *scenario* refers to a single mode of operation of the application [6]. When an application is executing at a given scenario, its computation and communication characteristics mostly remain invariable. Hence, each scenario can be modeled by a static dataflow model, such as a SDFG. We refer to a SDFG that models a single scenario of an application a *scenario graph*.

A sequence of operating modes of an application is modeled by a sequence of executions of the corresponding scenario graphs. These executions can also be pipelined in time. All possible scenario sequences of an application are represented by a finite-state machine (FSM), as defined in Definition 3.

Definition 3 (Finite-state machine (FSM)). *Given a set S of scenario graphs, a finite state machine $\mathbf{f}$ on S is a tuple $\mathbf{f} = (Q, q_0, \delta, \Sigma)$. Q is a set of states, $q_0 \in Q$ is an initial state, δ is a transition relation between two states, $\delta \subseteq Q \times Q$, and Σ is scenario labeling, $\Sigma : Q \rightarrow S$.*

Scenario graphs along with a FSM can be used to capture the dynamic behavior of an application [15] [8]. For example, let two SDFGs **A** and **B** represent two different scenarios of an application. Assume also that the application switches arbitrarily between these two scenarios. Figure 2 shows a FSM that captures this dynamic behavior. This modeling approach is referred to as *FSM-based Scenario-aware Dataflow (FSM-SADF)* [15] [8], as defined in Definition 4.

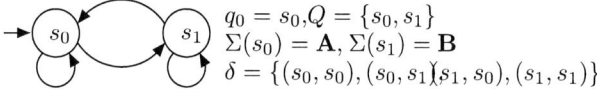
$q_0 = s_0, Q = \{s_0, s_1\}$
$\Sigma(s_0) = \mathbf{A}, \Sigma(s_1) = \mathbf{B}$
$\delta = \{(s_0, s_0), (s_0, s_1)(s_1, s_0), (s_1, s_1)\}$

Fig. 2.   Example of a FSM

Definition 4 (FSM-SADF). *An FSM-SADF model is a tuple $\mathbf{F} = (S, \mathbf{f})$, consisting of a set of scenario graphs S and a finite-state machine $\mathbf{f}$ on S.*

The FSM-SADF is a class of Scenario-aware Dataflow (SADF) model [6]. In FSM-SADF, scenario transitions are non-deterministically specified by a finite-state machine, instead of the stochastic model proposed in [6]. The FSM allows a worst-case timing analysis of FSM-SADF through the timing evolution of the time-stamp vector, as discussed in Section II-B. Efficient timing analysis techniques for FSM-SADF are presented in [15] and [7].

Next, in Section III, we show the applicability of FSM-SADF for modeling and analyzing a dynamic SDR application. The section shows how FSM-SADF captures the dynamism of the SDR application, which a single SDFG cannot do without introducing pessimistic assumptions.

III. LONG TERM EVOLUTION (LTE)

Long Term Evolution (LTE) is a recent standard in cellular wireless communication technologies. It aims at high bit rates: a downlink peak rate of up to 300 Mbit/s and an uplink of 150 Mbit/s [9]. Due to the high bit rates, and the resulting high workload, the complexity of LTE receivers is enormous. The complexity is further increased by dynamism (data-dependent variations) of frames. In this section, we focus on the dynamism of LTE's physical layer frames, as SDR deals with a software implementation of the baseband processing (physical layer processing) of radios.

LTE uses *adaptive modulation and coding (AMC)* that dynamically adjusts modulation schemes and transport block sizes to adapt to varying channel conditions [16]. Consequently, the workload of LTE's baseband processing changes dynamically. In this section, we present a variation-aware dataflow model for LTE baseband processing that captures this dynamic workload. We first start by discussing the source of dynamism in the physical layer processing of LTE in Section III-A. Then, we show how we model this dynamism using FSM-SADF dataflow in Section III-B.

A. Dynamism in LTE baseband processing

There are multiple sources of dynamism in LTE baseband processing that contribute to variable computation and communication requirements. These include variations in channel allocation of frames, modulation schemes and transport block sizes of upper layers. For the discussions of this paper, we limit ourselves to dynamism due to variations in channel allocations of frames. Nonetheless, the modeling concept can be equally applied to any complexity level of dynamism.

To show variations in channel allocations, we consider the downlink communication, which refers to the communication link from the base station (eNodeB) to the User Equipment (UE). Depending on the type of duplexing, there are two types of LTE physical layer frame structures. The downlink frame structure for Frequency Division Duplexing (FDD) is illustrated in Figure 3. A single frame is 10 milliseconds (ms) long. It consists of 10 *sub-frames* (1ms each) and each sub-frame consists of 2 *slots* (0.5ms each).

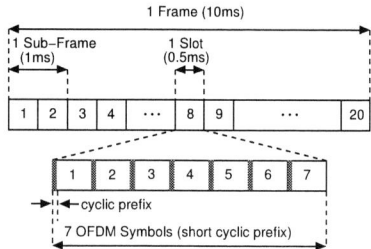

Fig. 3.   LTE frame structure for FDD

LTE employs Orthogonal Frequency Division Multiplexing (OFDM) for downlink data transmission. The transmission resource within a sub-frame is organized by a *resource grid*, as shown in Figure 4. The width of the resource grid (in time domain) equals twice the number of symbols per slot, N_{symb}^{DL}. The height of the grid (in frequency domain) equals the number of OFDM sub-carriers per resource block, N_{sc}^{RB}, multiplied by the number of resource blocks per sub-frame, N_{RB}^{DL}. N_{RB}^{DL} is determined by the downlink transmission bandwidth, while N_{symb}^{DL} and N_{sc}^{RB} are determined by the OFDM subcarrier spacing and the type of OFDM cyclic prefix used.

978-1-4577-0671-4/11 $26.00 © 2011 IEEE

In practice, N_{RB}^{DL}, N_{sc}^{RB} and N_{symb}^{DL} are fixed once the system is configured. In the rest of this paper, we consider a bandwidth of 20MHz, a subcarrier spacing of 15KHz and normal cyclic prefix. Hence, $N_{RB}^{DL} = 100$, $N_{symb}^{DL} = 7$ and $N_{sc}^{RB} = 12$, as shown in Figure 4.

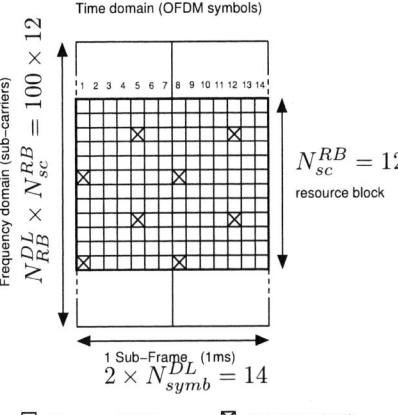

Fig. 4.   Resource grid of a sub-frame

The time-frequency unit for resource allocation of the resource grid is a *resource element*. Resource elements of the resource grid are allocated to different data and control channels. Resource elements of the first OFDM symbol (the first column of the grid) are allocated to the *Physical Control Format Indicator Channel* (PCFICH) and partly to the *Physical Downlink Control Channel* (PDCCH). PCFICH contains information regarding the resource allocation of PDCCH. PD-CCH can be allocated resource elements upto the third column of the resource grid. PDCCH, in turn, tells the locations of data channels, such as the *Physical Downlink Shared Channel* (PDSCH). PDSCH can be located between the second and the fourteenth columns of the resource grid.

Decoding a sub-frame consists of a number of tasks whose data dependency is captured by a directed graph, as shown in Figure 5. Some major tasks of the graph include *OFDM demodulation (dem)*, *channel estimation (est)*, *multiple-input and multiple-output summation (mimo)*, *OFDM demapping (dmp)* and *channel decoding (dec)*. The input-output data granularity of these tasks is an OFDM symbol (a column of the resource grid), that is about 4800 bytes. Hence, these tasks have to be carried out for each of the 14 symbols that constitute a sub-frame.

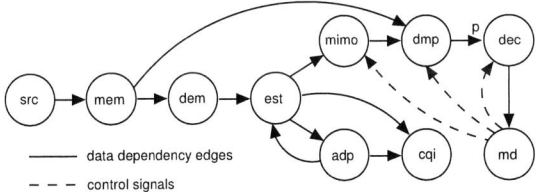

Fig. 5.   A directed task graph of LTE's baseband processing

However, depending on the type of channel a symbol is allocated to, the properties of these tasks, such as functionality, execution time and communication rates, vary. For instance, task *dec* has a different execution time for a symbol that carry a control channel (192 time-units) than a data channel (an average of 75 time-units). In addition, its input data rate can vary between 11, 12 or 13 symbols while decoding a data channel. This is because the control channel is always between

the first and the third symbols, leaving the remaining symbols for data channels.

Consequently, the execution time and the input-output data rates of tasks may change every symbol. This gives rise to the dynamic behavior of LTE's baseband processing. The challenge is now how to capture this dynamism in dataflow models for temporal analysis. Dataflow modeling and analysis of LTE, and other radios, that abstract from variable execution times and communication rates result in inefficient, if not invalid, implementations. This is because static worst-case conditions have to be considered for all operating conditions. Let us see, for instance, what a static SDFG model for the LTE baseband processing looks like.

All tasks of Figure 5, except *dec*, fortunately have fixed execution times and input-output token rates. Thus, the modeling effort simplifies to finding a fixed execution time for task *dec* and its input port rate p. The requirement for the selection of these two parameters is that the production time of tokens by actor *dec* must be conservative to (not earlier than) the actual production time of data by task *dec*. Symbols that carry control channels have to be decoded as soon as they arrive. This requires $\mathcal{R}(p) = 1$ and $\mathcal{X}(dec) = 192$. This configuration also ensures that the decoding of a data channel, which is carried out in a chunk of 11, 12 or 13 symbols, is also conservative at a sub-frame level. This SDFG of LTE's baseband processing is shown in Figure 6, where all port rates equal to one and execution times are shown by numbers written inside the actors.

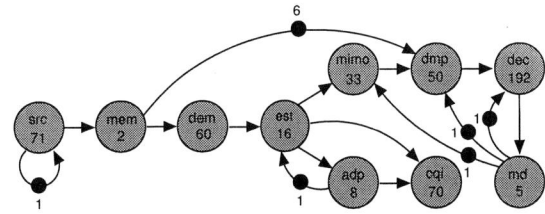

Fig. 6.   A SDFG of LTE's baseband processing

Due to the static nature of the SDFG, the execution time of *dec* is fixed to 192 time-units for all symbols, though it is on average 75 time-units for data channels. In addition, actor *md* (the mode detection task) is executed for every symbol, even though it is only needed for the first symbol of the resouce grid. As a result, the timing analysis of this SDFG gives a pessimistic throughput result, as shown later in Section V. This fact necessitates a variation-aware dataflow model that gives more accurate throughput results. In the next section, we present a dataflow model for LTE that captures variable execution times and input-ouput data rates through a set of scenario graphs.

B. Dataflow Model of LTE baseband processing

For the LTE's baseband processing, we identify five different modes of operation, depending on the type of symbol it is processing. When operating at a given mode of operation, the execution times and input-output rates of the tasks remain static. Therefore, each mode of operation can be modeled by a SDFG. In addition, the possible transitions between these five modes of operation are also known at design time. Hence, the transitions can be described by a finite-state machine (FSM), making FSM-SADF a suitable tool to model this application.

We present the FSM-SADF model for LTE's baseband processing in two steps: first in this section, without modeling the control signals (shown in Figure 5), and later in Section IV, with the modeling of these controls signals.

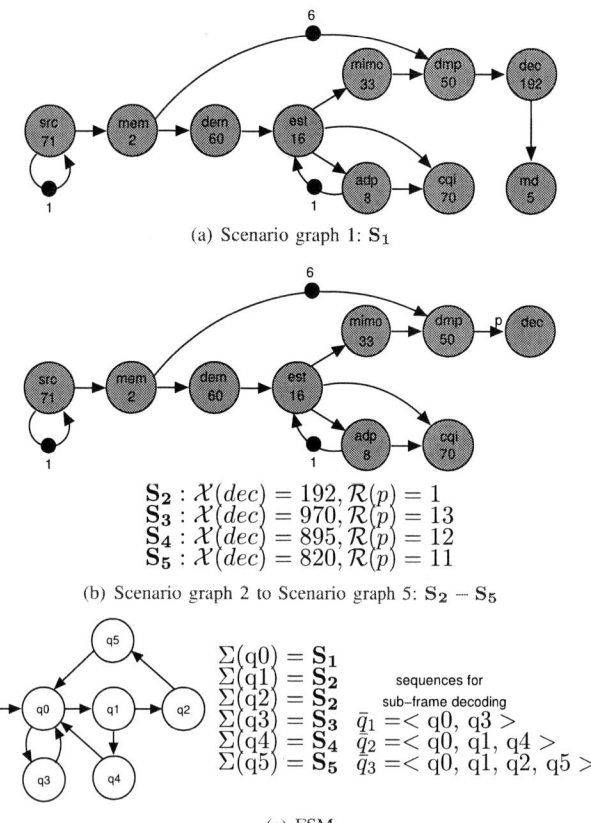

(a) Scenario graph 1: $\mathbf{S_1}$

$\mathbf{S_2} : \mathcal{X}(dec) = 192, \mathcal{R}(p) = 1$
$\mathbf{S_3} : \mathcal{X}(dec) = 970, \mathcal{R}(p) = 13$
$\mathbf{S_4} : \mathcal{X}(dec) = 895, \mathcal{R}(p) = 12$
$\mathbf{S_5} : \mathcal{X}(dec) = 820, \mathcal{R}(p) = 11$

(b) Scenario graph 2 to Scenario graph 5: $\mathbf{S_2} - \mathbf{S_5}$

$\Sigma(q0) = \mathbf{S_1}$
$\Sigma(q1) = \mathbf{S_2}$
$\Sigma(q2) = \mathbf{S_2}$ sequences for
$\Sigma(q3) = \mathbf{S_3}$ sub–frame decoding
$\Sigma(q4) = \mathbf{S_4}$ $\bar{q}_1 = <q0, q3>$
$\Sigma(q5) = \mathbf{S_5}$ $\bar{q}_2 = <q0, q1, q4>$
                  $\bar{q}_3 = <q0, q1, q2, q5>$

(c) FSM

Fig. 7. FSM-SADF model of LTE receiver

A compact representation of the five scenario graphs that correspond to each mode of operation is shown in Figure 7(a) and Figure 7(b). Except for those explicitly shown in Figure 7(b), all port rates are 1 and execution times are indicated by numbers written inside the actors.

The first scenario graph, $\mathbf{S_1}$, models the decoding of the first symbol, which has the control format channel (PCFICH) and part of the control channel (PDCCH). At the end of the execution of $\mathbf{S_1}$, the *mode detection (md)* actor determines the scenario sequence to decode the remaining 13 symbols. The three possible sequences are: 1) executing $\mathbf{S_3}$ to decode all the 13 symbols for the data channel (PDSCH), 2) executing $\mathbf{S_2}$ to decode the second symbol for the control channel (PDCCH), followed by $\mathbf{S_4}$ to decode the remaining 12 symbols for the data channel (PDSCH), and 3) executing $\mathbf{S_2}$ twice to decode the second and third symbols for the control channel (PDCCH), followed by $\mathbf{S_5}$ to decode the remaining 11 symbols for the data channel (PDSCH). The FSM in Figure 7(c) shows the three scenario sequences $\bar{q}_1, \bar{q}_2$ and $\bar{q}_3$ to decode one complete sub-frame.

The timing analysis of the FSM-SADF model can be carried out by executing these scenario sequences. However, before we present the timing analysis, we first need to discuss an important modeling aspect that is not captured by Figure 7. It is mentioned earlier that the type of scenario sequence for a given sub-frame is determined by actor *md* of $\mathbf{S_1}$. Hence, actor *md* should run to completion before scenario graphs $\mathbf{S_2}$ and $\mathbf{S_3}$ start execution. Scenario-independent actors in these graphs can, in fact, start ahead of the completion of actor *md*. The dashed edges in the directed task graph of Figure 5 show

actors that have dependency with actor *md*. The dashed edges represent data dependencies that exist across scenarios: from actor *md* of $\mathbf{S_1}$ to actors *mimo, dmp* and *dec* of $\mathbf{S_2}$ and $\mathbf{S_3}$. We refer to such types of data dependencies that exist between scenarios as *scenario dependencies*.

The presence of these scenario dependencies in the transitions $\delta_{q0q1} = (q0, q1)$ and $\delta_{q0q3} = (q0, q3)$ are not modeled by the FSM-SADF, shown in Figure 7. This has a serious consequence, as it may lead to invalid timing analysis. Section IV next presents a modeling technique for data dependencies that exist across scenario graphs.

IV. MODELING SCENARIO DEPENDENCIES

The temporal behavior of an FSM-SADF model is analyzed by executing the possible scenario sequences specified by the FSM. For a given sequence, the scenario graph of each state is executed for one complete iteration. Figure 8 illustrates an example of a sequence of iterations $\langle \cdots, k, k+1, \cdots \rangle$ of two scenarios. As shown in the figure, these iterations are possibly pipelined in time.

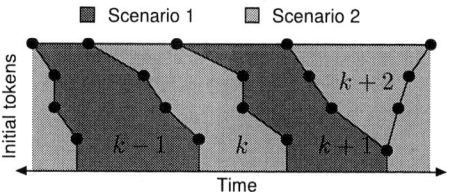

Fig. 8. Example of sequence of iterations

The end of iterations is marked by the production times of initial tokens, represented by the black dots in Figure 8. These initial tokens are the set of all initial tokens of the scenario graphs. Part of the initial tokens that are common between two scenario graphs (iterations) represent their data dependencies. This is because the starting times of actors that consume these common initial tokens is determined by the production times of the initial tokens in the previous iteration. Hence, to avoid earlier starting of actors, all data dependencies between iterations should be captured through common initial tokens between scenario graphs.

However, it is not possible to model data dependencies using common initial tokens unless both the source and destination actors of channels that carry the initial tokens exist in both scenario graphs. It requires, otherwise, channels that extend across the two scenario graphs, which FSM-SADF does not allow. This results in scenario dependencies, as discussed in Section III-B for the FSM-SADF model of Figure 7.

A scenario dependency is a data dependency from a source actor of a given scenario graph to a non-empty set of destination actors of a set of scenario graphs. We refer to the source actor as the *master actor* and the destination actors as the *slave actors*. A master actor and its slave actors can possibly belong to the same scenario graph. Slave actors must not fire before the master actor completes all of its firings of an iteration, i.e. as many firings as its entry in the repetition vector. Figure 9(a) illustrates the scenario dependency of the FSM-SADF model shown in Figure 7. Actors *mimo, dmp* and *dec* must receive configuration information from actor *md* before they start execution.

Therefore, to determine the correct firing times of slave actors in subsequent iterations, the completion time of the master actor must be recorded. This can be achieved by introducing a new actor for every master actor. We refer to this new actor as the *time-stamp* actor. The time-stamp actor

978-1-4577-0671-4/11 $26.00 © 2011 IEEE

is a SDFG actor with a single self-edge, as illustrated in Figure 9(b). The self-edge has exactly one initial token that carries the time-stamp of the completion of the master actor. The same time-stamp actor is introduced in all parent scenario graphs of the slave actors, as shown in Figure 9(b), so that the firing times of the slave actors cannot be earlier than the completion of the master actor in preceding iterations.

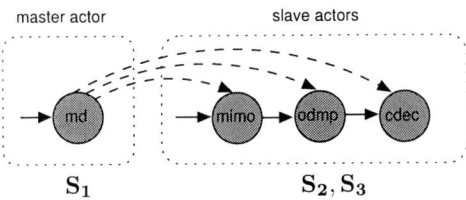

(a) Scenario dependency between master and slave actors

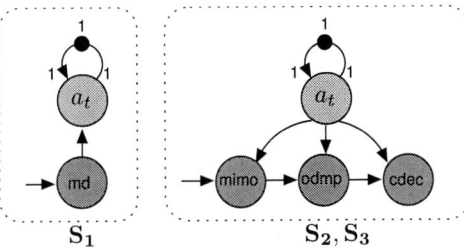

(b) time-stamp actor a_t where $\mathcal{X}(a_t) = 0$

Fig. 9. Modeling scenario dependency

Any number of scenario dependencies can be modeled with this technique. All sorts of scenario dependencies are also possible. For instance, a master actor of a scenario dependency can be a slave actor in another scenario dependency. An actor can also be a slave for multiple scenario dependencies. In addition, some of the slave actors of a scenario dependency can be on the same scenario graph as the master actor. In the last case, a channel that connects the time-stamp actor with a slave actor needs initial tokens to avoid deadlock.

A generalized formal presentation of the scenario dependency modeling technique is presented next, in Algorithm 1. The algorithm also covers the assignment of port rates ($\mathcal{R}$) and initial tokens ($\mathcal{I}$) of channels that connect time-stamp actors with master and slave actors.

Table I defines notation for a given FSM-SADF model $\mathbf{F} = (S, \mathbf{f})$. We use $\mathbb{P}$ to denote powerset.

TABLE I
NOTATION FOR AN FSM-SADF MODEL $\mathbf{F} = (S, \mathbf{f})$

Notation	Description
$\mathcal{A}(\mathbf{s})$	the set of actors of scenario graph $\mathbf{s} \in S$
$repVector(a)$	the repetition vector of an actor $a \in \mathcal{A}(\mathbf{s})$
$\mathcal{C}(\mathbf{s})$	the set of channels of scenario graph $\mathbf{s} \in S$
$srcPort(c)$	the source port of channel $c \in \mathcal{C}(\mathbf{s})$
$dstPort(c)$	the destination port of channel $c \in \mathcal{C}(\mathbf{s})$
$M_{\mathbf{F}}$	the set of all master actors in $\mathbf{F}$
$S_{\mathbf{F}}$	the set of all slave actors in $\mathbf{F}$
$T_{\mathbf{F}}$	the set of all time-stamp actors in $\mathbf{F}$
$O_{\mathbf{F}}$	the set of all ordinary actors in $\mathbf{F}$
$A_{\mathbf{F}}$	the set of all actors in $\mathbf{F}$, or equivalently $A_{\mathbf{F}} = M_{\mathbf{F}} \cup S_{\mathbf{F}} \cup T_{\mathbf{F}} \cup O_{\mathbf{F}}$
$\mathcal{G}(a_m)$	parent scenario graph of $a_m \in M_{\mathbf{F}}$
$\mathcal{S}(a_m)$	slave actors of a_m, $\mathcal{S} : M_{\mathbf{F}} \rightarrow \mathbb{P}(A_{\mathbf{F}})$
$\mathcal{SG}(a_m)$	the set of parent scenario graphs of slave actors of $a_m \in M_{\mathbf{F}}$, $\mathcal{SG} : M_{\mathbf{F}} \rightarrow \mathbb{P}(S)$

For every master actor $a_m \in M_{\mathbf{F}}$, there exists one unique time-stamp actor as defined in Definition 5.

Definition 5 (Time-stamp mapping). *Time-stamp mapping is a bijective function, denoted as $\mathcal{T} : M_{\mathbf{F}} \leftrightarrow T_{\mathbf{F}}$. The corresponding master actor of $a_t \in T_{\mathbf{F}}$ is given as $\mathcal{T}^{-1}(a_t)$.*

Each time-stamp actor $a_t \in T_{\mathbf{F}}$ represents a uni-directional data dependency from one scenario to a non-empty set of scenarios. The set of all such data dependencies constitutes the scenario dependency of the FSM-SADF model, as defined in Definition 6.

Definition 6 (Scenario dependency). *Scenario dependency of $\mathbf{F}$ is defined as a function $\mathcal{D} : T_{\mathbf{F}} \rightarrow S \times \mathbb{P}(S)$ such that for any $a_t \in T_{\mathbf{F}}, \mathcal{D}(a_t) = (\mathcal{G}(a_m), \mathcal{SG}(a_m))$, where $a_m = \mathcal{T}^{-1}(a_t)$.*

The Scenario dependency $\mathcal{D}$ defines all data dependencies between scenarios that cannot be modeled with existing common initial tokens between scenario graphs. To model these data dependencies, the original FSM-SADF model $\mathbf{F}$ is transformed into a new dependency-aware model, as given in Algorithm 1. The transformation involves two main steps. For every time-stamp actor $a_t \in T_{\mathbf{F}}$, (1) add the time-stamp actor in the parent scenario graph of the master actor ($\mathcal{G}(a_m)$), and (2) add the same time-stamp actor to all scenario graphs of slave actors of the master actor ($\mathcal{SG}(a_m)$).

Algorithm 1 Model scenario dependencies of an FSM-SADF model $\mathbf{F} = (S, \mathbf{f})$ from a given scenario dependency $\mathcal{D}$

1: ModelScenarioDependency($\mathcal{D}$)
2:   $T_{\mathbf{F}} :=$ domain($\mathcal{D}$) //set of time-stamp actors
3:   **for all** $a_t \in T_{\mathbf{F}}$ **do**
4:     $a_m := \mathcal{T}^{-1}(a_t)$ //get master actor
5:     $(\mathbf{s}, S_s) := (\mathcal{G}(a_m), \mathcal{SG}(a_m))$
6:     $\mathcal{A}(\mathbf{s}) := \mathcal{A}(\mathbf{s}) \cup \{a_t\}$ //add time-stamp actor to graph
7:     $c :=$ new channel $a_m \rightarrow a_t$ where $\mathcal{I}(c) := 0$
8:     $\mathcal{R}(srcPort(c)) := 1$
9:     $\mathcal{R}(dstPort(c)) := repVector(a_m)$
10:     $\mathcal{C}(\mathbf{s}) := \mathcal{C}(\mathbf{s}) \cup \{c\}$
11:     **for all** $\mathbf{s}_s \in S_s$ **do**
12:       $\mathcal{A}(\mathbf{s}_s) := \mathcal{A}(\mathbf{s}_s) \cup \{a_t\}$
13:       **for all** $a_s \in \mathcal{S}(a_m) \wedge a_s \in \mathcal{A}(\mathbf{s}_s)$ **do**
14:         $c :=$ new channel $a_t \rightarrow a_s$ where $\mathcal{I}(c) := 0$
15:         $\mathcal{R}(srcPort(c)) := repVector(a_s)$
16:         $\mathcal{R}(dstPort(c)) := 1$
17:         **if** $\mathbf{s}_s = \mathbf{s}$ **then**
18:           $\mathcal{I}(c) := repVector(a_s)$
19:         **end if**
20:         $\mathcal{C}(\mathbf{s}_s) := \mathcal{C}(\mathbf{s}_s) \cup \{c\}$
21:       **end for**
22:     **end for**
23: **end for**

Timing analysis of FSM-SADF whose scenario dependencies are not properly modeled, may give erroneous throughput results. This is shown next, in Section V, that discusses the throughput analysis of the LTE dataflow model.

V. THROUGHPUT ANALYSIS

Throughput requirements of SDRs, such as frame arrival rate, come from standards. We regard SDRs as hard real-time applications, as they must comply with standards and process frames at least at the rate of their throughput requirement. Thus, it is essential to analyze the worst-case throughput (WCT) of a SDR to ensure that the throughput requirement is satisfied in all operating conditions. The focus of this

978-1-4577-0671-4/11 $26.00 © 2011 IEEE

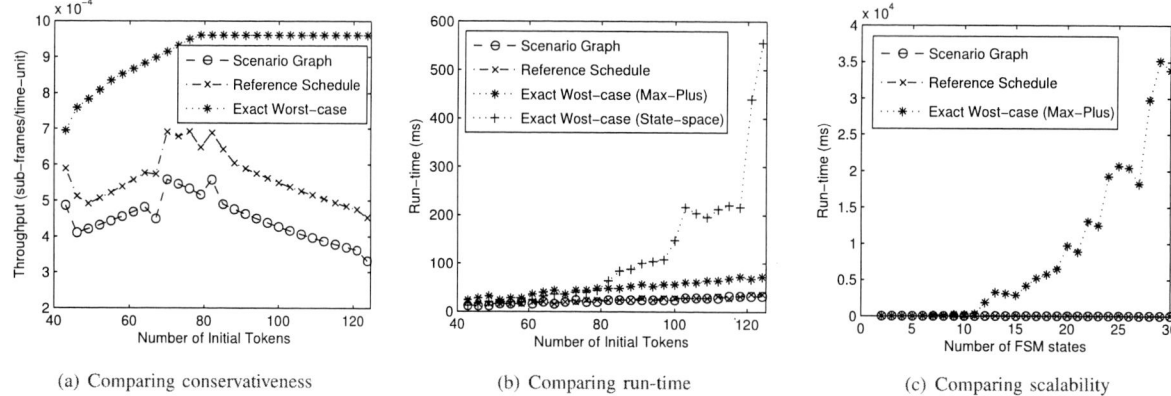

(a) Comparing conservativeness   (b) Comparing run-time   (c) Comparing scalability

Fig. 10. Conservativeness, run-time and scalability of various throughput analysis techniques

section is the WCT computation of FSM-SADF models. First we introduce four terminologies for our discussion: *validity*, *conservativeness*, *run-time* and *scalability*.

Throughput of an FSM-SADF is expressed in *number of iterations per time-unit*. The actual WCT of an FSM-SADF is the minimum throughput that may appear in the execution of the model. Any computed WCT is said to be *valid* if and only if it is less than or equal to the actual WCT. *Conservativeness* refers to how close a valid computed WCT is to the actual WCT. To obtain a less conservative (less pessimistic) result, a detailed analysis of the FSM-SADF model has to be carried out. However, this may cost us in terms of *run-time*, which is the length of time the computation takes to run to completion. Another important aspect is how the run-time of a given technique scales with increase in graph size, which we refer to as *scalability*. Run-time and scalability are important properties since design-space exploration (DSE) in scenario-based design flows [8] is a long iterative process that involves large graph sizes.

The throughput analysis of the FSM-SADF model of LTE's baseband processing, shown in Figure 7, is presented in this section. The analysis is carried out using the publicly available dataflow tool, SDF3 [17]. Table II shows computed WCTs according to five different computation techniques that are presented in detail in [10], [15] and [7]. Figure 10 shows the relative conservativeness, run-time and scalability of the scenario-based techniques (methods 2 to 5).

TABLE II
WCT COMPUTATION OF THE FSM-SADF MODEL OF FIGURE 7 WITH AND
WITHOUT SCENARIO DEPENDENCY MODELING (SDM)

($\times 10^{-4}$ sub-frames/time-unit)

Method	1	2	3	4	5
Name	Static SDFG	Scenario graph	Reference schedule	State-space	MaxPlus
Without SDM	2.6	9.7	10.4	14.5	14.5
With SDM	2.6	5.2	6.6	8.9	8.9

The scenario-based WCT computation techniques (methods 2 to 5) are based on analyzing the timing evolution of the time-stamp vectors of the constituent scenario graphs, as discussed in Section II-B. In addition, the scenario transitions specified by the FSM is also considered by some of them. Our main observations from Table II and Figure 10 are presented next.

A. Conservativeness

Methods 4 and 5 are based on computing the exact finishing time of iterations (time-stamp vectors). As a result, they give

the exact WCT of the FSM-SADF model. Therefore, they can be used as a reference for comparing the relative conservativeness of the throughput computation methods. Methods 2 and 3 have similarity as they are based on approximations on the finishing time of iterations [15]. The approximations are delay-period linear upper bounds on the time-stamp vectors, as discussed in Section II-B. However, as shown in Figure 10(a), method 3 is less conservative than method 2. This is because method 2 does not consider the scenario transitions specified by the FSM [15].

B. Validity

According to Table II, the actual WCT is 8.9×10^{-4} sub-frames/time-unit, since this result is based on the exact WCT computation techniques that also consider scenario dependencies. Table II, as expected, shows that timing analysis without modeling scenario dependencies may give invalid WCT, as the results of methods 2 to 5 are all greater than the actual WCT.

C. Pessimism of static dataflow

Method 1 is based on a single SDFG that has fixed execution times and port rates, assuming worst-case conditions. Such a static SDFG for the LTE baseband processing is given in Figure 6. The throughput of this SDFG is computed using the state-space exploration technique discussed in [10]. Table II shows that the static SDFG analysis gives a very pessimistic result. The other four scenario-based techniques improve this result by 2 to 3.4 times more (from 2.6 to 5.2 for method 2 upto 8.9 for methods 4 and 5).

D. Run-time and scalability

Methods 2 and 3 are based on delay-period approximations of the the time-stamp vectors. Hence, they are not significantly affected by neither lengthy time-stamp vectors nor large FSM states. As a result, they have a very low run-time that scale very well with increase in initial tokens and FSM state sizes, as shown in Figure 10(b) and 10(c).

Method 4, on the other hand, is based on a state-space exploration technique that is applied directly on the time-stamp vectors, considering all scenario sequences specified by the FSM. This enables it to give an exact WCT. However, its run-time exponentially grows with increasing initial tokens, as shown in Figure 10(b). As a result, it is also omitted from Figure 10(c).

Method 5 employs a compact representation of initial tokens using *throughput graphs* [7]. The throughput graph is a directed graph that has $V = |Q| \cdot |\gamma|$ vertices and $|V|^2$ edges, where $|Q|$ is the number of FSM states and $|\gamma|$ is the number of initial tokens in the FSM-SADF. The throughput computation on this graph gives an exact WCT but has an order of complexity $O(V^3)$. However, as shown in Figure 10(c), its run-time is in the order of tens of seconds that makes it practical for real-life applications, such as SDRs.

E. Summary

This section shows that there is a trade-off in conservativeness, run-time and scalability between different WCT computation techniques of FSM-SADF. Method 2 and 3 trade accuracy for lower run-time that make them useful for long and iterative DSE algorithms. Method 3 can be preferred to method 2, as it is less conservative and has an equivalent run-time and scalability. Method 4 and 5 give the exact WCT, at the cost of run-time. Method 4, however, is barely scalable and could be cumbersome to use it in scenario-based design flows. On other hand, the run-time of method 5 is in the order of tens of seconds that makes it practical for analyzing dynamic SDR applications.

VI. RELATED WORK

Synchronous Dataflow (SDF) [5] is the first dataflow-based model of computation to gain broad acceptance in DSP design tools due to its analyzablity as compared to other directed graph techniques such as computational graphs, petri nets and synchronous languages [3]. With SDF, it is possible to obtain a periodic schedule that can be implemented with bounded buffer size. However, the expressiveness of SDF is limited. and hence it cannot express applications' dynamism without over-allocation. For instance, it is shown in [8] that an SDF-based design-flow may lead to upto 66% in resource over-allocation, as compared to scenario-based techniques.

There are various proposed extensions of SDF to improve its expressiveness. Dynamic dataflow models such as Dynamic dataflow (DDF) [18], Boolean dataflow (BDF) [19],Integer dataflow (IDF) [20] and Core function dataflow (CFDF) [21] are expressively Turing-complete. However formal properties such as deadlock-freedom is an undecidable property for these dataflow models.

[22] and [23] suggest some dynamic dataflow models for software-defined radio (SDR) applications. However, they do not discuss the timing analysis of these dynamic dataflow models for predictable SDR design. The dataflow models proposed in these works include Scalable SDF (SSDF) [24], Parameterized Cyclo-static dataflows (PCSDF) [25], Cyclo-dynamic dataflow (CDDF) [26] and Mixed-mode vector-based dataflow (MMVBDF).These dataflow models, in one way or the other, can model applications' dynamism. However, one dataflow can be more expressive than the other, while still being analyzable. For example, Scenario-aware dataflow (SADF) [6] is more expressive than SSDF, PCSDF and MMVBDFs. SSDF allows integer multiples of token rates for an actor. SADF can model each rate with a separate scenario. PSDF and MMVBD are less expressive than SADF, since they require the parameterized consumption and production rates for a channel to be equal and does not support rates of 0 [6].

FSM-SADF is a version of SADF, where the transitions between scenarios are specified by a finite-state-machine (FSM). It has been shown in [8] that FSM-SADF can express dynamism in multimedia codecs. Efficient performance analysis techniques for FSM-SADF graphs are also presented in [15] and [7]. This paper extends these works by showing the applicability of FSM-SADF for SDRs.

VII. CONCLUSIONS

Software-defined radios (SDRs) are real-time streaming applications with throughput requirements. Dataflow modeling of SDRs for timing analysis is challenging due to their dynamically changing data processing workload. In this paper, we address the challenge of dataflow modeling of dynamic SDRs such that their timing behavior can be accurately analyzed to guarantee real-time requirements. The basis of our modeling approach is splitting the dynamic data processing behavior of a SDR into a group of static mode of operations. Each static mode of operation is then modeled by a Synchronous Dataflow (SDF), which we refer to as *scenario*. This work shows the applicability of this approach by modeling Long Term Evolution (LTE), which is a recent standard in cellular communications. Our results show that the worst-case throughput computation by scenario-based analysis is at least two times more accurate than a state-of-the-art SDF analysis technique. Our investigation also shows that existing timing analysis techniques of SDF scenarios have very low run-time that scales very well with increase in graph size. This makes SDF scenarios suitable in practice for modeling and analyzing SDRs as well as similar dynamic applications.

REFERENCES

[1] O. Gustafsson *et. al*, "Architectures for cognitive radio testbeds and demonstrators an overview," in *CROWNCOM*, 2010.

[2] F. Jondral, "Software-defined radio: basics and evolution to cognitive radio," *EURASIP Journal on Wireless Comm. and Netw.*, 2005.

[3] S. Sriram and S. S. Bhattacharyya, *Embedded Multiprocessors: Scheduling and Synchronization*, 2009.

[4] C. van Berkel, "Multi-core for Mobile Phones," in *DATE*, 2009.

[5] E. Lee and D. Messerschmitt, "Synchronous dataflow," *IEEE Proceedings*, 1987.

[6] B. Theelen *et. al*, "A scenario-aware data flow model for combined long-run average and worst-case performance analysis," in *MEMOCODE*, 2006.

[7] M. Geilen and S. Stuijk, "Worst-case performance analysis of synchronous dataflow scenarios," in *CODES/ISSS*, 2010.

[8] S. Stuijk *et al.*, "A predictable multiprocessor design flow for streaming applications with dynamic behaviour," in *DSD*, 2010.

[9] D. Martn-Sacristn, "On the way towards fourth-generation mobile: 3gpp lte and lte-advanced," *EURASIP Journal on Wireless Comm. and Netw.*, 2009.

[10] A. Ghamarian *et. al*, "Throughput analysis of synchronous data flow graphs," in *ACSD*, 2006.

[11] S. Stuijk, M. Geilen, and T. Basten, "Throughput-buffering trade-off exploration for cyclo-static and synchronous dataflow graphs," in *Computers, IEEE Transactions on*. USA: IEEE, 2008.

[12] C. Lee *et. al*, "A systematic design space exploration of MPSoC based on synchronous data flow specification," *Journal of Signal Processing Systems, Springer*, 2010.

[13] A. Bonfietti *et. al*, "An efficient and complete approach for throughput-maximal SDF allocation and scheduling on multi-core platforms," in *DATE*, 2010.

[14] F. Baccelli *et. al*, *Synchronization and Linearity: An Algebra for Discrete Event Systems*. John Wiley Sons, 1993.

[15] M. Geilen, "Synchronous dataflow scenarios," *ACM Transactions on Embedded Computing Systems*, 2011.

[16] "3GPP TS 36.211 V8.6.0: Physical Channels and Modulation," 2009.

[17] "SDF3 - Synchronous Dataflow for Free," http://www.es.ele.tue.nl/sdf3/.

[18] J. Buck, "A dynamic dataflow model suitable for efficient mixed hardware and software implementations of dsp applications," in *CODES*, 1994.

[19] E. Lee, "Consistency in dataflow graphs," *IEEE Transactions on Parallel and Distributed Systems*, 1991.

[20] J. Buck, "Static scheduling and code generation from dynamic dataflow graphs with integer-valued control streams," in *Asilomar*, 1994.

[21] W. Plishker *et. al*, "Functional dif for rapid prototyping," in *IEEE/IFIP*, 2008.

[22] H. Berg, C. Brunelli, and U. Lucking, "Analyzing models of computation for software defined radio applications," in *SOC*, 2008.

[23] C. Hsu *et. al*, "A mixed-mode vector-based dataflow approach for modeling and simulating lte physical layer," in *DAC*, 2010.

[24] S. Ritz *et. al*, "High level software synthesis for signal processing systems," in *ASAP*, 1992.

[25] B. Bhattacharya and S. Bhattacharyya, "Parameterized dataflow modeling for dsp systems," *IEEE Transactions on Signal Processing*, 2001.

[26] P. Wauters *et. al*, "Cyclo-dynamic dataflow," *IEEE Transactions on Signal Processing*, 1996.

978-1-4577-0671-4/11 $26.00 © 2011 IEEE

A Hybrid Model of Speculative Execution and Scout Threading for Auto-Memoization Processor

Tomoki IKEGAYA*, Ryosuke ODA*, Tatsuhiro YAMADA*,
Tomoaki TSUMURA*, Hiroshi MATSUO* and Yasuhiko NAKASHIMA†

*Nagoya Institute of Technology, Gokiso, Showa, Nagoya, Japan
Email: camp@matlab.nitech.ac.jp

†Nara Institute of Science and Technology, 8916-5, Takayama, Ikoma, Nara, Japan
Email: nakashim@is.naist.jp

Abstract—We have proposed an auto-memoization processor based on computation reuse, and merged it with speculative multi-threading based on value prediction into a parallel speculative execution. In the parallel speculative execution model, speculative cores do not work when the target instruction region is not suitable for computation reuse. This paper proposes a new parallel speculative execution model where the idle speculative cores execute scout threads for reducing cache miss penalties. The scout thread is based on value prediction, and can handle an instruction region which accesses the addresses with several strides. It also can reduce execution cycles by raising computation reuse ratio. The result of the experiment with SPEC CPU95 FP suite benchmarks shows that the new hybrid model of parallel speculative execution and scout threading improves the maximum speedup from 40.6% to 41.3%, and the average speedup from 15.0% to 19.1%.

I. INTRODUCTION

So far, various speed-up techniques for microprocessors have been proposed. The performance of microprocessors had been controlled by the gate latencies, and it had been relatively easy to speed-up microprocessors by transistor scaling. However, the interconnect delay has been going major, and it has become difficult to achieve speed-up only by higher clock frequency. Therefore, speed-up techniques based on ILP (Instruction-Level Parallelism), such as super-scalar or SIMD instruction sets, have been counted on.

Recently, multi-core processors equipped with two or more cores attract a great deal of attention. They are now in wide use from generic processors for PCs to embedded processors [1]. The UltraSPARC T2 [2] with eight cores, and the TILE64 [3] with 64 cores is available now, and many core processors such as the TILE-Gx processor [4] with 100 cores is planned to be shipped.

A program generally forms a poset, or a lattice. It has a length along time axis, and has a width (i.e. parallelism) orthogonal to time axis. Traditional speed-up techniques mentioned above are all based on some parallelism in different granularity. In other words, their approaches aim to increase performance by shrinking the width of the program lattice.

On the other hand, we have proposed an auto-memoization processor based on computation reuse [5], [6], [7]. In contrast

to traditional speed-up techniques for microprocessors, memoization or computation reuse tries to shrink the vertical length of the program lattice. As a speedup technique, memoization has no relation to parallelism of programs. It depends upon value locality, especially input values of functions or loops. Therefore, memoization has a potential for breaking through the stone wall, against which the speedup techniques based on ILP have come up.

We also have proposed a model called parallel speculative execution. It predicts the inputs for a reusable loop iteration, and additional shadow cores execute the iteration speculatively. The shadow cores register the results of the speculative executions onto the reuse table. If the value prediction for inputs was correct, the registered outputs can be reused by the main core and execution time will be reduced.

However, only the loops whose input values change monotonously can benefit from parallel speculative execution, and the speculative cores are idle without such loops. In this paper, we propose a new hybrid model of traditional parallel speculative execution and scout threading [8] for the auto-memoization processor. While being idle, the speculative cores execute scout threads and try to conceal memory access latencies without disturbing computation reuse.

II. RELATED WORKS

Studies for extracting ILPs with speculative executions based on value prediction have been proposed by Lipasti et. al. [9] or Wang et. al. [10] Many speculative multi-threading (SpMT) models also have been proposed. They have multiple processors or cores, and run threads speculatively using predicted value sets. In an SpMT model, a speculative thread will generally squashed when its input values are overwritten by main thread.

Roth et. al. [11] has proposed *register integration*. It is a mechanism for reusing the results of squashed instructions by writing back the past register mapping. It is shown that the model can provide performance improvements of up to 11.5%.

Some hybrid methods of computation reuse and value prediction have been studied. Wu et. al. [12] have proposed a speculative multi-threading supported by computation reuse.

In the model, the compiler identifies computation region for reuse or value prediction. At runtime, if a region cannot be reused, the processor predicts the outputs of the region, and speculatively execute its following instructions using the predicted values. Hence, if the value prediction fails, the speculative executions should be squashed, and it costs additional hardware and overhead for the squash.

Molina et. al. [13] have proposed a combination model of speculative thread and non-speculative thread. The execution results of speculative thread are stored into the FIFO called a *look ahead buffer*, and non-speculative thread picks up instructions from the FIFO. If current source operands and the stored operands are same, the non-speculative thread reuses the execution results and skips execution.

In contrast to them, the parallel speculative execution model we have proposed is a non-symmetric SpMT model based on the value prediction, and uses computation reuse technique. Our model has two advantages over [12]. The one is that there is no need to be assisted by compiler for computation reuse. The other is that there is no need to squash speculative executions. Molina's model [13] is similar to our model. However, our model can reuse some computation regions which require memory read as their inputs.

III. Research Background

In this section, we describe about an auto-memoization processor and a parallel speculative execution model as the background of our study.

A. Auto-Memoization Processor

Computation reuse is a well-known speed-up technique in the software field. It is storing the input sequences and the results of some computation regions, such as functions, for later reuse and avoiding recomputing them when the current input sequence matches one of the past input sequences. It is called **memoization** [14] to apply computation reuse to computation regions in programs.

Memoization is originally a programming technique for speed-up, and brings good results on expensive functions. However, it requires rewrite of target programs, and the traditional load-modules or binaries cannot benefit from memoization. Furthermore, the effectiveness of memoization is influenced much by programming styles. Rewriting programs using memoization occasionally makes the programs slower. Memoization costs a certain overhead because it is implemented by software.

On the other hand, the auto-memoization processor, which we have proposed, makes traditional load-modules faster without any software assist. There is no need to rewrite or recompile target programs. The processor detects functions and loop iterations as reusable regions dynamically, and memoizes them automatically.

Fig. 1 shows the two types of memoizable instruction regions. A region between a callee label and its associated `return` instruction is detected as a memoizable function. A region between a backward branch instruction and its

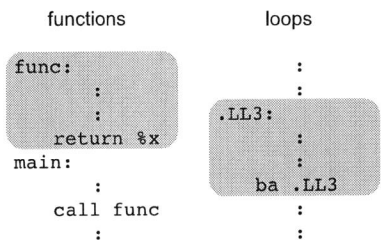

Fig. 1. Memoizable instruction regions.

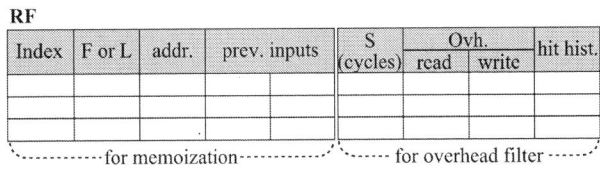

Fig. 2. Structure of MemoTbl

branch target is detected as a memoizable loop iteration. This processor detects these memoizable regions automatically and memoizes them.

The auto-memoization processor consists of the memoization engine, **MemoTbl** and **MemoBuf**. The MemoTbl is a set of tables for storing input/output sequences of past executed computation regions. The MemoBuf works as a write buffer for MemoTbl. The brief structure of MemoTbl is shown in Fig.2.

Entering to a memoizable region, the processor refers to MemoTbl and compares current input set with former input sets which are stored in MemoTbl. If the current input set matches with one of the stored input sets on MemoTbl, the memoization engine writes back the stored outputs, which are associated with the input set, to cache and registers. This omits the execution of the region and reduce the total execution cycles.

If the current input set does not match with any past input sets on MemoTbl, the processor stores the inputs and the outputs of the region into MemoBuf while executing the region. The input set consists of the register/memory values which are read in the region; the output set consists of the values which are written in the region and return value of function. Reaching the end of the region, the memoization engine stores the content of MemoBuf into MemoTbl for future reuse.

The MemoTbl consists of four tables. They are
RF: for start addresses of instruction regions.
RB: for input data sets of instruction regions.
RA: for input address sets of instruction regions.
W1: for output data sets of instruction regions.

978-1-4577-0671-4/11 $26.00 © 2011 IEEE   23

The RF, RA, and W1 are implemented with RAM. On the other hand, the RB is implemented with CAM (Content Addressable Memory) array, so that input matching can be done fast by associative search.

Each RF line corresponds to a reusable computation region. One RF line has two groups of fields, the one is for computation reuse and the other is for the overhead filter which will be explained later in III-C. The fields for computation reuse store whether the region is a function or a loop (*F or L*), the start address of the region (*addr.*), and previous two input sets for predicting next input set for parallel speculative execution (*prev. inputs*). The fields for overhead filter store the execution cycles of the region (*S*), its previous reuse overhead (*Ovh*), and its hit/miss pattern (*hit hist.*).

The brief execution mechanism of the auto-memoization processor is as follows. When the auto-memoization processor detects a function or a loop iteration, it first searches its start address through the RF table for deciding the inputs of the reusable region are stored or not. Then, the input matching for computation reuse starts.

The processor reads the value of program counter and registers, and searches their values from the RB. If one of the RB lines matches, the processor gets its index and reads RA using the index. The RA line has the address for the input of the region which should be tested next. Next, the processor gets input value from the cache or main memory using the address, searches the input value through the RB again, and repeats it. If all inputs of a reusable block have matched with one of the stored input set on the MemoTbl, the processor can get the output set from W1 by using the index for W1 (called 'W1 pointer') stored in the terminal RA entry. The detail of this execution mechanism is shown in [5], [6].

Meanwhile, accessing MemoTbl causes overhead inevitably. Through input matching, searching RB, referring RA, and reading caches cost a certain time. When input matching has succeeded, outputs of the reusable block should be written back from W1. This also costs some cycles. We call these two kinds of overheads '**reuse overheads.**'

B. Parallel Speculative Execution

As a matter of course, memoization can omit the execution of an instruction region only if the current input values for the region match completely with the input values which are used in former execution. Hence, a loop iteration whose inputs include its iterator variable never benefits from memoization.

Meanwhile, many of microprocessor companies are switching to multi-core designs today. There is a story going around that processors with hundreds of cores may be delivered in another decade [15]. But how we can use these many-core processors effectively is still under review between researchers.

Speculative multi-threading (SpMT) is an answer to this question, but it is not so easy to deal with cross-thread dependence violation and thread squash. We installed some SpMT cores called **SpC** (speculative cores) to our auto-memoization processor. These cores help the unsuitable regions for memoization mentioned above. Fig.3 shows the structure of the

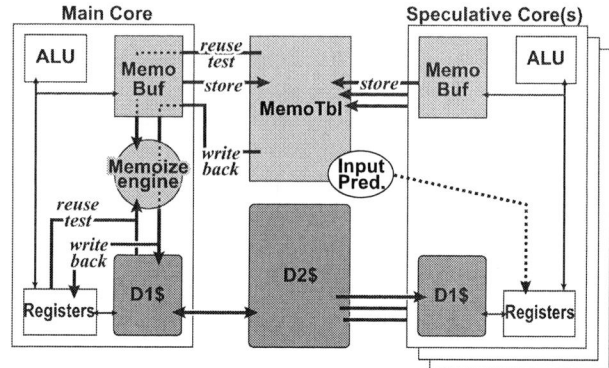

Fig. 3. Structure of a parallel speculative execution.

auto-memoization processor with three SpCs.

Each SpC has its own MemoBuf and a first-level data cache. The second-level data cache and MemoTbl is shared between all cores. While the main core executes a memoizable computation region, SpCs execute the same region using predicted inputs, and stores the results into MemoTbl. The inputs are predicted by stride prediction using the last two input sets stored in RF *prev. inputs* field. If the input prediction was correct, the main core can omit intended execution by reusing the result of SpC. Unlike as other SpMT methods, even if the input speculation proves to be incorrect later, the main core need not to pay a cost for any back-out management, it only fails reuse test and executes the region as usual. This extension can omit the execution of instruction regions whose inputs show monotonous increase/decrease. These SpCs not only omit some executions, but also work as a cache prefetch technique [16], [17].

C. Overhead Filter

For some reusable regions, reuse overhead may outweigh the eliminated execution cycles by reuse. This will go for some regions which have many input values to be tested, and all tiny regions. Hence, the auto-memoization processor should estimate the effect of reuse, and avoid memoizing unsuitable instruction regions. This can reduce useless input matching and contribute to good performance.

For the reusability estimation, we installed a small logic onto MemoTbl. This logic estimates how much cycles will be reduced by memoizing a block, and how much overhead will cost for its reuse. It is important how to decide which instruction region should be suitable for parallel speculative execution. When the results of speculative executions for an instruction region are frequently reused, the instruction region is supposed to be suitable. Shift registers, shown as *hit hist.* fields in Fig.2, are used for recording these reuse frequency. The reuse overhead of an instruction region can be calculated from these frequency values.

Assume that M represents the number of successful reuses about a certain region for recent T times tries ($0 \leq M \leq T$). The value of M can be gotten from *hit hist.* field in RF. With the execution cycles S of the region, which can be also got from RF, the reduced cycle can be represented as

$$M \cdot (S - Ovh^R - Ovh^W) \qquad (1)$$

where Ovh^R and Ovh^W represent search/writeback overheads for the region respectively.

Ovh^R also costs when input matching fails and reuse cannot be applied. This overhead can be calculated as follows.

$$(T - M) \cdot Ovh^R \qquad (2)$$

Here, if the *loss* (2) is larger than the *gain* (1), the computation region cannot be suitable for reuse. Now, we define the difference between (1) and (2) as *Gain* (3). An additional small logic calculates whether (3) goes positive or negative, and decides the suitability of computation regions.

$$Gain = M \cdot (S - Ovh^W) - T \cdot Ovh^R \qquad (3)$$

IV. SCOUT THREADING BY SPECULATIVE CORES

In this section, we will propose a new parallel speculative execution model, where the SpCs execute scout threads.

A. Outline

On the traditional auto-memoization processor, the overhead filter described in III-C stops registering input/output values of an instruction region, if applying computation reuse to the region will deteriorate the whole performance. The previous input values for value prediction, which are shown as *prev. inputs* in Fig. 2, will also not be updated for such worthless regions.

The *prev. inputs* for value prediction are initialized before the associated region is executed, and updated when the input values of the region are registered from MemoBuf to MemoTbl at the end of the region. The value predictor, which is shown as *Input Pred.* in Fig. 3, reads these previous input values, predicts next input values from them by stride prediction, and passes the predicted values to SpCs. Now, if the overhead filter prohibits registering input/output values onto MemoTbl, *prev. inputs* are not updated and the value predictor cannot generate correct input values. When value prediction fails, SpCs cannot issue speculative execution. We call this that SpCs are *idle*.

In this paper, we propose a new parallel speculative execution model, in which SpCs execute some scout threads for cache prefetching when they are idle. This hybrid model of parallel speculative execution and scout threading can conceal some memory access latencies of the main core and reduce execution cycles. At input-predictable instruction regions, SpCs issue parallel speculative execution. On the other hand, the input values are not predictable, the instruction region will not benefit from memoization, and SpCs execute scout threads.

Scout threading not only reduces some cache miss penalties but also can raise the hit-rate of computation reuse. The parallel speculative execution typically brings cache misses because it precedes the main core execution. The cache miss delays registering the result of parallel speculative executions

```
186 C
187      DO  600  I3 =2,N−1
188      DO  600  I2 =2,N−1
189      DO  600  I1 =2,N−1
190 600  R(I1 ,I2 ,I3 )=V(I1 ,I2 ,I3 )
191  >      −A(0 )*(U(I1 ,   I2 ,   I3   ))
192  >      −A(1 )*(U(I1 −1,I2 ,   I3   ) + U(I1 +1,I2 ,   I3   )
193  >          + U(I1 ,   I2 −1,I3   ) + U(I1 ,   I2 +1,I3   )
194  >          + U(I1 ,   I2 ,   I3 −1) + U(I1 ,   I2 ,   I3 +1))
195  >      −A(2 )*(U(I1 −1,I2 −1,I3   ) + U(I1 +1,I2 −1,I3   )
196  >          + U(I1 −1,I2 +1,I3   ) + U(I1 +1,I2 +1,I3   )
197  >          + U(I1 ,   I2 −1,I3 −1) + U(I1 ,   I2 +1,I3 −1)
198  >          + U(I1 ,   I2 −1,I3 +1) + U(I1 ,   I2 +1,I3 +1)
199  >          + U(I1 −1,I2 ,   I3 −1) + U(I1 −1,I2 ,   I3 +1)
200  >          + U(I1 +1,I2 ,   I3 −1) + U(I1 +1,I2 ,   I3 +1))
201  >      −A(3 )*(U(I1 −1,I2 −1,I3 −1) + U(I1 +1,I2 −1,I3 −1)
202  >      ·    + U(I1 −1,I2 +1,I3 −1) + U(I1 +1,I2 +1,I3 −1)
203  >          + U(I1 −1,I2 −1,I3 +1) + U(I1 +1,I2 −1,I3 +1)
204  >          + U(I1 −1,I2 +1,I3 +1) + U(I1 +1,I2 +1,I3 +1))
205 C
```

Fig. 4.   A part of 107.mgrid program code.

onto MemoTbl, and this leads to the low reuse hit-rate of *input-predictable* regions because the result registration by SpCs may be too late for reuse test by the main core. Now, consider that there is a loop A, which is not suitable for computation reuse, and A contains another loop B in it and the loop B is suitable for reuse. SpCs in the new hybrid model will execute scout threads for the outer loop A, and will issue parallel speculative execution for the inner loop B. This can reduce cache misses at the inner loop B and may raise the reuse hit-rate for the region B.

Fig. 4 shows a part of the program code of 107.mgrid from SPEC CPU95 FP benchmark suite. The code calculates $R = V - AU$, where R, V, U are 3-dimensional arrays. On the traditional parallel speculative execution model, the innermost loop at line 189 benefits from parallel speculative execution and computation reuse, but the outer two loops at line 187 and 188 do not, because they have too many input values. In this case, SpCs will execute scout threads for the outer loops, and it will lead to higher reuse hit-rate for the innermost loop, because the cache misses with first some iterations of the innermost loop will reduced by the scout threading.

B. Execution Model

For scout threading, SpCs should be able to execute the instruction region speculatively. Input values for the speculative execution are available if value prediction succeeds. Consequently, higher the hit-rate of value prediction is, more times SpCs can execute speculatively. As mentioned in III, value prediction depends on *prev inputs* fields in RF. However, on the traditional parallel speculative execution model, when the overhead filter detects that the *Gain* shown in (3) is negative, the processor stops not only applying computation reuse to the region, but also updating *prev inputs* fields.

On the new hybrid model, the processor does not stop updating *prev inputs* fields even if *Gain* is negative. This will increase the opportunities of speculative execution. However, the result of speculative execution should not to be stored onto MemoTbl, because *Gain* is negative and the result of the instruction region will not be suitable for reuse. Hence, the processor does not issue parallel speculative execution when *Gain* is negative, but executes a kind of scout thread. The scout thread is composed of *load* instructions in the instruction region. This can avoid storing useless entries onto MemoTbl, and leads to high cache hit-rate without disturbing computation reuse.

Scout thread includes only *load* instructions, and it will consume not so much energy. The targets of scout threading are the instruction regions, which are memoizable and will not benefit from computation reuse. Hence, running time of scout threads will be short and it will cause little additional energy consumption.

V. IMPLEMENTATION

This section describes an implementation of the new hybrid model.

A. Switching between Speculative Execution and Scout Threading

As mentioned in the previous section, *prev inputs* field in RF keeps being updated, even if *Gain* is negative and the instruction region is not suitable for computation reuse. Now, the processor needs to select whether SpCs should issue parallel speculative execution or execute scout threads for the region. This decision depends on whether the value of *Gain* is positive or negative. A 1-bit-width field is installed onto each RF line, and this field holds the previous result of $Gain <> 0$ for the associated instruction region. SpC dynamically select which should be issued, parallel speculative execution or scout threading, by checking the field.

Through parallel speculative execution, the processor needs to store input/output values onto MemoBuf and MemoTbl, and all units in SpCs should work. On the other hand, through scout threading, only units for prefetching should work. Consequently, SpCs' behavior should vary depending on whether parallel speculative execution or scout threading is now active. Therefore, 1-bit flag is installed to each SpC, and it manages the current execution mode of associated SpC.

B. How to Generate Scout Threads

For scout threading, each SpC should pick up *load* instructions from the instruction region, and execute only them. The auto-memoization processor is based on SPARC-V8, and it has fixed 32-bit width SPARC ISA, and the opcode has fixed 8-bit width. The SPARC ISA has many types of *load* instructions; five *load floating-point* instructions, three *load floating-point from alternate space* instructions, eight *load integer* instructions and eight *load integer from alternate space* instructions [18]. However, the opcodes of them are all in

```
              :
1c1d4:   sethi   %hi(0x1d000), %o1
1c1d8:   sll     %l1, 2, %l0
1c1dc:   inc     %l1
1c1e0:   ld      [ %o1 ], %f3
1c1e4:   fdivs   %f3, %f4, %f2
1c1e8:   fstod   %f2, %f2
1c1ec:   fadds   %f4, %f4, %f4
1c1f0:   std     %f2, [ %fp + −8 ]
1c1f4:   ldd     [ %fp + −8 ], %o2
1c1f8:   mov     %o3, %o1
1c1fc:   st      %f4, [ %fp + −116 ]
1c200:   call    1c05c
1c204:   mov     %o2, %o0
1c208:   cmp     %l1, 0x19
1c20c:   sethi   %hi(0x1d000), %o3
1c210:   fadds   %f0, %f0, %f0
1c214:   ld      [ %o3 + 8 ], %f2
1c218:   fdivs   %f2, %f0, %f0
1c21c:   st      %f0, [ %l2 + %l0 ]
1c220:   ble     1c1d4
              :
```

Fig. 5.   A part of FFT assembly code.

`11xxx0xx` format, and they can be easily distinguished from other instructions.

The processor recognizes all types of *load* instructions and *save/restore* instructions by instruction decoder, and issues them. In SPARC binaries, there is a *save* instruction at the beginning of a function, and a *restore* instruction at the end of a function. On the SPARC architectures, register windows are managed with *save/restore* instructions. To guarantee correct function call behavior, the new hybrid model is designed to issue not only *load*s but also *save* and *restore*.

On the traditional model, nest structures of loops are kept in MemoBuf and parallel speculative executions are issued for each nested loop. Fig. 5 shows a part of FFT benchmark program from Stanford benchmark suite. With this program, the loop from `1c1d4` to `1c220` will be applied parallel speculative execution. The function at `1c05c` is also applied parallel speculative execution because it is called at `1c200`.

On the other hand, through scout threading on the new hybrid model, function calls are ignored. For example in Fig. 5, the input values for the function at `1c05c` cannot be defined correctly without executing from `1c1d4` to `1c200`. Another case is that a branch instruction controls whether a function is called or not. Consequently, almost all instructions in an instruction region should be executed for correctly handling function calls inside the region. This will increase latencies of scout threads and power consumption by scout threading.

VI. PERFORMANCE EVALUATION

A. Simulation Environments

We have developed a single-issue SPARC-V8 processor simulator with auto-memoization structures and SpCs, which

TABLE I
SIMULATION PARAMETERS

MemoBuf	64 kBytes
MemoTbl CAM	128 kBytes
Comparison (register and CAM)	9 cycles/32Bytes
Comparison (Cache and CAM)	10 cycles/32Bytes
Write back (MemoTbl to Reg./Cache)	1 cycle/32Bytes
D1 cache	32 KBytes
line size	32 Bytes
ways	4 ways
latency	2 cycles
miss penalty	10 cycles
D2 cache	2 MBytes
line size	32 Bytes
ways	4 ways
latency	10 cycles
miss penalty	100 cycles
Register windows	4 sets
miss penalty	20 cycles/set

TABLE II
RATIO OF REDUCED D2 CACHE MISS PENALTIES

	(M) Traditional	(S) Proposed
102.swim	55.1%	96.9%
104.hydro2d	16.7%	56.4%
125.turb3d	12.4%	67.1%
146.wave5	13.0%	42.6%
Ave. of 4 programs	24.3%	65.8%

can issue both speculative execution and scout threading. This section discusses the performance of the new model proposed in this paper. The simulation parameters are shown in TABLE I. The cache structure and the instruction latencies are based on SPARC64 processors [19]. The on-chip CAM for RB in MemoTbl is modeled on MOSAID DC18288 [20]. The latencies of the CAM are defined on the assumption that the clock of the processor is 10-times faster than the CAM.

B. Results with SPEC CPU95 FP

We evaluated the hybrid model by parallel speculative execution and scout threading. Workloads are all benchmark programs in SPEC CPU95 FP suites and are executed with 'train' dataset. All benchmark programs are compiled by gcc version 3.0.2 with '-msupersparc -O2' options, and linked statically.

The evaluation results are shown in Fig. 6. We have evaluated following three models,

(M) No-memoization model
(P) Traditional model of parallel speculative execution
(S) Hybrid model parallel speculative execution and scout threading

and Fig. 6 shows the execution cycles of these models. Each bar is normalized to the number of executed cycles of (M) the model without memoization.

The legend in Fig. 6 shows the breakdown items of total cycles. They represent the executed instruction cycles ('**exec**'), the comparison overhead between CAM and the registers ('**test(r)**'), the comparison overhead between CAM and the caches ('**test(m)**'), the writeback overhead ('**write**'), the first-level and shared second-level data cache miss penalties ('**D\$1**', '**D\$2**'), and the register window miss penalty ('**window**') respectively.

As we can see in Fig. 6, (S) the new hybrid model is rewarded with good results. For 102.swim, 104.hydro2d, 125.turb3d and 146.wave5, *D\$1* increases and *D\$2* decreases. This means that the hit-rate of shared second-level data cache for the main core increases due to the scout threading by SpCs. The reduction of cache miss penalty cycles of the second-level

data cache for these four workloads are shown in TABLE II. On the other hand, *exec*, *test(r)*, *test(m)* and *write* for these four workloads do not differ from them of (P). This means that scout threading does not disturb computation reuse. The average of reduced cache miss penalty cycles for all SPEC95 FP workloads increases from 15.0% to 19.1%.

For 107.mgrid, 141.apsi and 145.fpppp, we cannot see the benefit of new hybrid model, because there occurred little cache misses. However, notice that *exec* is slightly reduced for 107.mgrid. This means that the hit-rate of computation reuse increased. As mentioned in IV-A, this should occur in some inner instruction regions of nested structures. We verified that the hit-rate of computation reuse for 107.mgrid increases 1.9% on the new hybrid model.

Now, let us roughly discuss the energy consumption of the new hybrid model. On the traditional model (P), the running time of SpCs is about 23% of whole execution cycles. On the new hybrid model (S), the running time for scout threads is about 26% besides it. However, for executing scout threads, not all units in SpC, but only clock and prefetching unit should work, and this will not lead to much increase of energy consumption. We estimated the energy consumption of the new hybrid model, and it is found that the increase is about 13% over the traditional model (P).

In conclusion, the performance of the new hybrid model (S) is better than the traditional model (P) as a whole. The model (S) improves the maximum speedup from 40.6% to 41.3%, and the average from 15.0% to 19.1%.

VII. CONCLUSIONS

In this paper, we have proposed a hybrid model of traditional parallel speculative execution and scout threading for auto-memoization processor. In the model, idle SpCs execute scout threads and conceal some of memory access latencies without disturbing computation reuse. This prefetching sometimes also raises the hit-rate of computation reuse.

Through an evaluation with SPEC CPU95 FP suite benchmark programs, it is found that the new hybrid model improves the maximum speedup ratio from 40.6% to 41.3%, and the average speedup ratio from 15.0% to 19.1%. The hit-rate of computation reuse also rises a little with 107.mgrid.

One of the our future works is merging this model with other low-overhead models such as [6] we had proposed.

ACKNOWLEDGMENT

This research was partially supported by the Kayamori Foundation of Informational Science Advancement.

978-1-4577-0671-4/11 $26.00 © 2011 IEEE

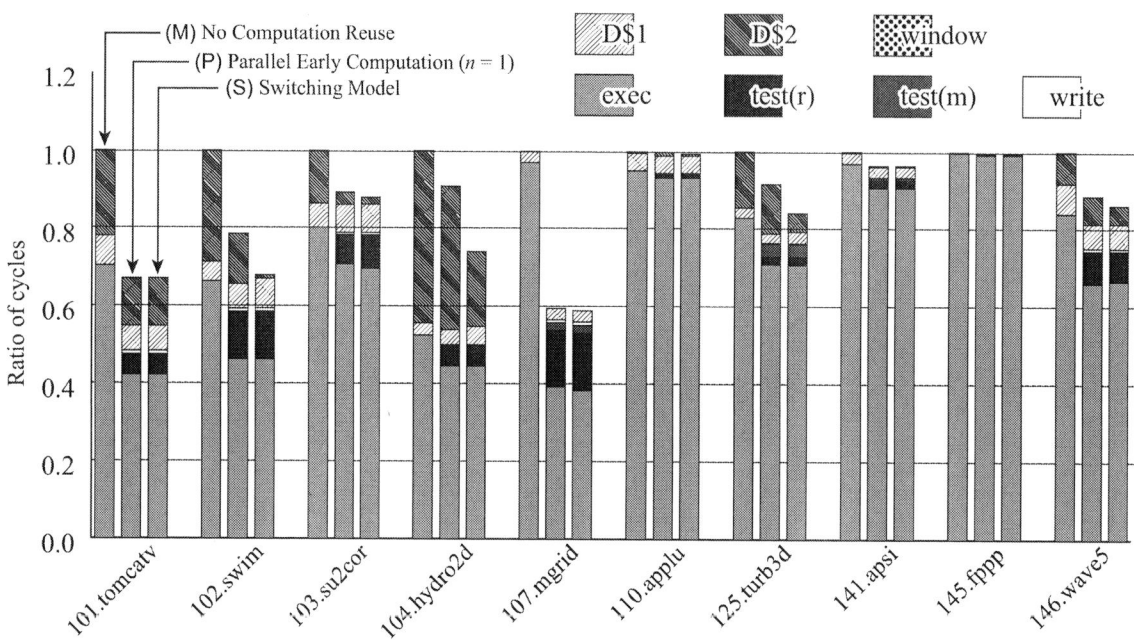

Fig. 6. Ratio of execution cycles (SPEC CPU95 FP).

REFERENCES

[1] ARM Ltd, *The ARM Cortex-A9 Processors*, Sep 2007.
[2] M. Shah, J. Barreh, J. Brooks, R. Golla, G. Grohoski, N. Gura, R. Hetherington, P. Jordan, M. Luttrell, C. Olson, B. Saha, D. Sheahan, L. Spracklen, and A. Wynn, "UltraSPARC T2: A Highly-Threaded, Power-Efficient, SPARC SOC," A-SSCC 2007, Tech. Rep., 2007.
[3] Tilera Corporation, *Product Brief: TILE64 Processor*, 2007.
[4] ——, *TILE-Gx Processor Family Product Brief*, 2009.
[5] T. Tsumura, I. Suzuki, Y. Ikeuchi, H. Matsuo, H. Nakashima, and Y. Nakashima, "Design and evaluation of an auto-memoization processor," in *Proc. Parallel and Distributed Computing and Networks*, Feb. 2007, pp. 245–250.
[6] Y. Kamiya, T. Tsumura, H. Matsuo, and Y. Nakashima, "A Speculative Technique for Auto-Memoization Processor with Multithreading," in *Proc. 10th Int'l. Conf. on Parallel and Distributed Computing, Applications and Technologies (PDCAT'09)*, Dec. 2009, pp. 160–166.
[7] T. Ikegaya, T. Tsumura, H. Matsuo, and Y. Nakashima, "A Speed-up Technique for an Auto-Memoization Processor by Collectively Reusing Continuous Iterations," in *Proc. 1st Int'l. Conf. on Networking and Computing (ICNC'10)*, Nov. 2010, pp. 63–70.
[8] S. Chaudhry, P. Caprioli, S. Yip, and M. Tremblay, "High-Performance Throughput Computing," *IEEE Micro*, vol. 25, pp. 32–45, May. 2005.
[9] M. H. Lipasti and J. P. Shen, "Exceeding the dataflow limit via value prediction," in *29th MICRO*, Dec. 1996, pp. 226–237.
[10] K. Wang and M. Franklin, "Highly accurate data value prediction using hybrid predictors," in *30th MICRO*, Dec. 1997, pp. 281–290.

[11] A. Roth and G. S. Sohi, "Register integration: A simple and efficient implementation of squash reuse," in *33rd MICRO*, Dec. 2000.
[12] Y. Wu, D. Chen, and J. Fang, "Better exploration of region-level value locality with integrated computation reuse and value prediction," in *28th ISCA*, 2001, pp. 98–108.
[13] C. Molina, A. González, and J. Tubella, "Trace-level speculative multi-threaded architecture," in *ICCD*, 2002.
[14] P. Norvig, *Paradigms of Artificial Intelligence Programming*. Morgan Kaufmann, 1992.
[15] S. Y. Borkar, P. Dubey, K. C. Kahn, D. J. Kuck, H. Mulder, S. S. Pawlowski, and J. R. Rattner, "Platform 2015: Intel processor and platform evolution for the next decade," Intel Corp., White Paper, 2005.
[16] J. A. Brown, H. Wang, G. Chrysos, P. H. Wang, and J. P. Shen, "Speculative precomputation on chip multiprocessors," in *Proc. of the 6th Workshop on Multithreaded Execution, Architecture, and Compilation (METAC-6)*, 2002.
[17] I. Ganusov and M. Burtscher, "Future execution: A hardware prefetching technique for chip multiprocessors," in *Proc. Int'l Conf. on Parallel Architectures and Compilation Techniques (PACT'05)*, 2005, pp. 350–360.
[18] D. L. Weaver and T. Germond, Eds., *The SPARC Architecture Manual Version 9*. Prentice-Hall, Inc., 1994.
[19] *SPARC64-III User's Guide*, HAL Computer Systems/Fujitsu, May 1998.
[20] MOSAID Technologies Inc., *Feature Sheet: MOSAID Class-IC DC18288*, 1st ed., Feb. 2003.

Customizable Datapath Integrated Lock Unit

Pekka Jääskeläinen, Erno Salminen, Otto Esko and Jarmo Takala
Tampere University of Technology
Department of Computer Systems
Tampere, Finland
Email: {pekka.jaaskelainen, erno.salminen, otto.esko, jarmo.takala}@tut.fi

Abstract— *Multicore Application-Specific Instruction-Set Processors (MCASIP) offer an interesting alternative for implementing parallel applications in MPSoCs. Flexible MCASIP architecture templates allow matching the instruction and task level parallelism provided by the processor to the requirements of the application at hand.*

The processing throughput provided by shared memory (SM) multicores is commonly limited by the SM bandwidth. Synchronizing the execution of multiple threads using lock variables residing in the SM further adds to the bottleneck. In this paper we present a technique to reduce the SM contention in the case of MCASIPs where application-specific hardware customization can be used. The proposed solution is to use customized Datapath Integrated Lock Units (DILU) that enable the implementation of light weight synchronization primitives which minimize SM traffic.

The paper presents an experiment with a 48-core MCASIP which shows that the SM impact of the proposed fast barrier based on DILU in comparison to a basic SM polling one is up to 64% smaller. The size of the DILU hardware is negligible.

I. INTRODUCTION

Multicore Application-Specific Instruction-Set Processors (MCASIP) extend the design space of single core ASIPs to multicores. The additional customization point of core count enables matching the task level parallelism provided by the processor to the requirements of the application at hand. However, improving the performance by adding more cores to the design is often hindered by the overheads due to the need to synchronize the execution of multiple threads running in multiple cores.

Locks, barriers, and semaphores are basic synchronization primitives used to orchestrate the execution of a multithreaded program. Lock variables typically reside in shared memory and atomic *Read-Modify-Write (RMW)* instructions are needed for manipulating them without corruption. A lock performs mutual exclusion to ensure critical sections that manipulate shared data structures are executed only by one thread at a time.

A common synchronization primitive heavily used in system code is the *spin lock*. It performs busy wait until the *lock variable* is marked "free" (usually by writing 0 to the variable). It can be implemented with a loop that "spins" until it manages to write 1 the target lock variable before other cores do. The atomicity of the lock acquiring operation can be implemented with RMW instructions in the instruction set.

Another common synchronization primitive used especially in massively parallel programs is the *barrier*. Barriers are used to synchronize the control flow of all threads co-operating in the execution of the multithreaded program. The semantics of the barrier is to wait at the barrier call site until all other threads have reached the barrier. After all threads have reached the barrier, the execution of the threads continues freely. Simplest barrier implementations use counter variables that are protected with locks. They count how many threads have reached the barrier and block the waiting threads until the counter reaches the total number of threads.

Unfortunately, the polling during spinning the lock variables causes spurious traffic to the shared memory hierarchy causing unnecessary slowdown to other threads. The need to reduce shared memory contention due to synchronization led us to design the proposed hardware lock unit. The novel feature is to expose a simple address lock book keeping hardware to the programmer by integrating it to the processor datapath. Using this *Datapath Integrated Lock Unit (DILU)* the programmer is free to implement various atomic operations in software using the lock variable book keeping functionality for mutual exclusion.

Exposing the lock unit has the benefit of avoiding shared memory accesses altogether in some synchronization scenarios. The flexibility of DILU enables tailoring both the software and the hardware for implementing the synchronization primitives to match the varying synchronization demands of MCASIP designs. The results with a 48-core MCASIP prove the benefits of the flexibility. With a shared memory heavy workload the proposed fast barrier alternative which consumes two lock registers per barrier reduces shared memory contention up to 64% in comparison to a simpler version that consumes only one lock register.

The rest of the paper is organized as follows. The next section studies the related work, which is followed by a description of the context of the proposed work. Section IV describes the proposed solution that is evaluated in Section V. Finally, the paper is concluded in the last section.

II. RELATED WORK

Techniques to reduce overheads of software based synchronization implementations have been studied widely. For example, *Adaptive Backoff Techniques* use simple heuristics to compute a time to wait before polling the barrier variables again to reduce the traffic [1], [2], [3], [4]. These approaches usually rely on a cache coherent memory hierarchy to enable fast spinning on a local cached copy of the lock variable. However, cache coherence hardware brings additional chip area costs to the design which are avoided in embedded multicores with explicitly accessed local memories.

Another approach is to use dedicated synchronization hardware. One of the earliest works is presented by *Beckmann* and *Polychronopoulos* in [5] that supports barriers by means of a barrier register hardware. Each 1-bit register denotes whether each processor has reached the barrier or not. However, their work relies fully on hardware, while our proposal goes a step towards the software side, thus adding more flexibility.

The *Synchronization-operation Buffer (SB)* [6] reduces the spinning overheads by performing the polling independently from the processor core using dedicated hardware unit in the memory block. The unit sends notifications to the processor after a memory location changes its content to a desired value. Like our method, also SB avoids the need for coherent caches. However, SB monitors the shared memory bus for updates to the interesting variables, in contrast to DILU which separates itself completely from the shared memory.

Distributed Synchronization Controller (DSC) [7] is an approach close to ours. Both consume two lock registers per barrier. The main difference is that in DSC the monitoring synchronization traffic and updating the synchronization registers happens in hardware while DILU generalizes the concept of lock registers and pushes the main synchronization implementation logic to software side. Similar ideas are used in the *test-and-set registers* of *Intel's Single-chip Cloud Computer (SCC)* [8]. However, in case of SCC, the lock registers are mapped to memory and their number is limited to one per core which reduces the options for fast synchronization primitive implementation.

DILU resembles SCC, DSC and SB in their idea of isolating the synchronization logic to an independent hardware block. The distinctive feature of DILU is that it makes the hardware as simple as possible and exposes the lock register manipulation operations to the instruction set of the processor. DILU does not use a shared bus for monitoring or communicating the synchronization activities, but an arbitrated access to a set of synchronization registers. Moving complexity from hardware to software provides more flexibility for implementing the synchronization software library.

III. CONTEXT OF THE WORK

The experiments for this paper have been conducted using an MCASIP design environment called TTA-based Co-design Environment Multicore (TCEMC) [9]. TCEMC allows the co-design of application-specific multicores based on a customizable single core template. Using TCEMC it is easy to experiment with custom instructions such as the lock unit instructions proposed in this paper and to produce multicores with an arbitrary number of cores.

The overview picture of the TCEMC MCASIP architecture template is shown in Fig. 1. The template is homogeneous in its single core instruction set architecture. However, the memory architecture is optimized for distributed execution of threads using fast core local memories. As shown in Fig. 1, each core has access to a local memory and the shared

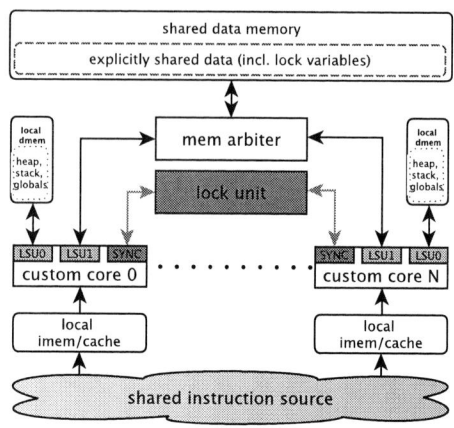

Fig. 1: TCEMC MCASIP template, the Local Default Data Memory (LDDM) version. In LDDM each tailored homogeneous core (number of cores can be customized) uses a private disjoint local memory for thread stacks and heaps. Shared memory is strictly for shared data which includes the possible synchronization variables. The proposed lock unit is accessed via the SYNC function units integrated to each core. LSU stands for *Load-Store Unit*.

memory. The address spaces of the two memories are disjoint and accessed with separate load-store units (and instructions). This configuration allows the stack, the heap and the global data of each thread to reside in the fast local memory while thread scheduler book keeping data and shared application data can be stored in the slower shared memory. The programs are either hand-optimized or compiler-optimized to explicitly use the local memories for speeding up the execution. In other words, caches are omitted from the microarchitecture and the shared memory accesses are assumed to be always very slow. This enables implementations with tens of cores with minimal complexity increase to the memory system.

Fig. 1 also includes the proposed lock unit which can be used to reduce shared memory accesses due to synchronization. The lock unit is accessible from each core's datapath using a *SYNC* function unit.

IV. DATAPATH INTEGRATED LOCK UNIT

The key concept in the proposed lock unit is datapath integration. In practice it means that the basic operations to handle locks are presented to the programmer as processor instructions. The goal in the hardware design was to produce a simple enough lock unit that can be easily generated automatically according to the needs of the MCASIP design at hand while providing enough flexibility to support various synchronization primitive implementations in software.

A. Lock Unit Hardware Design

The *Lock Unit (LU)* hardware consists of a customizable sized *lock register* file. The number of registers, connected cores and address bits are configurable. Each lock register stores the status of a shared memory address currently being locked. Unlocked shared memory addresses do not consume

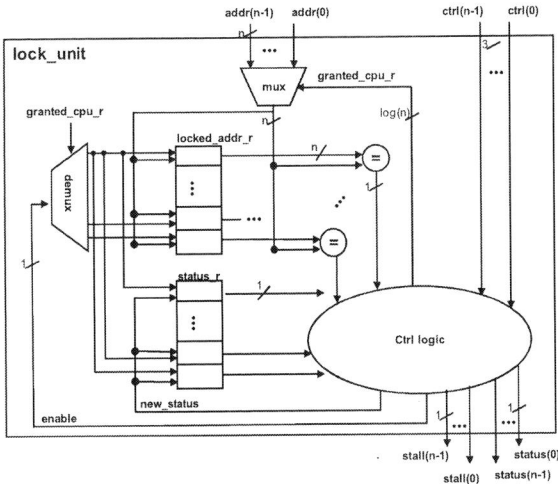

Fig. 2: The hardware design of the lock unit.

```
// spinlock_fast

lock_lu(lock_var* A) {
    while (!TRY_LOCK(A));
}

wait_lu(lock_var* A) {
    while (!READ_LOCK(A));
}
```

```
// spinlock_basic:

lock_var global_lock;

spin_lock(lock_var* A) {
  try_lock:
      lock_lu(&global_lock);

      if (*A == 1) {
          unlock_lu(&global_lock);
          goto try_lock;
      } else {
          *A = 1;
          unlock_lu(&global_lock);
      }
}
```

Fig. 3: Two spin lock implementations using the lock unit instructions. The *fast* version uses the lock unit operations solely while the basic version uses a single global lock register to guard accesses to all lock variables. *wait_lu()* will be used in a later barrier example to implement spin waiting on a lock register without lock acquisition.

any registers. LU ensures there is at most one lock register reserved per address at a time.

The hardware design of the LU is shown in Fig. 2. The inputs from the cores are shown on the top and outputs on the bottom. Each core provides an address and a 3-bit command. LU can stall the core until it can be served, and after that, LU gives the status of the completed operation.

The current implementation of the LU needs two cycles for each operation: register update and status reporting. During the status cycle it uses round-robin to arbitrate which core gets the next access to the lock register file.

Most of the chip area of the LU hardware is consumed by the register file and the comparators that find the slot to be reserved or released. In addition, the comparators are used to prevent double-locking of an address. The lock register file is split to two blocks in the picture: locked address and associated status bit. The lock status(i) is 1 in case the address register (i) contains a valid locked address and 0 otherwise. The control logic is rather small. It controls the stalling of the cores and produces the lock status based on the comparator outputs and the lock status bit.

The HW unit supports 4 basic operations: *read*, *lock*, *unlock*, and *wait until unlocked*. *Read* and *unlock* operations are non-blocking, while the *wait until unlocked* and *lock* can be implemented in both ways. The instruction set presented in this work utilizes only the non-blocking operations.

B. Synchronization Software

Three instructions are used to access the lock unit. All of them are non-blocking and take a memory address of a shared memory lock variable as an operand. The instructions use the address as a key to query the lock status, to attempt to acquire a lock, and to free a lock. Using these lock instructions it is possible to implement variations of synchronization functions that use a varying number of lock registers in the lock unit with the tradeoff of required shared memory accesses.

The instruction set in the *SYNC* function unit included in all cores is as follows:

got_lock := TRY_LOCK A	Tries to acquire a lock at address A.
UNLOCK A	Unlocks the lock at address A.
status := READ_LOCK A	Reads the lock status of the address A.

Two alternative implementations for both the spin lock and the barrier were implemented for this paper: a basic one that reduces lock register usage and one that minimizes shared memory accesses by relying more on the lock unit registers.

The spin lock implementations are shown in Fig. 3. *spin-lock_basic* uses a single global lock unit register to guard accesses to all lock variables residing in the shared memory. First, function *lock_lu()* is called and it blocks until the global lock is acquired. If the actual lock variable contained in shared memory address A is locked (contains value 1), the global lock is unlocked and the lock acquisition is retried by jumping back. This consists the "spin loop" which is retried until the lock variable is 0 and can thus be locked by the core. Depending on the guarded critical section length and a potential "spin backoff algorithm" used, this version generates heavy traffic to the shared memory bus during its spin wait.

On the other extreme, *spinlock_fast* in *lock_lu()* assumes that lock information can be stored fully into the lock unit registers, thus it does not access shared memory at all during mutual exclusion. All spinning is done by using the instructions of the lock unit. Thus, the number of these lock primitives that can be "alive" at the same time in a program is limited by the number of lock registers in the DILU. Another restriction is that the program should not rely on the shared memory value of the lock as that value is not updated by the spin lock implementation at all.

The programmer or the compiler is responsible for deciding which synchronization function version to use in which occasion. A middle ground implementation between the *basic* and *fast* that would reduce contention on a single global lock register would be to use two or more lock registers that each guard an even share of lock variable memory addresses.

```
barrier_fast(barrier_t* b) {
  int executing;

  lock_lu(&b->count_l);
  b->running -= 1
  executing = b->running;

  if (executing == -1) {
  /* init barrier */
      executing =
          b->running =
              b->total -1;
      lock_lu(&b->barrier_l);
  }
  unlock_lu(&b->count_l);
  if (executing == 0)
      unlock_lu(&b->barrier_l);
  else /* spin wait */
      wait_lu(&b->barrier_l);
}
```

```
lock_var global_lock;

barrier_basic(barrier_t* b) {

  int executing;

  lock_lu(&global_lock);
  b->running -= 1
  executing = b->running;

  if (executing == -1) {
  /* init barrier */
      b->running =
          b->total - 1;
  }
  unlock_lu(&global_lock);

  /* spin wait */
  while (b->running > 0) {}
}
```

```
for (int round = 0; round < ROUNDS; ++round) {
  if (core_id == 0) {
      for (int delay = 0;
          delay < smAccesses;
          ++delay, ++shared_value) {}
  }
  barrier(&b);
}
```

Fig. 5: The microbenchmark that "stress tests" the shared memory overheads of the barrier alternatives.

Fig. 4: Two barrier implementations using the lock unit instructions.

TABLE I: Required lock registers and shared memory read (r) and write (w) accesses for the alternative lock unit based synchronization implementations. The shared memory accesses are per thread participating in the synchronization. C is a variable depending on the critical section length or the thread imbalance in case of the barrier. L is the number of locks or barriers in use at the same time during the program execution.

| lock style | # of shared memory accesses | | | Required # of lock regs |
	acquire	release	spin wait	# of lock regs
basic	$r + w$	w	$C \times r$	≥ 1
fast	0	0	0	$1 \times L$
barrier	init	reach	wait others	# of lock regs
basic	w	$r + w$	$C \times r$	≥ 1
fast	w	$r + w$	0	$2 \times L$

The barrier implementation alternatives are shown in Fig. 4. *barrier_basic* uses a counter variable in shared memory. The counter is used to count how many threads are still to reach the barrier. Updates to it are protected with a single global lock register. After acquiring the lock, a thread decrements the counter to denote that it has arrived. If it was the first one, the counter goes to -1 and the barrier must be set up. After releasing the global lock, it spin waits until the counter goes to zero.

The fast version, *barrier_fast*, uses a counter variable similarly, but consumes two lock unit registers per barrier, named *count_l* and *barrier_l*. The first lock guards the counter variable updates and the latter records the whole barrier status. The first steps are similar to the basic algorithm except the barrier initialization locks also the *barrier_l*. It will be unlocked by the last thread reaching the barrier, i.e. when the count of threads still to reach the barrier goes to 0. This version accesses the shared memory when each thread reaches the barrier the first time (read, decrement, write). However, after that it spins on the second lock register by calling *wait_lu()* which minimizes the shared memory traffic while waiting for the other threads.

Table I summarizes the shared memory and lock unit register costs of the alternative implementations. Locks can be implemented without shared memory access at all. In both cases, the biggest difference is during spinning.

V. EVALUATION

The effect of using the proposed alternatives to the final performance of the application depends on various factors. First, it is affected by the synchronization operation per program operation ratio of the program. Furthermore, the shared memory pressure of the program is one important factor. In case the program performs frequent accesses to the shared memory, it is hindered more by the avoidable traffic caused by the synchronization.

A. MCASIP Hardware

For the evaluation of the proposed lock unit we designed a 48-core MCASIP using TCEMC. It should be noted that the results should be reproducible with any multicore architecture as DILU is designed to be generic enough to be integrated to any multicore's datapath.

All the cores in the designed MCASIP use the DILU for synchronization operations and have a relatively simple single core architecture with an integer ALU, an integer multiplier and a register file with 16 32-bit registers. The memory configuration of the MCASIP was as shown in the Figure of Section III. That is, each core had its own fast private data memory large enough for local stack and data in addition to an arbitrated uncached shared memory. The shared memory access time is from 4 to 96 cycles, depending on how many cores are competing for access.

The design was synthesized to an Altera Stratix II FPGA with a clock rate of 50 MHz. As expected, the additional area overhead of DILU was very low. DILU with four lock registers consumed about 1% (925 ALM) of the total logic utilization of the multicore (72 kALM).

B. Benchmark program

The scalability of the proposed barrier alternatives was measured by implementing a synthetic microbenchmark that performs a tight barrier synchronized loop and executing it in the MCASIP programmed to the FPGA. The benchmark loop is shown in Fig. 5.

The benchmark represents a "stress test" where the computation to synchronization ratio is extremely low and where the progress is heavily limited by concurrent shared memory accesses from different cores. Artificial shared memory traffic was generated by adding an update to a counter residing in shared memory to the first core. This forces the other cores to wait that time in the barrier spin wait loop. While being a

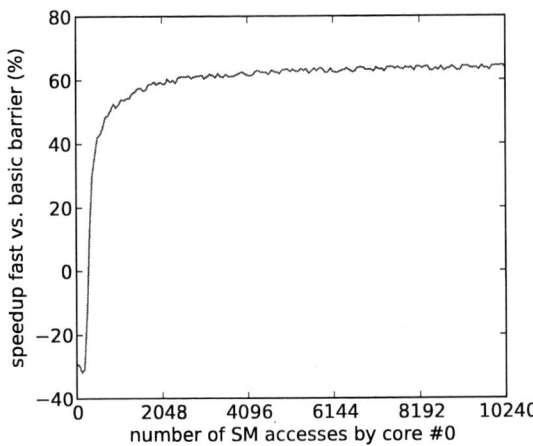

Fig. 6: The speedups obtained by using the *fast* barrier in comparison to the *basic* barrier version. The run times were measured as wall clock time with the FPGA implementation of the 48-core MCASIP.

synthetic example, the case represents the worst case of *thread imbalance* where one shared memory heavy thread takes more time to complete than the others and the completed threads cause it to slow down due to the SM traffic from barrier spin waiting.

In this benchmark, the number of threads equals the number of cores, thus the context switch overheads were ruled out. If there were more threads per core, the barrier call would induce a thread context switch to allow the waiting threads to reach the barrier. This would reduce the total "unsuccesful" spin waiting for the time of each context switch, possibly reducing SM noise. However, the context switch time and its effect to the performance depends at least on a) the size of the thread context to store/restore b) the location of the context data (SM or a local storage) c) the thread scheduler overhead d) if a pre-emptive scheduler is used, the number of pre-emptions, etc. Therefore, we decided to simplify this benchmark to emphasize the effect of the synchronization overheads alone.

The results illustrated in Fig. 6 show that the speedup from using the fast barrier that eliminates shared memory polling is drastic. The speedup increases up to about 3000 shared memory accesses after which it stabilizes to around 64%. For less than 320 accesses the *basic* barrier is faster as the additional barrier software complexity of the *fast* barrier dominates the shared memory access reduction benefits.

VI. CONCLUSION

A flexible *Datapath Integrated Lock Unit (DILU)* was proposed. DILU provides a middle ground solution between hardware based and software based synchronization implementations. The proposed hardware unit is simple enough to be customized according to the synchronization needs of the application at hand by varying the number of lock registers in the unit. Therefore, it is suitable to be used as a building block in co-design of MCASIPs. The unit does not require a coherent fast local memory to implement very low overhead spin lock and barrier synchronization, making it useful for multicore designs with simplified memory hierarchies.

The flexibility and scalability of the approach was shown with a 48-core MCASIP design. The results showed that in case of workloads with high shared memory access demands, the faster of the proposed barrier alternatives places about 60% less stress to the shared memory in comparison to the basic version with shared memory spinning.

In the future we plan to study compiler techniques to automatically optimize the usage of the different synchronization implementation alternatives and to further optimize the hardware implementation of the DILU.

ACKNOWLEDGMENT

This work has been supported by Academy of Finland. The authors wish to thank Mr. Lasse Lehtonen for his help in verifying and synthesizing the first implementation of the lock unit.

REFERENCES

[1] J. M. Mellor-Crummey and M. L. Scott, "Algorithms for scalable synchronization on shared-memory multiprocessors," *ACM Trans. Comput. Syst.*, vol. 9, pp. 21–65, February 1991. [Online]. Available: http://doi.acm.org/10.1145/103727.103729

[2] D. S. Nikolopoulos and T. S. Papatheodorou, "The architectural and operating system implications on the performance of synchronization on ccNUMA multiprocessors," *Int. J. Parallel Program.*, vol. 29, pp. 249–282, June 2001. [Online]. Available: http://portal.acm.org/citation.cfm?id=608719.608775

[3] T. Anderson, "The performance of spin lock alternatives for shared-memory multiprocessors," *Parallel and Distributed Systems, IEEE Transactions on*, vol. 1, no. 1, pp. 6 –16, jan 1990.

[4] A. Agarwal and M. Cherian, "Adaptive backoff synchronization techniques," in *Proceedings of the 16th annual international symposium on Computer architecture*, ser. ISCA '89. New York, NY, USA: ACM, 1989, pp. 396–406. [Online]. Available: http://doi.acm.org/10.1145/74925.74970

[5] C. J. Beckmann and C. D. Polychronopoulos, "Fast barrier synchronization hardware," in *Proceedings of the 1990 ACM/IEEE conference on Supercomputing*, ser. Supercomputing '90. Los Alamitos, CA, USA: IEEE Computer Society Press, 1990, pp. 180–189. [Online]. Available: http://portal.acm.org/citation.cfm?id=110382.110433

[6] M. Monchiero, G. Palermo, C. Silvano, and O. Villa, "Efficient synchronization for embedded on-chip multiprocessors," *IEEE Trans. Very Large Scale Integr. Syst.*, vol. 14, pp. 1049–1062, October 2006. [Online]. Available: http://dx.doi.org/10.1109/TVLSI.2006.884147

[7] C. Yu and P. Petrov, "Low-cost and energy-efficient distributed synchronization for embedded multiprocessors," *Very Large Scale Integration (VLSI) Systems, IEEE Transactions on*, vol. 18, no. 8, pp. 1257 –1261, aug. 2010.

[8] T. G. Mattson, M. Riepen, T. Lehnig, P. Brett, W. Haas, P. Kennedy, J. Howard, S. Vangal, N. Borkar, G. Ruhl, and S. Dighe, "The 48-core scc processor: the programmer's view," in *Proceedings of the 2010 ACM/IEEE International Conference for High Performance Computing, Networking, Storage and Analysis*, ser. SC '10. Washington, DC, USA: IEEE Computer Society, 2010, pp. 1–11. [Online]. Available: http://dx.doi.org/10.1109/SC.2010.53

[9] P. Jääskeläinen, E. Salminen, C. Sánchez de La Lama, J. Takala, and J. Matrinez, "TCEMC: A co-design flow for application-specific multicores," in *2011 International Conference on Embedded Computer Systems: Architectures, Modeling and Simulation 2011 (SAMOS XI)*, July 2011.

978-1-4577-0671-4/11 $26.00 © 2011 IEEE

Exploring Instruction Caching Strategies for Tightly-Coupled Shared-Memory Clusters

Daniele Bortolotti, Francesco Paterna, Christian Pinto,
Andrea Marongiu, Martino Ruggiero and Luca Benini

Dipartimento di Elettronica, Informatica e Sistemistica (DEIS) - University of Bologna
Viale Risorgimento, 2 - 40135 Bologna, Italy

{daniele.bortolotti, francesco.paterna, christian.pinto,
a.marongiu, martino.ruggiero, luca.benini}@unibo.it

Abstract—Several Chip-Multiprocessor designs today leverage tightly-coupled computing *clusters* as a building block. These clusters consist of a fairly large number N of simple cores, featuring fast communication through a shared multibanked L1 data memory and ≈ 1 Instruction-Per-Cycle (IPC) per core. Thus, aggregated I-fetch bandwidth approaches $f * N$, where f is the cluster clock frequency. An effective instruction cache architecture is key to support this I-fetch bandwidth. In this paper we compare two main architectures for instruction caching targeting tightly coupled CMP clusters: (i) private instruction caches per core and (ii) shared instruction cache per cluster. We developed a cycle-accurate model of the tightly coupled cluster with several configurable architectural parameters for exploration, plus a programming environment targeted at efficient data-parallel computing. We conduct an in-depth study of the two architectural templates based on the use of both synthetic microbenchmarks and real program workloads. Our results provide useful insights and guidelines for designers.

I. INTRODUCTION

To keep the pace of Moore's law, several Chip-Multiprocessors (CMP) platforms are embracing the many-core paradigm, where a large number of simple cores are integrated onto the same die. Current examples of many-cores include GP-GPUs such as NVIDIA *Fermi* [1], the *HyperCore Architecture Line (HAL)* [3] processors from Plurality, or ST Microelectronics *Platform 2012* [5]. While there is renewed interest in Single Instruction Multiple Data (SIMD) computing, thanks to the success of GP-GPU architectures, strict instruction scheduling policies enforced in current GP-GPUs are being relaxed in the most recent many-core designs to exploit data parallelism in a flexible way. Single Program Multiple Data (SPMD) parallelism can thus be efficiently implemented in these designs, where processors are not bound to execute the same instruction stream in parallel to achieve peak performance.

All of the cited architectures share a few common traits: their fundamental computing tile is a tightly coupled cluster with a shared multibanked L1 memory for fast data access and a fairly large number of simple cores, with ≈ 1 Instruction Per Cycle (IPC) per core. Key to providing I-fetch bandwidth for a cluster is an effective instruction cache architecture design. Due to the lack of sophisticated hardware support to hide L2/L3 memory latency (e.g. prefetch buffers), the simple processors embedded in many-cores may indeed experience prolonged stalls on long-latency I-fetch.

The main contribution of this work is the analysis and comparison of the two main architectures for instruction caching targeting tightly coupled CMP clusters: (i) private instruction

caches per core and (ii) shared instruction cache per cluster. We developed a cycle-accurate model of the target cluster with the two cache organizations, and with several configurable architectural parameters for exploration. The private template achieves higher speed, due to its simpler design, but the smaller L1 memory space seen by each core may induce a lower hit ratio. Moreover, the co-existence of multiple program copies in the system may require more bandwidth to main memory in case of multiple concurrent misses. In contrast, the shared template can offer a lower miss ratio and better memory utilization (less copies) at the cost of increased hardware complexity and thus lower speed.

To efficiently analyze and assess pros and cons of the two architectures we also developed a programming environment targeted at efficient data-parallel computing and based on the popular OpenMP programming model. The compilation toolchain and runtime support have been tailored to the target cluster, thus allowing effective benchmarking. We first characterize the two architectural templates by using synthetic microbenchmarks, useful to stress specific corner cases and to assess the best and worst operating conditions for the two cache architectures. Then we further validate the two approaches with several kernels from representative applications from the image processing and recognition domain, parallelized with OpenMP.

The results of this in-depth study provide useful insights and guidelines for designers. The rest of the paper is organized as follows: In Sec. II we discuss related work to ours. The target tightly-coupled cluster and the two cache architectures are described in Sec. III, while the programming framework, compiler and runtime support are discussed in Sec. IV. We describe our experimental setup and results in Sec. V, while Sec. VI concludes the paper and provides future research directions.

II. RELATED WORK

The organization of memory hierarchy is one of the most important and critical phases in architecture design. This aspect has become more relevant with the advent of modern manycores. Dealing with massively parallel systems, instruction caching plays a fundamental role since it must provide the required bandwidth to all cores and software tasks, complying with tight constraints in terms of size and complexity [6].

The Fermi-based General Purpose Graphic Processing Units (GPGPU) comprises hundreds of Streaming Processors (SP) organized in groups of Streaming Multiprocessors (SM) [1].

978-1-4577-0671-4/11 $26.00 © 2011 IEEE

The numbers of SMs and SPs per device vary by device. GPGPUs employ massively multi-threading in order to hide the latency of main memory. The GPU achieves indeed efficiency by splitting application workload into multiple groups of threads (called warps) and multiplexing many of them onto the same SM. When a warp that is scheduled attempts to execute an instruction whose operands are not ready (due to an incomplete memory load, for example), the SM switches context to another warp that is ready to execute, thereby hiding the latency of slow operations such as memory loads. All the SPs in an SM execute their threads in lock-step, according to the order of instructions issued by the per-SM instruction unit. SPs within the same SM share indeed one single instruction cache [2].

Plurality's HyperCore Architecture Line (HAL) family includes 16 to 256 32-bit RISC cores, a shared memory architecture, and a hardware-based scheduler that supports a task-oriented programming model [3]. HAL cores are compact 32-bit RISC cores, which execute a subset of the SPARC v8 instruction set with extensions. The memory system is composed by a single shared memory which operates also as instruction cache. The shared memory holds indeed program, data, stack, and dynamically allocated memory. Each core has two memory ports: an instruction port that can only read from memory, and a data port that can either read or write to memory. Both ports can operate simultaneously, thus allowing an instruction fetch and a data access by each individual core at each clock cycle. The processors do not have any private cache or memory, avoiding coherency problems. However, conflicting accesses cannot be avoided causing latency increasing for not-served requests [4].

STMicroelectronics Platform 2012 (P2012) is a high-performance architecture for computationally demanding image understanding and augmented reality applications [5]. P2012 architecture is composed by several processing clusters interconnected by a Network on Chip (NoC). Each computing cluster features a shared instruction cache memory.

All of the cited platforms have adopted different instruction cache architectures, meaning that there is still not a dominant paradigm for instruction caching in the manycore scenario. Clearly, a detailed design space exploration and analysis are needed to evaluate how micro-architectural differences in L1 instruction cache architectures may affect the overall system behavior and IPC.

III. SHARED L1 CLUSTER ARCHITECTURE

This section provides description of the building blocks of both private and shared instruction cache architectures. To help system designers to compare different L1 instruction cache architectures, we have developed a flexible L1 instruction cache architecture system. The proposed templates, written in SystemC [8], can be used either in stand-alone mode or plugged into any virtual platform, we integrated them in an accurate virtual platform environment specifically designed for embedded MPSoC design space explorations [7]. Our enhanced virtual platform is highly modular and capable of simulating at cycle-accurate level an entire shared L1 cluster including cores, L1 instruction caches, shared L1 tightly coupled data memory, external (L3) memories and system interconnections.

A. Architectural components

Processing Elements: Our shared L1 cluster consists of a configurable number of 32-bit ARMv6 processor (ARM11 family [10]). An accurate model of separate instruction and data buses path (i.e. Harvard Architecture) is fundamental for the purpose of this work. There are several ARMv6 instruction set simulator already available, Skyeye [11], SoClib [12] and SimSoc [9] are just a few representative examples. We chose the one in [9] as our base ISS. To obtain timing accuracy, after modifying its internal behavior to perform concurrent load/store and instruction fetch, we wrapped the ARMv6 ISS in a SystemC module.

L1 Instruction Cache Module: The Instruction Cache Module has a core-side interface for instruction fetches and an external memory interface for refill. The inner structure consists of the actual memory (TAG + DATA) and the cache controller logic managing the requests. The module is configurable in its total size, associativity, line size and replacement policy (FIFO, LRU, random).

Logarithmic interconnect: The logarithmic interconnect module has been modeled, from a behavioral standpoint, as a parametric, Mesh-of-Trees (MoT) interconnection network to support high-performance communication between processors and memories in L1-coupled processor clusters resembling the hardware module described in [13], shown in Fig. 1. The module is intended to connect processing elements to a multi-banked memory on both data and instruction side. Data routing is based on address decoding: a first-stage checks if the requested address falls within the intra-cluster memory address range or has to be directed off-cluster. To increase module flexibility this stage is optional, enabling explicit L3 data access on the data side while, on the instruction side, can be bypassed letting the cache controller take care of L3 memory accesses for lines refill. The interconnect provides fine-grained address interleaving on the memory banks to reduce banking conflicts in case of multiple accesses to logically contiguous data structures. The last $log_2(interleaving\ size)$ bits of the address determine the destination. The crossing latency consists of one clock cycle. In case of multiple conflicting requests, for fair access to memory banks, a round-robin scheduler arbitrates access and a higher number of cycles is needed depending on the number of conflicting requests, with no latency in between. In case of no banking conflicts data routing is done in parallel for each core, thus enabling a sustainable full bandwidth for processors-memories communication. To reduce memory access time and increase shared memory throughput, read broadcast has been implemented and no extra cycles are needed when broadcast occurs.

L1 Tightly Coupled Data Memory: On the data side, a multi-ported, multi-banked, Tightly Coupled Data Memory (TCDM) is directly connected to the logarithmic interconnect. The number of memory ports is equal to the number of banks to have concurrent access to different memory locations. Once a read or write requests is brought to the memory interface, the data is available on the negative edge of the same clock cycle, leading to two clock cycles latency for conflict-free TCDM access. As already mentioned above, if conflicts occur there is no extra latency between pending requests, once a given bank is active, it responds with no wait cycles. Banking factor (i.e. ratio between number of banks and cores) can be configured to explore how this affects banking conflicts.

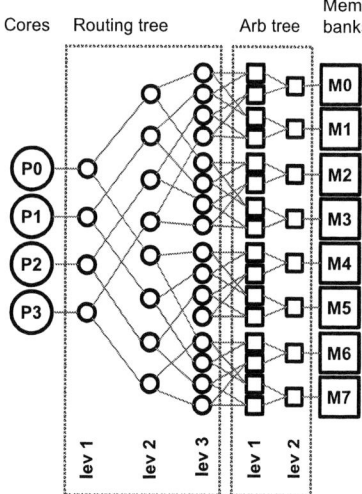

Fig. 1. Mesh of trees 4x8: empty circles represent routing switches and empty squares represent arbitration switches (banking factor of 2)

Synchronization: To coordinate and synchronize cores execution, we modeled two different synchronization mechanisms. The first one consists of *HW semaphores* mapped in a small subset of the TCDM address range. They consist of a series of registers, accessible through the data logarithmic interconnect as a generic slave, associating a single register to a shared data structure in TCDM. By using a mechanism such as a hardware *test&set*, we are able to coordinate access: if reading returns '0', the resource is free and the semaphore automatically locks it, if it returns a different value, typically '1', access is not granted. This module enables both single and two-phases synchronization barriers, easily written at the software level. Theoretically all cores can be resumed at the same time (reading broadcast the value of the semaphore), but there is no guarantee that this happens because of execution misalignment. To get tight execution alignment, we developed two *fast synchronization primitives* based on a *HW Synchronization Handler Module* (SHM). This device acts as an extra slave device of the logarithmic interconnect and has a number of hardware registers equal to the number of cores, where each register is mapped in a specific address range. When a write operation is issued to a given register, a synchronization signal is raised to the corresponding core suspending its execution after one cycle, when the synchronization signal is lowered the execution is resumed. The SHM is programmable in different ways from the software level via APIs. Writing to the OP_MODE register, different synchronization mechanisms can be enabled: if operating in SYNC_MODE, synchronization signals are lowered when all cores have executed the `sync()` API (writing to their respective register, increasing an HW counter inside the SHM), obtaining a cycle-accurate execution alignment. When operating in TWO_PH_MODE, a simple state machine inside the SHM distinguishes cores behavior between master and slaves enabling a two-phases barrier. When the master reaches a `master_wait_barrier()` primitive, it is suspended until all slaves have reached the `slave_enter_barrier()`. After that, the master is awakened and is the only core executing until the `master_release_barrier()` primitive

is reached, reactivating all slaves exactly in the same clock cycle. These APIs and the underlying HW mechanism offered by the SHM are fundamental for the OpenMP library described in Section IV-B.

B. Private Instruction Cache Architecture

All the previously described architectural elements are combined together to form the private instruction cache architecture as shown in Fig. 2. The cluster is made of 16 ARMv6 cores, each one has its own private instruction cache with separate line refill paths while the L1 data memory is shared among them. An optional DMA engine can be used to carry out L3 to TCDM data transfers. Access to the off-cluster L3 memory is coordinated by the L3 BUS, requests are served in a round-robin fashion. On the data side all cores are able to perform access to TCDM, L3 memory and eventually to HW semaphores or SHM. The logarithmic interconnect is responsible of data routing based on address ranges as already described in the previous section. Default configuration for the private instruction cache architecture and relevant timings are reported in TABLE I.

TABLE I
DEFAULT PRIVATE CACHE ARCHITECTURE PARAMETERS AND TIMINGS

PARAMETER	VALUE
ARM v6 cores	16
$I\$_i$ size	1 KB
$I\$_i$ line	4 words
t_{hit}	= 1 cycle
t_{miss}	$\geq$ 59 cycles
TCDM banks	16
TCDM size	256 KB
L3 latency	50 cycles
L3 size	256 MB

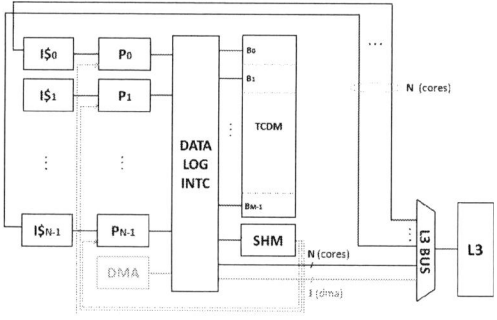

Fig. 2. Cluster with private L1 instruction caches

C. Shared Instruction Cache Architecture

Shared instruction cache architecture is shown in Fig. 3. From the data side there is no difference between the private architecture except for the reduced contention for data requests to L3 memory (line refill path is unique in this architecture). Shared cache inner structure is represented in Fig. 4. A slightly modified version of the logarithmic interconnect described in III-A (the first stage of address deconding is disabled) connects processors to the shared memory banks operating line interleaving (1 line consists of 4 words). A round robin scheduling guarantees fair access to the banks. In case of two or more processors requesting the same instruction, they

978-1-4577-0671-4/11 $26.00 © 2011 IEEE

are served in broadcast not affecting hit latency. In case of concurrent instruction miss from two or more banks, a simple MISS BUS handles line refills in round robin towards the L3 BUS.

TABLE II
SHARED CACHE ARCHITECTURE PARAMETERS AND TIMINGS

PARAMETER	VALUE
$I\$$ size	16 KB
$I\$$ line	4 words
t_{hit}	≥ 1 cycle
t_{miss}	variable

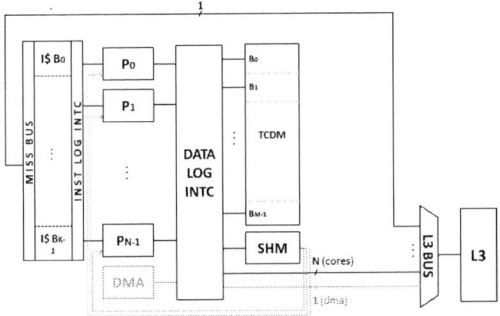

Fig. 3. Cluster with shared L1 instruction cache

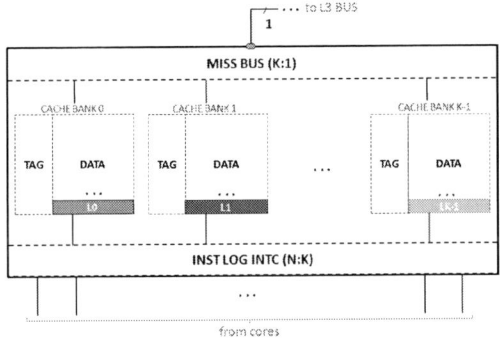

Fig. 4. Shared instruction cache architecture

IV. SOFTWARE INFRASTRUCTURE

In this section we briefly describe the software infrasctructure: first compiler and linking strategies used to compile and allocate all the data needed for the execution of all benchmarks. In the second part we will introduce our custom implementation of the openMP library, developed to run on the proposed target architectures.

A. Compiler and Linker

Before describing compiling and linking strategies applied for our benchmarks, it is of primary importance to introduce the memory map seen by all processors in the architecture. Fig. 5 shows the global memory map of one cluster, in which it is possible to distinguish two memory regions: the **L3 memory region** (256 MB) and the **TCDM memory region** (256KB). The first is the off-chip memory used to store the

executable of the applications, and data too big to be stored in the on-chip data memory. The TCDM region, mapping the shared data cache, is in turn divided in three sub-regions: *LOCAL_SHARED*, *STACK* and *HEAP*.

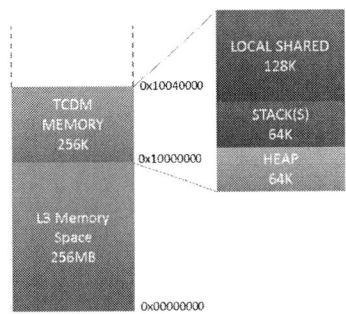

Fig. 5. Cluster global memory map

The *LOCAL_SHARED* region is intended to maintain variables of static size (known at compile time) explicitly defined to be stored in the TCDM memory. To force the allocation of a variable in the on-chip data memory we combined the use of a linker script and gcc attributes. We defined a new section in the ARM binary, namely *.local_shared*, used to contain variables to be stored in this region of the memory map.

The *STACK* region is defined to maintain the stack of all 16 processors, with 4K assigned to each of them. Each processor calculates its own stack top at simulation startup using a combination of linker script and an assembly boot routine.

Finally, the *HEAP* region is used for dynamically allocated structures. The allocation is allowed through the *shmalloc* function, provided by the MPARM applications support (app-support).

B. Custom OpenMP Library

To parallelize our benchmarks we used a custom implementation of the OpenMP APIs for parallel programming, adapted to run on our MPARM based architecture. The OpenMP programming paradigm is based on two different parallel constructs: **parallel for** and **parallel sections**. The first allow the exploitation of SIMD or SPMD parallelism, the iterations of the for cycle are divided in chunks and assigned to the available cores. The second describes task parallel sections of a program, each core can execute a different portion of code so a different task. To tailor these two constructs to the target architecture it is necessary to consider that our software infrastructure has no operating system. In our implementation all cores execute the same binary file as a single process running on each processor, and the work performed is differentiated according to the processor id.

The Master-Slave mechanism on which OpenMP is based is realized using the two-phase barriers described in Section III-A. We also had to modify the compiler (GCC 4.3) to transform OpenMP annotations in a correct binary form for the MPARM architecture. The compiler has to create all the structures needed to run a certain application and to differentiate the the work to be performed by all processors, using appsupport functions.

Our OpenMP runtime has a thin software layer based on a set of shared structures used by the processors to synchronize, share data and control the different parallel regions of the applications. All these structures are stored in on-chip TCDM memory using both statically (LOCAL_SHARED) and dynamically allocated structures (*shmalloc*), some are also protected by a lock which is implemented via the hardware semaphores described in Section III-A.

V. EXPERIMENTAL RESULTS

As already outlined in previous sections, we considered a cluster made of 16 ARMv6 cores connected through a low latency logarithmic interconnect to a multiported, multibanked 256 KB TCDM memory. On the instruction side, private and shared architectures differ in the cache architecture. An off-cluster (L3) 256 MB memory is accessible through the data logarithmic interconnect or through the line refill path. Our investigations focus on varying the total instruction cache size, and hereafter the L3 memory latency.

A. Microbenchmarks

In this section we present the results of three synthetic benchmarks intended to characterize both architectures and to highlight interesting behaviors. The synthetic benchmarks were written using Assembler language in order to have complete control of the software running on top of the architectural templates. They consist of a set of iterated ALU or MEMORY instructions performed to highlight a specific behavior. All the synthetic benchmarks share the common structure shown in Listing 1 below.

```
        mov r6, N_LOOP
        mov r5, #0
_loop:  cmp r5, r6
        blt _body
        b _end
_body:  ...
        add r5, r5, #1
        b _loop
_end:   ...
```

Listing 1. Synthetic benchmarks structure

The performance metrics considered here are the *cluster IPC* ($\overline{IPC}$, $0 < \overline{IPC} \leq 16$) and its average value, calculated as the number of instructions executed by all the processors divided by the number of cluster execution cycles.

Cold misses: The body of this benchmark consists of only ALU operations (i.e. `mov r0, r0`) leading to a theoretical $\overline{IPC} = 16$ (and avarage IPC = 1) for both architectures. The plot in Fig. 6 shows on the Y-axis the cluster average IPC while X-axis reports how many times the loop is executed. Increasing N_LOOP both architectures tend to the theoretical value, but the private architecture starts from a lower IPC due to the heavy impact of cold misses serialization (16 cores contending for L3 access).

Conlict free TCDM accesses: This benchmark adds the effect of TCDM access. As already mentioned before, in case of conflict free access, TCDM latency is two cycles leading to a single cycle stall between two consecutive instruction fetches. The loop is iterated a fixed number of times (4K in order to lower cold misses effect) and has a variable number of memory operations inside its body. We are considering a banking factor of 1, allowing every core to access a different bank without conflicts. The plot in Fig. 7 shows on the Y-axis the average cluster IPC while on X-axis varies the percentage of memory instructions over the number of instructions the loop is made of. Both architectures are affected in the same

way, with IPC tending to the asymptotic value value of $\frac{1}{2}$[1] (and cluster IPC respectively to 8 because of any conflict leading to misalignement). Private architectures as an initial lower IPC due to the cold misses effect discussed in the previous paragraph.

Conflicts on TCDM accesses: This benchmark adds another aspect of TCDM accesses: conflicts. Conflicting accesses to the same bank increase TCDM latency thus affecting IPC. In this scenario we considered a realistic ratio between memory and ALU operations of 20%. As before, the loop is iterated 4K times to reduce cold misses effect. The plot in Fig. 8 shows on the Y-axis the cluster IPC while on the X-axis varies the percentage of memory accesses creating conflicts on the same bank. It is interesting to notice that, while there are no conflicts on TCDM, the shared architecture performs better the private one because of its intrinsic lower miss cost, in presence of TCDM conflicts the execution misalignment penalizes the shared cache architecture increasing the average hit time. It is important to underline that just a single conflict creates execution misalignment.

To explain the sharp reduction of the IPC for the shared cache due to TCDM conflicts, let us consider a simple program consisting of 16 instructions. Before any TCDM conflict occur, the execution is perfectly aligned leading to synchronous instruction fetching. The conflicting access in TCDM leads to a single-cycle misalignment among all cores in the next instruction fetch. As shown in Fig. 9, assuming a cache line is made of 4 32-bit words, there will be 4 groups of 4 processors accessing the same line (i.e. bank) but requesting instructions at different addresses. When this situation arises, the average hit time increases from 1 cycle (concurrent access) to 4 cycles (conflicting requests are served in a round-robin fashion). This particular case clearly shows how this architecture is sensitive to execution misalignment.

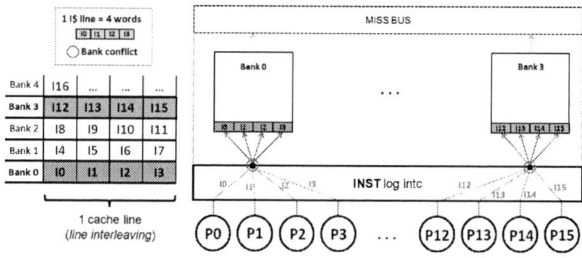

Fig. 9. Misalignement in instruction fetching for shared cache

This phenomenon can stand out in an even worse case when the 4 blocks of instructions that are fetched by processors reside in the same bank (situation depicted in Fig. 10). This leads to the worst-case for instruction fetch, increasing average hit time from 1 to 16 cycles.

B. Real Benchmarks

In this section we compare the performance of the private and shared I\$ architectures by using two real applications,

---

[1]A program consisting of only ALU (1 cycle) or MEMORY (2 cycles for TCDM access) operations gives a per-core IPC equal to $\frac{N_{alu}+N_{mem}}{1 \cdot N_{alu}+2 \cdot N_{mem}}$. Increasing N_{mem}/N_{alu} ratio, leads to an asymptotic value value of $\frac{1}{2}$. Cluster IPC, in this case of perfectly aligned execution, is $\overline{IPC} = 16 \cdot IPC_i$ and its average is equal to the IPC of a single core.

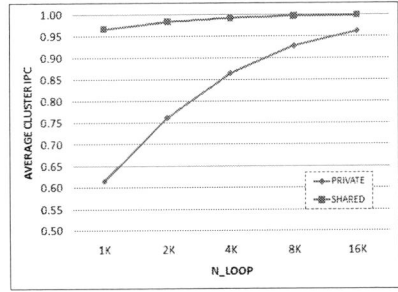
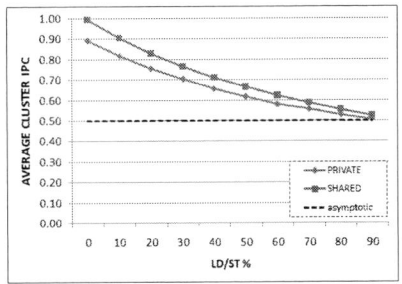
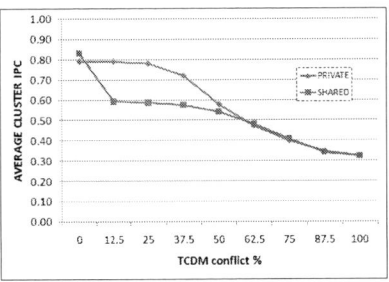

Fig. 6.   Private vs. Shared architectures IPC with only ALU operations

Fig. 7.   Private vs. Shared architectures IPC with conflict free TCDM accesses

Fig. 8.   Private vs. Shared architectures IPC with conflicting TCDM accesses

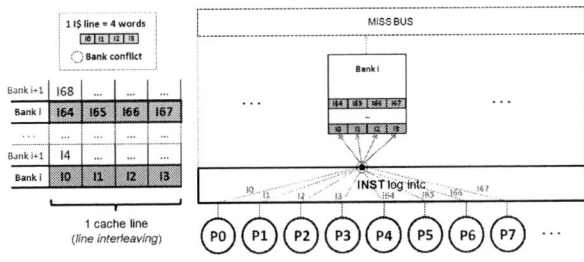

Fig. 10.   Worst case for instruction fetching in shared cache

namely a JPEG decoder and a Scale Invariant Feature Transform (SIFT), a widely adopted algorithm in the domain of image recognition. In particular, our aim is to evaluate the behavior of the two target architectures when considering different types of parallelism at the application level. We thus parallelized our benchmarks with OpenMP, and considering three different examples.

The first example expresses data-parallelism at the application level. Thus we focused on the two data-independent computational kernels in JPEG: Dequantization (DQTZ) and Inverse Discrete Cosine Transform (IDCT). With this parallelization scheme all processors execute the same instructions, but over different data sets.

In the second example we adopt pipeline parallelism in the same JPEG application, where each of the four stages of JPEG – Huffman DC, Huffman AC, Dequantization, IDCT – is wrapped in an independent task assigned to a processor. To keep all the processors busy we execute four pipelines in parallel.

The third example considers three main kernels from SIFT: up-sampling, Gaussian convolution, and difference of Gaussians, all leveraging data-parallelism. In relation with the JPEG data parallel application, SIFT is composed of more complex computational steps that can stress the cache capacity causing more misses.

In what follows we carry out two main experiments. We evaluate the performance by (i) varying the cache size and (ii) the L3 latency.

Figure 12 shows the results for the first experiment. Here we consider a fixed latency of 50 cycles for the L3 memory, and we vary the cache size. Focusing on the plots on the top part of the figure, for each of the three benchmarks and for the two architectures we show execution time, normalized to the slowest value (i.e. the longest execution time for that

benchmark). Looking at the data parallel variant of JPEG it is possible to see that the shared cache architecture performs worse than the private cache architecture. We would expect the SIMD parallelism exploited by this application to be favoured on the shared cache architecture, so this finding is seemingly counterintuitive. The reason for this loss of performance is to be found in an increased average hit cost, due to the banking conflicts in the instruction cache described in the previous section (Fig. 8). So if we mathematically model the overall execution time of an application as

$$N_H \times C_H + N_M \times C_M \qquad (1)$$

where N_H and N_M represent the number of hits and misses, and C_H and C_M represent the average cost for a hit and for a miss, C_H may be higher than 1 for the shared cache.

To confirm this assumption we report average cache hit ratio (left Y-axis, solid lines) and cost (right Y-axis, dashed lines) in the plots on the bottom part of Fig. 12. It is possible to see that the average cache hit cost for the shared cache architecture is ≈ 2.4 cycles, while the number of misses is negligible (miss rate = 0.003%). As a consequence, the right-hand part of the formula above does not contribute to the overall execution time. To understand the absence of cold cache miss impact we analyzed the disassembled program code. The Dequantization kernel consists of a loop composed by a few tens of instructions, while the IDCT kernel loop contains roughly 200 instructions. Overall this results in a hundred misses, and no capacity misses are later experienced. Due to the SIMD parallelism all cores fetch the same instructions, thus only the first core executing the program incurs cold cache misses. Instruction fetch from the remaining cores always results in a hit. In the private cache architecture, on the contrary, each core individually experiences 104 misses for cache sizes of 32 and 64 Kbytes, while around 400 for 16 Kbytes. This results in a cluster miss rate (total number of misses over total number of instructions) of 0.05% for the private cache and 0.003% for the shared cache.

For the SIFT application the difference in the number of misses between the shared and the private architecture is major, as we can see in Figure 12. In this case, due to the high average miss cost, the shared architecture provides best results despite the high average cost of an instruction hit (more than 2.25 cycles). Indeed, the average cost of a miss is around 800 cycles for the private cache (any size), while for the shared cache it is around 300 cycles. Again, this is due to the fact that for the private cache multiple refills from different cores are serialized on the L3.

978-1-4577-0671-4/11 $26.00 © 2011 IEEE

For the JPEG pipelined application, the shared cache has a miss rate of 0.03% against 0.3% (64 Kbyte) of the private cache. Moreover in this case the shared cache has lower average costs for a hit (around 1.5 cycles) in respect with the other applications. The shared approach delivers 60% faster execution time for small cache sizes (16K), which is reduced to $\approx 10\%$ for bigger caches.

We must distinguish when an instruction is missed for the first time or not. In the first case we identify it as cold miss, while in the second case as a capacity miss. Table III shows the number of the capacity misses on the total number of misses in percentage for the private cache architecture across all the applications. The shared cache architecture can better exploit the total cache size, then it experiences no capacity miss.

Figure 13 shows the results for the second experiment, where we keep a fixed cache size of 32KB and change the latency of the L3 memory. The plots on the upper part show normalized execution time (to the slowest, as before), whereas the plots on the lower part show the average cost of a miss.

Overall, it is possible to see that for L3 latency values beyond 100 cycles the shared cache architecture always performs better than the private cache architecture.

Considering Eq. 1 again, C_M is the parameter which is mostly affected by the varying L3 latency. In particular, the term $N_M \times C_M$ linearly increases with the L3 latency as we can see on the lower part of Fig. 13. In the data parallel applications (first JPEG variant and SIFT), the average cost for a miss in the private cache architecture sharply increases with the L3 latency, whereas the same curve for the shared cache has a much smaller slope. This is due to the fact that private caches generate much more traffic towards the L3 memory (16 line-refill requests against a single refill needed by the shared cache). Then, despite the very low number of misses for the JPEG data parallel application, their contribution accounts for 50% of the overall execution time in Eq. 1.

Regarding the pipelined JPEG application, different from the other examples the average miss cost is slightly higher for the shared cache. However, the miss rate for the shared cache is around 0.02%, while for the private cache it is around 0.2%, thus the shared architecture achieves slightly faster execution times.

TABLE III
NUMBER OF CAPACITY MISSES ON TOTAL NUMBER OF MISSES IN PERCENTAGE FOR THE PRIVATE CACHE ARCHITECTURE ACROSS ALL THE APPLICATIONS AND THE CACHE SIZES (BYTES)

	JPEG PARALLEL	JPEG PIPELINED	SIFT
SIZE 16K	73%	88%	86%
SIZE 32K	5%	41%	84%
SIZE 64K	5%	4%	52%

C. Frequency Comparison

As a last experiment we want to investigate how faster the private cache design should be clocked to deliver a similar performance to that achieved with the shared cache architecture. We considered as baseline configuration an L3 latency of 150 cycles and an Instruction cache of 32 KB. To carry out this comparison increasing the frequency of the clock within the cluster, we kept constant the L3 latency: our default T_{clk} is 10 ns leading to 1500 ns. The plot in Fig. 11 shows on the Y-axis the ratio between shared and private execution

time for the benchmarks described in Section V-B, while on the X-axis varies the percentage of frequency speedup.

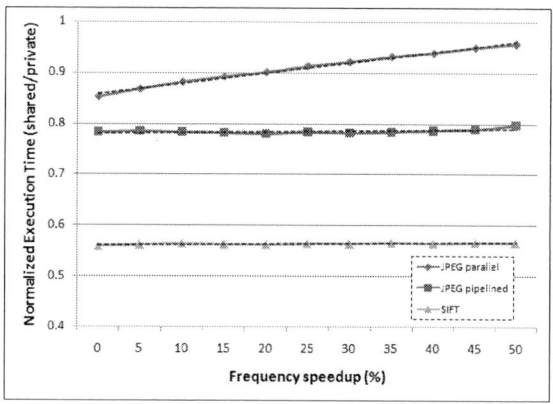

Fig. 11. Frequency comparison of private and shared cache architectures

Increasing cluster clock frequency has significant effect only for JPEG parallel while private architecture is quite insensitive for both SIFT and JPEG pipeline benchmarks. To explain such behavior we have to look at the execution time breakdown. A faster clock inside the cluster affects hit time and TCDM latency but has negligible effect on miss latency (dominated by L3 latency) and L3 data accesses. TABLE IV shows execution time breakdown for all benchmarks.

TABLE IV
EXECUTION TIME BREAKDOWN FOR JPEG PARALLEL, JPEG PIPELINED AND SIFT BENCHMARKS

	HIT & TCDM	MISS & L3 DATA
JPEG parallel	51.13%	48.87%
JPEG pipelined	14.72%	85.28%
SIFT	0.96%	99.04%

A $\approx 51\%$ of execution time affected by cluster clock frequency gives the performance improvement of the private architecture for JPEG parallel. The same is not true for JPEG pipelined and SIFT benchmarks. This behavior underlines cluster performance is not affected by clock frequency when the program running has the execution time dominated by L3 memory accesses.

VI. CONCLUSIONS

Key to providing I-fetch bandwidth for cluster-based CMP is an effective instruction cache architecture design. We analysed and compared the two most promising architectures for instruction caching targeting tightly coupled CMP clusters, namely private instruction caches per core and shared instruction cache per cluster. Experimental results showed that private cache performance can be significantly affected by the higher miss cost, on the other hand the shared cache has better performance, with speedup up to $\approx 60\%$. However, it is very sensitive to execution misalignment, which can lead to cache access conflicts and high hit cost.

ACKNOWLEDGMENT

This work was supported by projects PRO3D (FP7-ITC4-248776) and JTI SMECY (ARTEMIS-2009-1-100230), funded by the European Community.

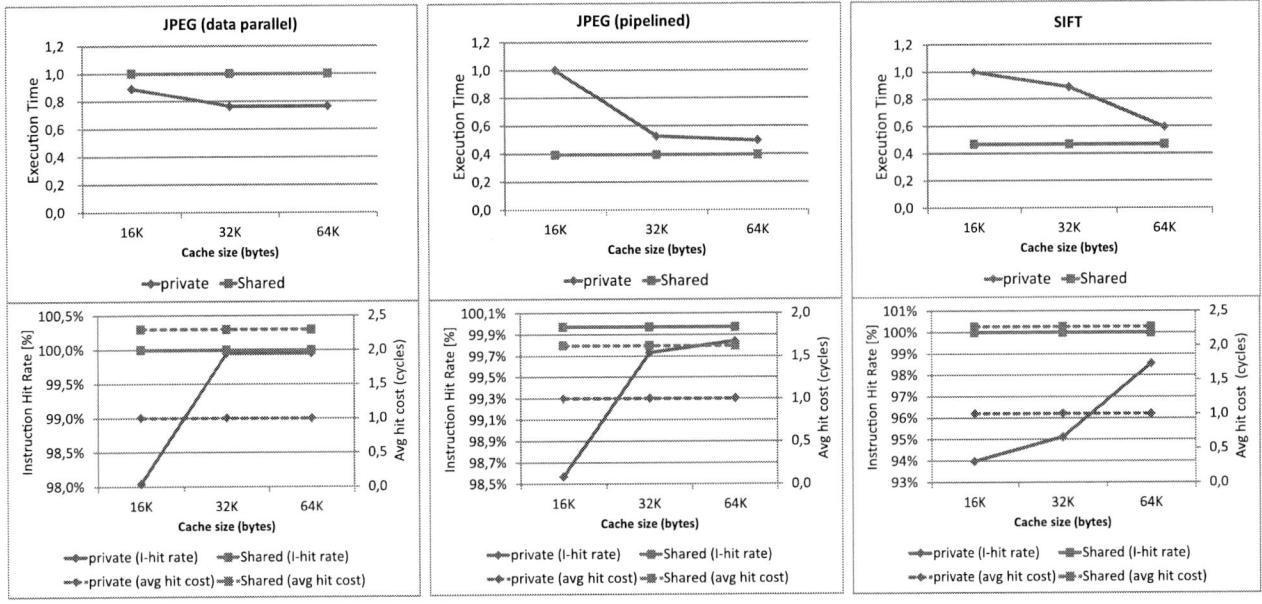

Fig. 12. Impact of varying the cache size for different benchmarks

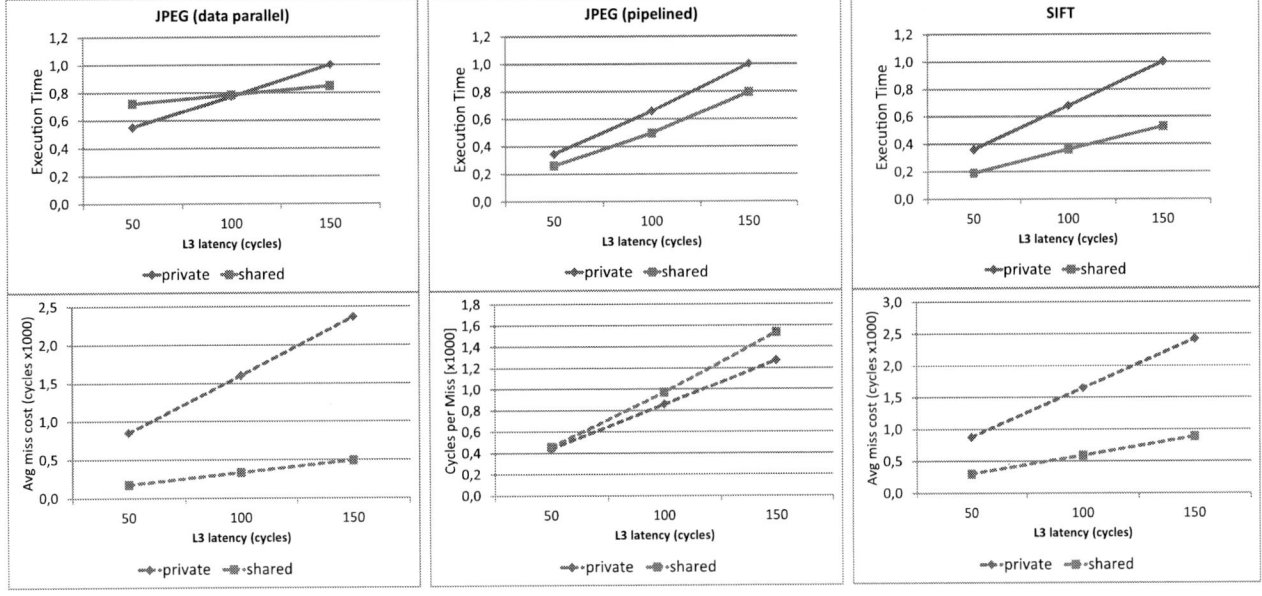

Fig. 13. Impact of varying the latency of L3 memory for different benchmarks

REFERENCES

[1] NVIDIA - Next Generation CUDA Compute Architecture: Fermi - WhitePaper
[2] H. Wong et al. - *Demystifying gpu microarchitecture through microbenchmarking* - In ISPASS, 2010.
[3] Plurality Ltd. - The HyperCore Processor
www.plurality.com/hypercore.html.
[4] Plurality Ltd. - HyperCore Software Developer's Handbook
www.plurality.co.il/software/documents/SDH-draft-1.5.pdf
[5] ST Microelectronics and CEA - *Platform 2012: A Many-core programmable accelerator for Ultra-Efficient Embedded Computing in Nanometer Technology*, 2010.
[6] M. M. S. Aly et al. - *Performance and Energy Trade-offs Analysis of L2*

on-Chip Cache Architectures for Embedded MPSoCs - In: GLSVLSI, 2010.
[7] The MPARM virtual platform
http://www-micrel.deis.unibo.it/sitonew/research/mparm.html
[8] SystemC Language - http://www.systemc.org/
[9] The SimSoc project
https://gforge.inria.fr/projects/simsoc/
[10] ARM 11 product page
http://www.arm.com/products/processors/classic/arm11
[11] SkyEye full system simulator
http://sourceforge.net/projects/skyeye/
[12] The SoClib platform - http://www.soclib.fr/
[13] A. Rahimi et al. - *A fully-synthesizable single-cycle interconnection network for shared-L1 processor clusters* - In: DATE, 2011.

978-1-4577-0671-4/11 $26.00 © 2011 IEEE

Static Analysis Method for Deadlock Detection in SystemC Designs

Mikhail Moiseev, Alexey Zakharov
Saint-Petersburg State Polytechnical University
Saint-Petersburg
Email: mikhail.moiseev@gmail.com,
zakharov@kspt.ftk.spbstu.ru

Ilya Klotchkov, Sergey Salishev
Intel Labs
Saint-Petersburg
Email: ilya.v.klotchkov@intel.com,
sergey.i.salishev@intel.com

Abstract— One of the goals of SystemC is high level system design verification at the early stage. Currently, simulation is widely used for this purpose. As the level of design parallelism grows, efficiency of simulation-based verification methods decreases. Thus different formal verification methods for SystemC are actively researched.

In this paper we present an approach to deadlock detection in SystemC designs based on static code analysis. Our approach to static analysis considers SystemC scheduler semantics. The developed approach has been implemented in Deadlock Analyzer tool. We demonstrate efficiency of our tool by applying it to dining philosophers, crossroads, producer-consumer cases and to a real-life model of video accelerator.

I. INTRODUCTION

System level modeling has become an essential part of SoC (System On Chip) design process. SystemC [1] is an IEEE standard for system-level software/hardware co-design. SystemC is very popular in the industry as system level modeling tool.

A typical SoC design consists of a hardware part and a software part running on a CPU. The system model will thus include models for software, hardware and their interfaces and is highly parallel. Due to high complexity of real system designs they might contain errors. Debug and verification of such designs is often done by simulation. However synchronization error detection in highly parallel programs is known to be difficult due to the exponential state space growth. Another approach to error detection is use of static methods. There are several static methods used in software development which are aimed at detecting synchronization errors. These methods allow to increase the system state space coverage comparing to simulation and thus to detect more errors in the system design at early stage.

In this paper we will focus on finding deadlocks in SystemC designs. Deadlocks are selected due to their high debug complexity and dramatic consequences for the system - in SoC deadlock usually means partial or full lockout and reset. Furthermore, limited on-chip observability makes deadlock detection in real silicon hard and unpredictable. We have selected static analysis because it allows sound deadlock detection at an expense of some false positives.

The contribution of this paper is a novel method for deadlock detection in SystemC designs based on the static code analysis. This method considers SystemC scheduler semantics. It supports immediate notifications, delta notifications and channel updates.

The rest of this paper is organized as follows: Section II presents program model. Section III describes analysis algorithms used for the SystemC code and section IV presents deadlock detection rules. Efficiency of the proposed approach is discussed in section V and the last section concludes the paper.

A. Related Work

The SystemC program is a parallel C++ program. So it is natural to apply formal methods used in C++ software verification for the synchronization error detection in SystemC. There are several main classes of formal methods that can be applied to SystemC [2].

Error detection boils down to checking some constraints on a program state space. These constraints can be predefined for certain types of errors or derived from user defined assertions. These assertions can be written using program model formalism or in property specification language [3], [4]. The latter approach has been successfully used for interface protocol verification.

The first step of error detection is extracting the program model. This can be done dynamically or statically. The run-time approach captures the program execution traces including events and constraint sensitive components of program state. In order to apply the run-time method to SystemC, the simulation kernel must be modified [5], [6]. This method can only capture errors that could potentially happen on the execution trace, so its accuracy heavily depends on the program test coverage [7].

Static approach extracts the program model directly from a program source code. Static methods can usually handle only a limited subset of SystemC language [8]. The SystemC kernel is usually abstracted at this stage by replacing it with annotations [9], [10] or with a simplier C/C++ stubs [11] to reduce the complexity of subsequent analysis. In the latter case it is possible to apply tools for C/C++ to the modified program directly.

A mixed method of program model extraction is implemented in PinaVM [12]. System design is extracted using runtime information at elaboration phase. This information

978-1-4577-0671-4/11 $26.00 © 2011 IEEE

allows to improve static analysis precision for virtual methods and pointers. The tool is based on LLVM for C/C++ language.

The program model formalism heavily depends on the error detection methods to be applied. Possible formalisms include Control Flow Graph, Dynamic Dependency Graph [5] Petri nets [13], [14], ISO LOTOS [9], [10], PROMELA [15], mCRL2 [16], etc. After the program model is extracted, a formal analysis method is applied to it. Model checking and static code analysis are of the most practical interest among different analysis methods.

Model checking methods are based on proving some exact model properties. These are popular for SystemC programs analysis [14], [5], [13], [7], [17]. The approach has limited applicability for large parallel systems due to the state space explosion [18] that leads to unfeasible memory and computational requirements.

Methods of static code analysis operate on program model approximations exhibiting modest resource requirements at the cost of false-positives. This approach has been used in [19] to extract certain characteristics of SystemC program for measuring its parallelism and testing complexity. The static code analysis is known to provide good results for C++ error detection. We conclude that the potential of static code analysis for SystemC is yet to be fully explored.

II. SystemC Program Model

The proposed deadlock detection method uses the formal program model, which is convenient for analysis algorithms. The program model is control flow graph in static single assignment form. It is automatically extracted from SystemC program source codes.

The program model contains SystemC-specific statements which are listed in table I.

TABLE I
SystemC-Specific Statements

Statement	Description
start()	Start simulation
sense(t, e)	Add event e to the sensitivity list of thread t
run(this, t, f)	Run method f in thread t
run_update(this, update)	Run update(this) method at update phase of delta-cycle
wait(e, time)	Wait for event e for less than time
notify(e, time)	Notify event e after time

Process creation statements, sensitivity list management statements and interprocess synchronization statements are all expressed with statements from table I. The representation of main SystemC statements is shown in table II.

The program model and analysis algorithms have some limitations:

- Only immediate and delta notifications are supported (time parameter in wait and notify statements can be −1 or 0 only, see table II).

TABLE II
Representation of SystemC Statements

Source Code	Model

SC_THREAD(f)
 sensitive << e1
 << e2;
	t = scThread();
	 run(t, f, this);
	 sense(e1, t);
	 sense(e2, t);

request_update();	run_update(this, this->update);
sc_start();	start();
wait();	wait(NO_EVENT, -1);
wait(e);	wait(e, -1);
wait(SC_ZERO_TIME, e);	wait(e, 0);
sc_time t(0, SC_NS); wait(t);	wait(NO_EVENT, t);
notify(e);	notify(e, -1);
notify(SC_ZERO_TIME, e);	notify(e, 0);
notify(0, SC_NS, e1);	notify(e, 0);
sc_time t(0, SC_NS); notify(t, e);	notify(e, t);

- Spawned processes are not supported. Statement sc_spawn, macros SC_FORK and SC_JOIN are not presented in program model.
- Method processes are not supported. Function next_trigger calls are not presented in program model.

III. Analysis Algorithms

The deadlock detection method is based on exploration of program processes' states. Processes' states are calculated by applying the following static analysis algorithms to the program model:

- Interval analysis algorithm,
- Points-to analysis algorithm,
- Lockset algorithm,
- Interprocess interaction algorithm,
- Variant separation algorithm.

A. Interval and Points-to Analyses

SystemC design is a parallel C++ program. Since synchronization statements may occur in the branch of if, switch or loop statement it is necessary to evaluate possible values of branch condition variables.

In order to obtain possible values of scalar variables *interval analysis* is used. SystemC events, ports and other essential entities are usually class members of sc_module. These objects are implicitly accessed via this pointer in the source code, so *points-to analysis* is required to deal with synchronization.

Both interval and points-to algorithms are specific implementations of well-known dataflow static analysis approaches [20].

The points-to algorithm uses representation of program state in the form of points-to relation. Each program object is a tuple (o, k), where o is an area of memory, k is an offset within this object. Points-to relation includes tuples $((o_1, k_1), (o_2, k_2))$, where (o_1, k_1) points to (o_2, k_2). The algorithm considers dynamically allocated objects, stack-allocated objects and special objects that represent uninitialized value of a pointer and `NULL` pointer.

Interval analysis algorithm operates on tuples in the form $((o, k), i)$, where i is an interval value which is an ordinary interval $[low, high]$ or a special interval that represents uninitialized value.

Static analysis algorithms discussed in this section are sound, context- and flow-sensitive. These features allow high precision and soundness of deadlock detection.

B. Elaboration Phase Analysis

Analysis of the elaboration phase involves interval and points-to analyses to extract structure of SystemC program, e.g. processes, their start methods, events, process sensitivity, possible values of global variables.

The elaboration phase covers analysis of a global initilizer and analysis of entry function `sc_main`. The global initializer is a block of statements representing initialization of global variables. Analysis of SystemC program entry function ends when `start` statement is met. The result of elaboration phase is a program state that contains:

- values of global variables and variables of `sc_main`,
- set of constructed processes,
- start method for each process,
- static sensitivity lists of processes.

C. Evaluation Phase Analysis

Evaluation phase analysis begins with a non-empty set of start points. A *start point* is a program point which is active in this evaluation phase. A program point is $\langle s, p, C, I \rangle$ tuple, where s – statement, p – process, C – call stack, I – cycle iteration counters. Start points in the first delta cycle are first statements of methods that are run in processes with empty call stack and cycle iteration counters. Start points of subsequent evaluation phases are finish points of previous evaluation or update phase which has delayed notification. A *finish point* $\mathcal{F}_i^j$ is a program point with statement s which is `wait` or `return` statement in a process p_j (i – number of finish point in the process). There is one program state for each start point.

The evaluation phase analysis contains interval analysis, points-to analysis and lockset analysis. A *lockset analysis* counts delayed notifications, immediate notifications and `request_update` calls in execution paths that are analyzed in this phase.

Let designate $L(s_k, e_m)$ as numbers of `notify` (e_m) statements and $U(s_k, p)$ as numbers of `request_update` calls in process p on all possible execution paths from the start point of this delta cycle to the statement s_k. The lockset algorithm consists of rules for program statements:

$[notify(e_m)]_k^p :$
$$\begin{cases} L(s_k, e_r) = L(s_n, e_r) + 1, & r = m \\ L(s_k, e_r) = L(s_n, e_r), & r \neq m \end{cases} , s_n \in Pred(s_k)$$

$[run\_update(...)]_k^p :$
$$U(s_k, p) = U(s_n, p) + 1, \quad s_n \in Pred(s_k)$$

$[\phi]_k^p :$
$$L(s_k, e_r) = \bigcup_{\forall s_n \in Pred(s_k)} L(s_n, e_r), \quad \forall r$$
$$U(s_k, p) = \bigcup_{\forall s_n \in Pred(s_k)} U(s_n, p)$$

$[if(...)]_k^p :$
$$\begin{cases} L(s_k, e_r)_{true} = L(s_n, e_r) \\ U(s_k, p)_{true} = U(s_n, p) \\ L(s_k, e_r)_{false} = L(s_n, e_r) \\ U(s_k, p)_{false} = U(s_n, p) \end{cases} , s_n \in Pred(s_k), \forall r,$$

ϕ – a program model statement, where program branches are joined, $Pred(s_k)$ – a set of statements which are direct predecessors of s_k. These rules are applied for program model statements according to their sequence in a process execution path.

After all execution paths are analyzed, evaluation phase analysis checks program points blocked on preceding evaluation phases. If there is an immediate notification for such a blocked point, this point is moved to a start point set and analysis of this evaluation phase continues (Fig. 1). Evaluation phase analysis continues until start point set is empty.

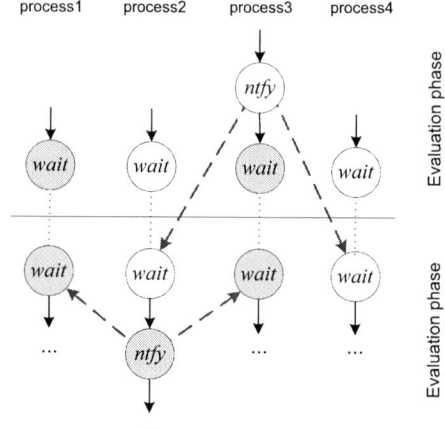

Fig. 1. Immediate notification analysis

Evaluation phase analysis determines set of finish points. For each finish point there are:

- a program state,
- a set of delayed notifications,
- a set of `request_update` calls.

D. Update phase analysis

Update phase analysis starts after the evaluation phase analysis if there are `request_update` calls in the evaluation phase of current delta cycle. For each finish point of execution, if paths have `request_update` call then corresponding `update` method is analyzed. The input state of an `update()` method analysis is a state in this finish point. It is not required to consider states of other finish points because `update` method may operate with current object data only. An example of update phase analysis in shown in Fig. 2 (`ru` denotes `request_update` call and `upd` – a first statement of `update`).

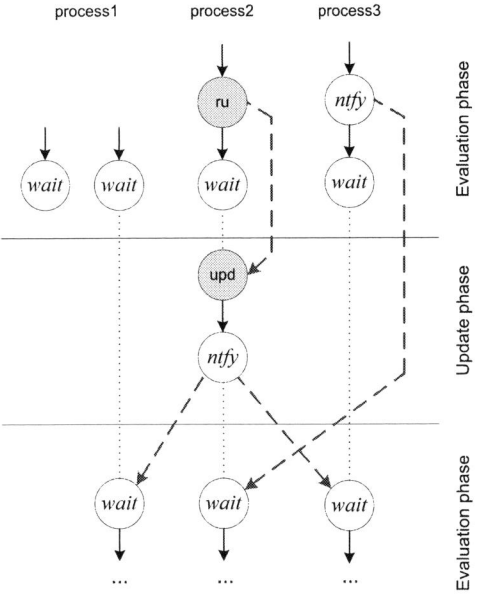

Fig. 2. Update phase analysis

Analysis of `update` methods can be performed in any order because there is no interaction between methods on update phase. Results of the update phase analysis are linked to corresponding finish points. For such finish point there are:
- modified program state,
- an additional set of delayed notifications.

E. Interprocess Interaction Algorithm

For each delta cycle static analysis algorithms work with individual program state on execution paths. At the end of delta cycle, all program states' changes are merged. Interprocess interaction algorithm determines actual values of shared variables in processes. An *actual value* of a variable is the last value assigned to this variable.

To determine actual values, a set of changed objects (Def) is used. This set is filled during evaluation and update phase analysis. If a value of variable is changed in an execution path to finish point $\mathcal{F}_i^j$, it is added to $Def(\mathcal{F}_i^j)$. If a variable leaves its scope, it is removed from corresponding Def.

The state merging is done for states of all finish points. The result value of a variable a in a finish point $\mathcal{F}_{i_1}^{j_1}$ is computed using the following formula:

$$val(a, \mathcal{F}_{i_1}^{j_1})' = \bigcup_{\forall i,j:j \neq j_1, a \in Def(\mathcal{F}_i^j)} val(a, \mathcal{F}_i^j) \cup$$
$$\bigcup_{[a \in Def(\mathcal{F}_{i_1}^{j_1})] \vee [\forall j \exists i: a \notin Def(\mathcal{F}_i^j)]} val(a, \mathcal{F}_{i_1}^{j_1}),$$

where $val(a, \mathcal{F}_{i_1}^{j_1})$ and $val(a, \mathcal{F}_{i_1}^{j_1})'$ – values of variable a before and after a finish point $\mathcal{F}_{i_1}^{j_1}$.

An example of interprocess analysis using the developed algorithm in shown in Fig. 3.

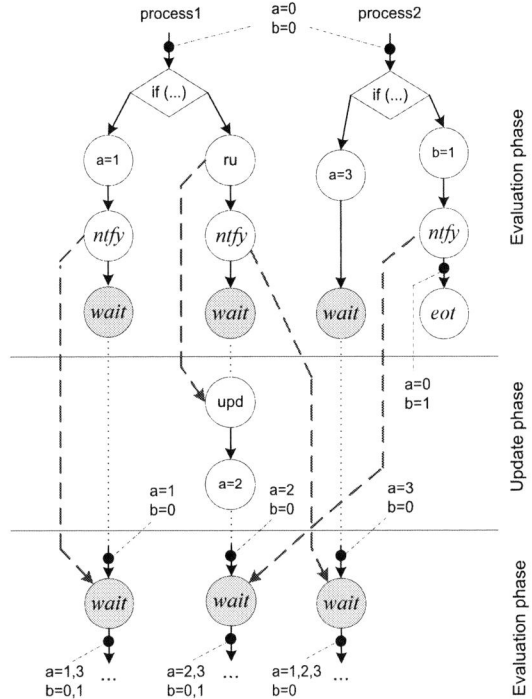

Fig. 3. Interprocess analysis

F. Variant Analysis

A delta notification may occur in an arbitrary number of execution paths of a process. For an event e all finish points at the end of delta cycle can be divided into following groups:
- finish points, all paths to which contain `notify(e, 0)` statement;
- finish points, all paths to which don't contain `notify(e, 0)` statement;
- finish points, some paths to which contain `notify(e, 0)` statement.

Both the first and the second groups unambiguously define whether process blocked on corresponding event has to be resumed or not. Finish points from the last group define notification ambiguously. Ambiguous notifications don't allow to determine a set of start points of the next delta cycle. To cope with this problem we use variant analysis.

Variant separation algorithm has the following steps:
- Determine a set of events $\{e_i\}$, $i = 1...n$ that have been notified in the delta cycle.
- For each finish point of a process calculate a notification vector $V = <v_1, v_2, ..., v_n>$, where $v_i = 1$ if there is notification for e_i, and $v_i = 0$ otherwise. If a finish point contains ambiguous notification, then it has several notification vectors.

- Create combinations of finish points. A combination C contains one finish point for each process.
- Calculate notification vector $V(C)$ for each combination C as $V(C) = V_1 \oplus V_2 \oplus ... \oplus V_K$, where V_l is a notification vector for a finish point and $\oplus$ is a bit OR operation.
- Combinations with the same notification vectors are joined into one variant.

Obviously, a variant has only unambiguous notification for all events. All variants are analyzed individually.

G. *Computational Complexity*

Computational complexity of the developed algorithms for one delta cycle is the following:

$$O(2^v \cdot k \cdot n),$$

v – number of ambiguous notifications (see 3.6), k – number of processes, n – number of statements analyzed in a delta cycle.

IV. DEADLOCK DETECTION

The basis of synchronization primitives in SystemC includes blocking statement `wait` and notification statement `notify`. All other synchronization primitives are based on `wait` and `notify`. So the only possible deadlock situation is infinite wait on an event that will never be notified.

There are several types of deadlocks in SystemC programs. A deadlock may lead to infinite blocking of all program processes or infinite blocking of some program processes. These two cases make it necessary to have different deadlock detection rules. The detection rules are based on an analysis of blocked points and are applied at the start of each delta-cycle.

Rule 1

A deadlock of the first type is possible if for each process there is at least one blocked point. In case these blocked points are consistent, all program processes are waiting for each other. The rule for detection of such deadlocks is presented below.

$$\forall j : \exists p_j \in P, \exists \mathcal{B}_i^j,$$

where P is a set of processes, $\mathcal{B}_i^j$ is a blocked program point.

Rule 2

A deadlock of the second type is possible if no notification exists for a `wait` statement. In practice waiting for an event rather long usually means a deadlock error. For a given design it is possible to determine what waiting time period corresponds to deadlock. To deal with deadlocks of this type, we consider time period while wait statement is being blocked. The detection rule is presented below.

$$\exists i, j : \exists \mathcal{B}_i^j, count(\mathcal{B}_i^j) > T,$$

where $\mathcal{B}_i^j$ is a blocked program point, $count(\mathcal{B}_i^j)$ is a number of delta-cycles the program is blocked in $\mathcal{B}_i^j$ and T is a threshold value.

Threshold values can be set globally (for all `wait` statements), or locally (for any subset of `wait` statements independently). The threshold value is provided by a developer, who determines it based on his knowledge of a particular system design.

V. EFFICIENCY EVALUATION

The approach presented in this paper has been implemented in Deadlock Analyzer tool. Source codes of the tool include more than 50 KLOC of Java code in total.

In this section we describe experiments being conducted in order to estimate an efficiency of Deadlock Analyzer. Since there is no available SystemC deadlock testbenches, we created our own testbench including artificial tests and one real-life model.

Test cases include well-known multi-process synchronization problems: dining philosophers, producer-consumer, crossroads. We created program modifications of the code with deadlock as well as without deadlock. There are 65 test programs in total (each program is about 100 LOC and contains 2-5 processes). All deadlocks were successfully detected, no false errors were found.

For scalability estimation we have conducted several experiments with dining philosophers. In these experiments number of philosophers (SystemC processes) varies from 2 to 100. The results of the experiments are shown in Fig. 4.

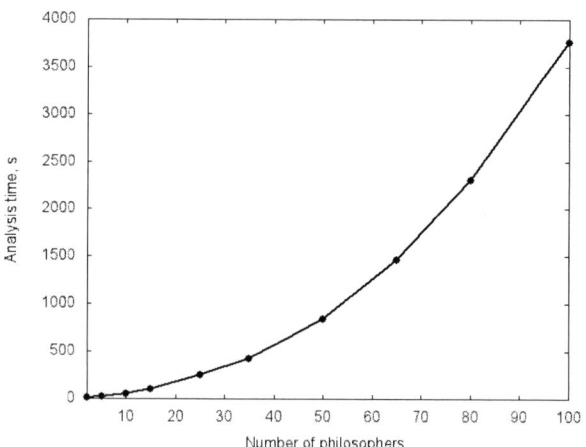

Fig. 4. Dining philosophers experimental results

A real-life example is a model of a video accelerator provided by Intel Labs. Figure 5 shows a top level of the accelerator and its modelled surrounding. The accelerator itself has deeply embedded microcontroller and high performance data crunching engine consisting of custom datapath, local SRAM memory and Direct Memory Access unit (DMA). An accelerator environment is represented in simplified form as a CPU and its memory. The system memory can be accessed by DMA independent of CPU and thus is represented as a separate unit.

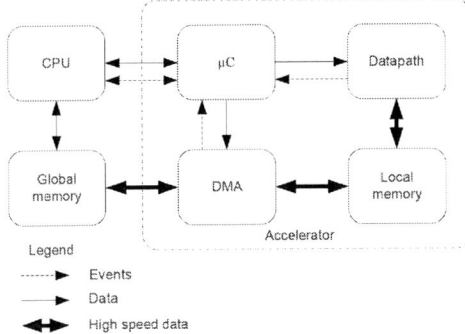

Legend
┄┄► Events
───► Data
◄──► High speed data

Fig. 5. Example system under test

Source codes of the video accelerator model contain more than 7800 LOC. This model contains one real deadlock, which was successfully found by our tool.

To estimate soundness of the deadlock detection we utilized mutation testing. We used two types of mutation for artificial deadlock injection:

- moving `notify` into the branch of `if`,
- insertion of unsatisfied `wait`.

Using simple bash script we generated 11 mutants of the video accelerator model. All injected deadlocks were successfully found in these test cases. In these experiments 3 false deadlocks were found.

We use Intel Xeon E5410 2.33 GHz processor with 12 GB of RAM for the described experiments. Analysis time for each mutant varies from 10 to 30 minutes.

VI. CONCLUSIONS

In this paper we have described a method for deadlock detection in SystemC programs using static analysis. The presented method has been implemented in Deadlock Analyzer tool, which we used to estimate precision and soundness of our algorithms. Developed algorithms yield correct results on the set of artificial tests: all the expected deadlocks were successfully detected, no false errors were found. Moreover, our analysis of a real-life video accelerator model developed at Intel Labs corroborates the ability of deadlock detection in industrial-level SystemC designs. Soundness of our method is confirmed by series of experiments with mutants of this model.

REFERENCES

[1] "IEEE standard SystemC language reference manual," *IEEE Std 1666-2005*, pp. 0_1 –423, 2006.

[2] M. Y. Vardi, "Formal techniques for SystemC verification," in *Proceedings of the 44th annual Design Automation Conference*, ser. DAC '07. New York, NY, USA: ACM, 2007, pp. 188–192. [Online]. Available: http://doi.acm.org/10.1145/1278480.1278527

[3] A. Habibi and S. Tahar, "Towards an efficient assertion based verification of SystemC designs," in *Proceedings of the High-Level Design Validation and Test Workshop, 2004. Ninth IEEE International*. Washington, DC, USA: IEEE Computer Society, 2004, pp. 19–22. [Online]. Available: http://portal.acm.org/citation.cfm?id=1255481.1256104

[4] L. Pierre and L. Ferro, "Enhancing the assertion-based verification of TLM designs with reentrancy," in *Formal Methods and Models for Codesign (MEMOCODE), 2010 8th IEEE/ACM International Conference on*, 2010, pp. 103 –112.

[5] E. Cheung, P. Satapathy, V. Pham, H. Hsieh, and X. Chen, "Runtime deadlock analysis of SystemC designs," in *High-Level Design Validation and Test Workshop, 2006. Eleventh Annual IEEE International*, 2006, pp. 187 –194.

[6] D. Tabakov and M. Vardi, "Monitoring temporal SystemC properties," in *Formal Methods and Models for Codesign (MEMOCODE), 2010 8th IEEE/ACM International Conference on*, 2010, pp. 123 –132.

[7] C. Helmstetter, F. Maraninchi, and L. Maillet-Contoz, "Full simulation coverage for SystemC transaction-level models of systems-on-a-chip," *Form. Methods Syst. Des.*, vol. 35, pp. 152–189, October 2009. [Online]. Available: http://portal.acm.org/citation.cfm?id=1644391.1644395

[8] K. Marquet, M. Moy, and B. Karkare, "A theoretical and experimental review of SystemC front-ends," in *Forum for Design Languages (FDL)*, 2010, B.1.4, C.3 OpenTLM (Projet Minalogic).

[9] C. Helmstetter and O. Ponsini, "A comparison of two SystemC/TLM semantics for formal verification," in *Formal Methods and Models for Co-Design, 2008. MEMOCODE 2008. 6th ACM/IEEE International Conference on*, 2008, pp. 59 –68.

[10] H. Garavel, C. Helmstetter, O. Ponsini, and W. Serwe, "Verification of an industrial SystemC/TLM model using LOTOS and CADP," in *Proceedings of the 7th IEEE/ACM international conference on Formal Methods and Models for Codesign*, ser. MEMOCODE'09. Piscataway, NJ, USA: IEEE Press, 2009, pp. 46–55. [Online]. Available: http://portal.acm.org/citation.cfm?id=1715759.1715765

[11] D. Grosse, H. Le, and R. Drechsler, "Proving transaction and system-level properties of untimed SystemC TLM designs," in *Formal Methods and Models for Codesign (MEMOCODE), 2010 8th IEEE/ACM International Conference on*, 2010, pp. 113 –122.

[12] K. Marquet and M. Moy, "PinaVM: a SystemC front-end based on an executable intermediate representation," in *Proceedings of the tenth ACM international conference on Embedded software*, ser. EMSOFT '10. New York, NY, USA: ACM, 2010, pp. 79–88. [Online]. Available: http://doi.acm.org/10.1145/1879021.1879032

[13] D. Karlsson, P. Eles, and Z. Peng, "Formal verification of SystemC designs using a petri-net based representation," in *Proceedings of the conference on Design, automation and test in Europe: Proceedings*, ser. DATE '06. 3001 Leuven, Belgium, Belgium: European Design and Automation Association, 2006, pp. 1228–1233. [Online]. Available: http://portal.acm.org/citation.cfm?id=1131481.1131824

[14] C.-N. Chou, C.-H. Hsu, Y.-T. Chao, and C.-Y. Huang, "Formal deadlock checking on high-level SystemC designs," in *Computer-Aided Design (ICCAD), 2010 IEEE/ACM International Conference on*, 2010, pp. 794 –799.

[15] C. Traulsen, J. Cornet, M. Moy, and F. Maraninchi, "A SystemC/TLM semantics in PROMELA and its possible applications," in *Proceedings of the 14th international SPIN conference on Model checking software*. Berlin, Heidelberg: Springer-Verlag, 2007, pp. 204–222. [Online]. Available: http://portal.acm.org/citation.cfm?id=1770532.1770552

[16] H. Hojjat, M. Mousavi, and M. Sirjani, "Process algebraic verification of SystemC codes," in *Application of Concurrency to System Design, 2008. ACSD 2008. 8th International Conference on*, 2008, pp. 62 –67.

[17] N. Blanc and D. Kroening, "Race analysis for SystemC using model checking," in *Proceedings of the 2008 IEEE/ACM International Conference on Computer-Aided Design*, ser. ICCAD '08. Piscataway, NJ, USA: IEEE Press, 2008, pp. 356–363. [Online]. Available: http://portal.acm.org/citation.cfm?id=1509456.1509540

[18] E. M. Clarke and B.-H. Schlingloff, *Model checking*. Amsterdam, The Netherlands, The Netherlands: Elsevier Science Publishers B. V., 2001, pp. 1635–1790. [Online]. Available: http://portal.acm.org/citation.cfm?id=778522.778533

[19] M. Holzer and M. Rupp, "Static code analysis of functional descriptions in SystemC," in *Proceedings of the Third IEEE International Workshop on Electronic Design, Test and Applications*. Washington, DC, USA: IEEE Computer Society, 2006, pp. 243–248. [Online]. Available: http://portal.acm.org/citation.cfm?id=1109224.1109387

[20] F. Nielson, H. R. Nielson, and C. Hankin, *Principles of Program Analysis*. Secaucus, NJ, USA: Springer-Verlag New York, Inc., 1999.

SAMOSA: Scratchpad Aware Mapping Of Streaming Applications

Zubair Wadood Bhatti,
Davy Preuveneers and Yolande Berbers
DistriNet, K.U.Leuven, Belgium
{zubairwadood.bhatti, davy.preuveneers,
yolande.berbers}@cs.kuleuven.be

Narasinga Rao Miniskar and Roel Wuyts
IMEC, Belgium
{miniskar, wuytsr}@imec.be

Abstract—Scratchpad memories have now emerged as an alternative to caches for energy constrained embedded systems. However, effectively mapping data on them while considering energy/timing trade-offs remains a challenge. We present SAMOSA as a technique for mapping streaming applications to scratchpad based MPSoCs. The contribution of this approach is a representation and transformation of the mapping problems — buffer dimensioning and allocation — to a constraint-based optimization problem. SAMOSA was used to explore energy-execution time trade-offs for mapping the H.264 decoder to a scratchpad-based MPSoC. Results show that scratchpad awareness has significant impacts on the energy-execution time trade-offs.

I. INTRODUCTION

With the recent widespread availability of mobile broadband internet and multi-core computing resources on hand held computers, the use of such devices for data intensive applications, such as multimedia is increasing. However, battery time remains a great concern for such devices. Viredaz et al. show in [26] that multimedia applications are one of the most power hungry activities on a mobile device and that the energy consumption of the memory sub-system for multimedia applications is significant. In this paper we focus on reducing the energy consumption of multimedia applications on such devices through better memory management.

In [2], the authors show that scratchpad memories (SPM) consume on average 40% less energy and 46% less area-time when compared to caches of the same capacity. Moreover, they provide much better timing predictability than caches. Therefore, SPMs are now included in many modern day embedded platforms such as ARM 10E, IBM Cell BE, GeForce GTX and Texas Instruments TMS370CX7X. Unlike caches that are managed by hardware and that select their contents on the principle of spatio-temporal locality, in SPM based systems the software is responsible for the allocation to scratchpads.

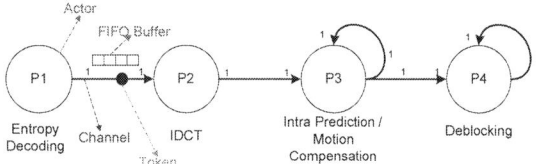

Fig. 1. Synchronous Dataflow Graph for H.264 decoder

One particular application domain that can benefit from scratchpad memories are data intensive streaming applications, such as multimedia players. These applications are often modeled with Synchronous Dataflow Graphs (SDFG) [17] as shown in Fig. 1. In an SDFG, *actors* communicate with each other by passing *tokens* over the logical links called *channels*. These logical channels need to mapped in the memories as *FIFO buffers*. Being a simple model of computation, SDFGs are easy to parallelize, schedule and analyze for timing and resource requirements. OpenDF [5] and StreamIt [24] are examples of stream programming languages whose programming models resemble SDFG.

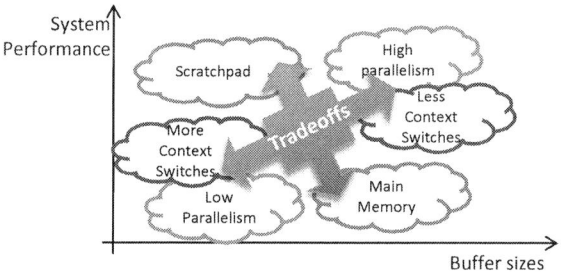

Fig. 2. Trade-offs in buffer dimensioning

Mapping applications described as synchronous dataflow graphs to scratchpad-based MPSoCs raises two important challenges, i.e. *buffer dimensioning* and *scratchpad allocation*:

- *Buffer dimensioning* is defined as, deciding the sizes of the different buffers between actors over intervals of time. Fig. 2 highlights some trade-offs in buffer dimensioning for a scratchpad based system.
  - From a scheduling perspective, a larger buffer provides better decoupling between a pair of actors. Better decoupling provides the freedom to construct schedules with fewer context switches and/or with more parallelism as shown in Fig. 3.
  - For a scratchpad based MPSoC, the execution time and energy consumption of an actor depends on the memory its reads and writes are addressed to. Therefore it is sometimes desirable to have smaller

978-1-4577-0671-4/11 $26.00 © 2011 IEEE

buffers, so that they could fit in the scratchpad.

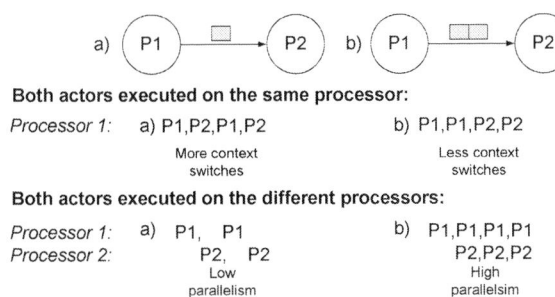

Both actors executed on the same processor:

Processor 1: a) P1,P2,P1,P2   b) P1,P1,P2,P2

More context   Less context
switches   switches

Both actors executed on the different processors:

Processor 1: a) P1, P1   b) P1,P1,P1,P1
Processor 2:   P2, P2   P2,P2,P2
Low   High
parallelism   parallelsim

Fig. 3.   Context switches and parallelism

- *Scratchpad allocation* is defined as, deciding which buffers are mapped onto the scratchpad memory and over what period of time.

  - Data access density of a buffer is the number of accesses made to the data in the buffer during a firing of an actor divided by the amount of data moved to/from the buffer. In uniprocessor system with only one execution path, allocating buffers with the highest access density to the scratchpad improves both the energy efficiency and the execution time of the application.

  - In a multiprocessor system, however, there are several execution paths of different lengths due to the way the application is scheduled on the different processors. In order to meet a given real-time deadline, it is sometimes necessary to allocate the buffers with lesser data access densities in the longest execution path to the faster scratchpad (as shown in Fig. 4) to reduce the length of the longest execution path.

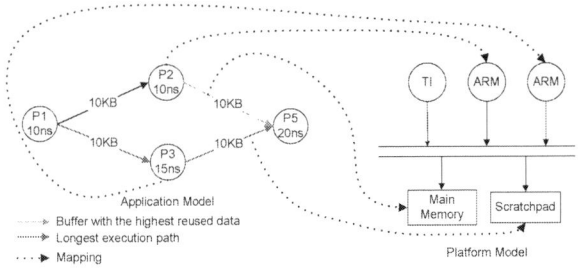

Fig. 4.   Trade-offs with scratchpad allocations

The two decisions are interdependent. Moreover, our results show that both these decisions can have significant impact on the timing and energy aspects of system performance. SAMOSA is an technique for mapping SDFGs to scratchpad based MPSoCs. Actors are scheduled over processors and tokens are mapped onto memories. The trade-off between execution time (or throughput) and energy is explored. The buffer dimensioning in SAMOSA is scratchpad aware and the buffer sizes are allowed to vary over time. The scratchpad

allocation in SAMOSA implicitly takes into account access densities to buffers and the lengths of the different execution paths on a multiprocessor system.

The proposed technique was used to find the energy-throughput trade-offs for the H.264 decoder (shown in Fig. 1) to a scratchpad-based MPSoC. The platform is similar to TI OMAP, with two StrongARM processors, a TI C64X+ processor, a scratchpad memory and a main memory. We observed that scratchpad sizes strongly influence the energy-throughput trade-off and that letting buffer sizes be functions of time produced mappings that were significantly more energy efficient when compared to fixed buffer sizes.

The paper is organized as follows. Section II gives an overview of the SAMOSA methodology. Section III explains the formats of the application model and platform model required as inputs for SAMOSA. Section IV explains how the application model is transformed to a model that is easier to schedule. Section V presents the constraint programming model for exploration and discusses the variables, constraints and the objective function. Section VI presents the results of our SAMOSA technique on the H.264 decoder. Section VII discusses related work in the domains of buffer dimensioning and scratchpad allocation. Finally, section VIII discusses the conclusion and future work.

II. OVERVIEW OF SAMOSA

SAMOSA explores schedules of tasks over multiple processors and schedules of data buffers in different types of memories. Buffer sizes are explored as a property of the task schedules, whereas scratchpad allocations are explored as a property of the schedules of data buffers in the memories. The objective of SAMOSA is to find schedules and scratchpad allocations such that the energy consumption is minimized while meeting the quality of service (timing/throuput) requirements. All aspects of the problem, including context switches, parallelism and data access densities, are transformed into the domain of energy consumption and execution time of task schedules and memory allocations. SAMOSA combines off-line compiled structural information about the application with dynamically profiled information of non-functional properties such as execution time and energy consumption to carry out this multi-objective optimization.

Fig. 5 presents an overview of SAMOSA. We start with an SDFG model of the application. Each actor of the SDFG is profiled for energy consumption and execution time on the different processors in the platform, with all possible memory mappings of its inputs and outputs. This profiling

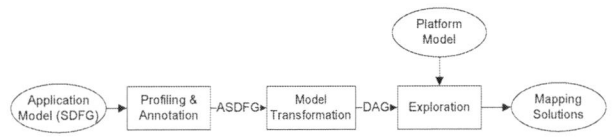

Fig. 5.   Overview of SAMOSA

978-1-4577-0671-4/11 $26.00 © 2011 IEEE

information is annotated on the SDFG to get an Annotated Synchronous Dataflow Graphs (ASDFG). In order to simplify the scheduling, we limit the concurrency of the ASDFG by unfolding it into a Directed Acyclic Taskgraph (DAG), as shown in Fig. 6. This DAG is then used for exploring the required schedules.

The scheduling exploration is realized with IBM ILOG where the mapping is modeled as a constraint based scheduling problem in Optimization Programming Language (OPL) [15]. The execution schematics of the DAG, along with the platform resources (processing, memory) are modeled as constraints. The solver is then asked to find a minimum energy schedule for a given deadline. A set of schedules that give a trade-off between energy and execution time is found by repeating the procedure for the different deadlines.

III. PLATFORM & APPLICATION MODELS

This section describes the models for the application and the platform that are used as an input for SAMOSA.

A. Platform Model

Most of the information about the platform required is extracted through detailed profiling of the application on the platform. A platform is described as a tuple *(Processors, Memories)*, where *Processors* and *Memories* represent the sets of processors and memories in the platform. A processor is defined as a tuple *(ProcessorID, CtxSwthTime, CtxSwthEnergy)*, where *ProcessorID* is a unique number for every processor, *CtxSwthTime* and *CtxSwthEnergy* are the time and energy consumed during a context switch. A context switch includes the loading and the initialization of the actor code before it can start processing data. A memory is defined as a tuple *(MemID, Size)*, where *MemID* is a unique identifier for every memory and *Size* is the size of the memory. The energy values and execution time of the context switch include those spent in the memories and interconnect along with those of the processor.

B. Application Model

An Annotated Synchronous Dataflow Graph (ASDFG) is a Synchronous Dataflow Graph annotated with profiling information. The ASDFG is described as a tuple (A, C), where A is the set of *Actors* and C is the set of *Channels*.

1) Actors: An Actor is assumed to be a deterministic piece of code. In each execution instance it consumes a fixed amount of data from each of its input channels and produces a fixed amount of data on its outputs. The execution time and energy consumption of the actor depends only on the type of processor it is executed on and the memories where its input/output is mapped onto. In case when these properties also depend on the input data, worst case assumptions are taken. The actors themselves are stateless. Any required state is explicitly represented as a self loop channel. Actors are described as a tuple *(ActorID, Ports, Modes)*, where *ActorID* is a unique number for every actor. *Ports* is the set of all input and output ports of the actor. *Modes* is the set of all possible execution configurations for the actor, i.e. one configuration for each possible combination of processors and memories.

2) Modes: A mode is a configuration in which an actor can execute. Consider the application and the platform shown in Fig. 4; the actor *P2* can be executed on either of the two processors. It has two ports each of which can be mapped to read/write the data either from/to the main memory or the scratchpad. Therefore, the actor *P2* has eight modes. A mode describes the execution configuration as a tuple *(ModeID, ProcessorID, PortMemIDs, ExecTime, Energy)*, where *ModeID* is a unique number for every mode. *ProcessorID* identifies the processor used in this configuration. *PortMemIDs* is the set of *MemID* that identifies the memories used for each port, i.e. one *MemID* for each port. *ExecTime* and *Energy* are the execution time and energy consumed if the actor is executed in the given mode. These energy and execution time values are obtained through profiling and include the time and energy spent in the processors, memories and interconnect.

3) Channels: A channel is defined as *(ChID, Source, Sink, InRate, OutRate)*, where *ChID* is a unique identifier for each channel. *Source* and *Sinks* are *ActorID*s of the source and sink actors connected to the channel. *InRate* and *OutRate* are the amounts of data produced or consumed by the source and sink actors respectively, each time they execute.

IV. MODEL TRANSFORMATION

A Synchronous Data Flow Graph is an auto-concurrent model of computation where parallelism is implicit. In order to derive an optimal schedule on a multiprocessor, all possible combinations of the different instances of actors need to be considered which is often not possible. A common approach is to first construct a Directed Acyclic Graph (DAG) where parallelism is explicit [21]. Fig. 6 illustrates the model transformation of a synthetic SDFG. In order to construct a DAG from an SDFG, a periodically admissible sequential schedule (PASS) needs to be computed first. To compute a PASS balance equations for each channel are formulated and an integral vector is found that solves the system of equations. The PASS is then unfolded by an 'unroll factor' and dependencies are added between all the different instances of the actors [14]. The 'unroll factor' depends on the target platform and can be specified by the developer. A DAG is defined as (T, E) where, T is a set of *Tasks* and E is a set of *Edges*.

a) Tasks: Tasks are instances of SDFG actors. Therefore they have all of the same properties and modes as the SDFG actor they belong to. It should be noted that if the tasks belonging to the same actor are executed one after another on the same processor, there is no context switch because the same actor is fired multiple times. Tasks are defined as *(TaskID, ActorID, Ports, Modes)*.

b) Edges: Edges are data transfers between the tasks. An edge is defined as *(EdgeID, Source, Sink, Size)*. It is possible that several edges in the DAG belong to the same channel in the SDFG. The *Size* of an edge is the greatest common divisor of *InRate* and *OutRate* of the SDFG channel it belongs to. The question of time dependant buffer sizes is now transformed

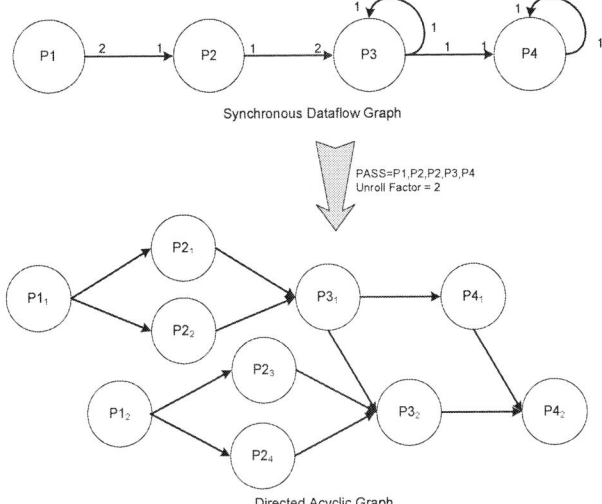

Fig. 6.   Model Transformation

into the question of the number of live edges in the DAG and their fixed sizes.

V. Scheduling Exploration

This section describes the scheduling exploration in SAMOSA. In [7], the authors show that scheduling a DAG onto multiple processors is an NP-Complete problem. In order to find a schedule for tasks on the processors and a schedule for the edges on the memories, we formulate the problem as a constraint based scheduling problem and solve it with a constraint solver [15]. In this section we describe the solution space, the constraints and the minimization objective of the problem.

A. Solution Space

We have two types of activities in our problem: tasks and edges. These activities use four types of resources: processors, memories, time and energy. The usage of these resources by the activities form the two schedules that define the solution.

The resources in our system are classified into two types; *Renewable* and *Nonrenewable*. Renewable resources are resources that can be returned to the system once a task is finished, such as processors and memories. The nonrenewable resources are permanently consumed, such as energy and time. We model usage of nonrenewable resources with the variables of types *Interval* and *Sequence*. The usage of renewable resources are represented as *Cumulative functions*.

1) Intervals: An Interval variable $\underline{a}$ has a start $s(\underline{a})$ and an end $e(\underline{a})$ when it is present; these variables can also be declared as 'optional' in which case they can also be absent $\perp$ i.e. they don't have a start or end. The domain of $\underline{a}$ is $dom(\underline{a})$: The size of interval $\underline{a}$ is $IntervalSize(\underline{a})$:

$$IntervalSize(\underline{a}) = e(\underline{a}) - s(\underline{a})$$

a) TaskTime: is an array of intervals (one for each task) that represent the time slot occupied by each task. These are not optional intervals. Therefore each task must be allocated at least one time slot. The size of this interval is not fixed and depends on the selected mode. $TaskTime[T]$ represents the interval that belongs to the task T.

b) TaskModeTime: is a array of optional intervals in time (one for each mode of every task). $TaskModeTime[M]$ represents the time interval that belongs to the mode M. If the mode M is selected:

$$Size(TaskModeTime[M]) = M.ExecTime$$

The dot operator '.' is used to represent tuple components. *M.ExecTime* is the task execution time under mode M. If the mode M is not selected:

$$TaskModeTime[M] = \{\perp\}$$

c) TaskModeEnergy: is an array of optional intervals in energy (one for each mode of every task). $TaskModeEnergy[M]$ represents the energy interval that belongs to the mode M. If the mode M is selected:

$$Size(TaskModeEnergy[M]) = M.Energy$$

where *M.Energy* is the energy consumption under mode M. If the mode M is not selected:

$$TaskModeEnergy[M] = \{\perp\}$$

d) EdgeTime: is an array of intervals that represents the lifetimes of the edges. $EdgeTime[E]$ represents the lifetime of the edge E.

e) EdgeMode: is a two dimensional matrix of optional intervals. The interval $EdgeMode[E, M]$ is present if the edge E is connected to the task of mode M and the mode M is selected see Section V-B3. This variable is responsible for the allocation of buffers to either the main memory or the scratchpad.

2) Sequences: A sequence is a total ordered set of interval variables. These interval variables can also have types assigned to them. An additional constraint of *noOverlap* on sequences can force all the intervals in a sequence to be non-overlapping. Optionally the intervals in a sequence can have *Types*. These types can be used to impose a minimum transition distance between the intervals of a sequence in the presence of a *noOverlap* constraint. This translates into

$$noOverlap(\pi, M) \Leftrightarrow \forall \underline{a}, \underline{b} \in \underline{A},$$
$$((\pi(\underline{b}) = \pi(\underline{a}) + 1) \Rightarrow (e(\underline{a}) + M(T(\pi, \underline{a}), T(\pi, \underline{b})) \leq s(\underline{b})))$$

where $\underline{A}$ is a set of intervals, π is a sequence in $\underline{A}$ with types T and M is a matrix with the minimum transition distances between the different types of intervals.

a) ProcessorTime: is an array of non-overlapping sequences, one for every processor. The domain of these sequences is the array of time intervals of task modes *TaskModeTime*, for all modes that use the particular processor. In the sequence every *TaskModeTime* interval has a *Type* that represents the SDFG actor to which the task belongs *ActorID*. A minimum transition distance that equals *CtxSwitchTime* of the processor is imposed whenever tasks belonging to different actors are executed consecutively on the same processor, as shown in Fig. 7.

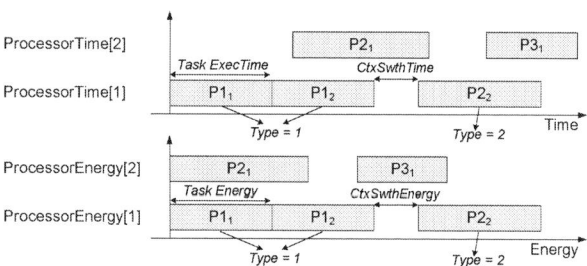

Fig. 7.   ProcessorTime & ProcessorEnergy sequences

b) ProcessorEnergy: is an array of non-overlapping sequences, one for every processor. The domain of these sequences is the array of energy intervals of task modes *TaskModeEnergy*, for all modes that use the particular processor. In the sequence every *TaskModeTime* interval has a *Type* that represents the SDFG actor to which the task belongs *ActorID*. A minimum transition distance that equals *CtxSwitchEnergy* of the processor is imposed whenever tasks belonging to different actors are executed consecutively on the same processor, as shown in Fig. 7.

3) Cumulative functions: Cumulative functions are used to represent the usage of renewable resources, such as the usage of different memories. These functions can be composed of elementary functions such as *Pulse(height,interval)* and *StepAtStart(height,interval)* shown in Fig. 8.

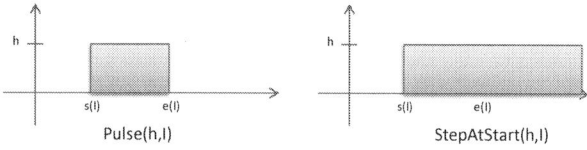

Fig. 8.   Elementary Functions

a) MemUsage: is a set cumulative functions that represents the memory usage of different memories over time. $MemUsage[Mem]$ is a model of the usage of the memory *Mem*.

$$MemUsage[Mem] =$$
$$\sum_{\{E \in Edges | EdgeMode[E,Mem] \neq \perp\}} Pulse(E.Size, EdgeTime(E))$$

B. Constraints

A valid mapping solution needs to satisfy several constraints. These constraints impose additional boundaries within the solution space, thus reducing the search space.

1) Tasks:

- All tasks need to finish before the deadline.

$$\forall_{T \in Tasks} e(TaskTime[T]) \leq Deadline$$

- One and only one *TaskModeTime* interval is present for every task in a valid schedule, and the *TaskTime* interval starts and ends with the selected *TaskModeTime* interval. This behavior is captured by the *alternative* constraint in OPL. An $alternative(\underline{a}, \{b1,..,bn\})$ constraint implies that, if the interval $\underline{a}$ is present then exactly one of $\{b1,..,bn\}$ is present. And that, $\underline{a}$ starts and ends with the chosen interval.

- If a task mode M is chosen the interval $TaskModeTime[M]$ is present, the corresponding interval $TaskModeEnergy[M]$ also has to be present and the sequences *ProcessorTime* and *ProcessorEnergy* should have the same order:

$$\forall_{M \in Modes},$$
$$presenceOf(TaskModeTime[M]) \Rightarrow$$
$$presenceOf(TaskModeEnergy[M]).$$

$$\forall_{M \in Modes, K \in Modes | M.ProcessorID = K.ProcessorID},$$
$$s(TaskModeTime[M]) \leq s(TaskModeTime[K]) \Rightarrow$$
$$s(TaskModeEnergy[M]) \leq s(TaskModeEnergy[K]).$$

2) Memory Size: At all times the maximum data allocated to a memory is less then the size of the memory:

$$\forall_{Mem \in Memories},$$
$$MemUsage[Mem] \leq Mem.Size$$

3) Edges:

- The edges enforce precedence constraints between the tasks and the life time of an edge starts at the end of its source task and ends at the end of its sink task:

$$\forall_{E \in Edges},$$
$$s(TaskTime[E.Sink]) \geq e(TaskTime[E.Source]),$$
$$s(EdgeTime[E]) = e(TaskTime[E.Source]),$$
$$e(EdgeTime[E]) = e(TaskTime[E.Sink]).$$

- The interval $EdgeMode[E,M]$ is present if the edge E is connected to the task of mode M and the mode M is selected:

$$\forall_{E \in Edges, M \in Modes},$$
$$presenceOf(TaskModeTime[M]) \Rightarrow$$
$$s(EdgeMode[E,M]) = s(EdgeTime[E]),$$
$$e(EdgeMode[E,M]) = e(EdgeTime[E]),$$
$$\neg presenceOf(TaskModeTime[M]) \Rightarrow$$
$$\neg presenceOf(EdgeMode[E,M]).$$

C. Objective function

The objective of the optimization is to minimize energy consumption while meeting the timing requirements and resource constraints. For our model we assume all energy spent in the processors and memories is used by tasks, in the form of *TaskModeEnergy* intervals. These intervals are aligned into *ProcessorEnergy* sequences, minimizing the total sum of the ends of these energy sequences will minimize the total energy consumption.

$$\text{minimize} \sum_{P \in Processors} ProcessorEnergy[P]$$

In order to explore the trade-off between energy consumption and execution time, a number of mapping solutions are calculated by incrementally varying the timing deadline.

VI. H.264 DECODER USE CASE

In this section we present the results of our technique for mapping an implementation of the H.264 decoder (Fig. 1) to the platform shown in Fig. 5. The H.264 decoder implementation takes a stream of H.264 and processes it frame by frame. Our TI OMAP 35xx like multiprocessor platform consists of two RISC processors (Strong ARM 1100x) operating at 200MHz, one VLIW processor (TI-C64X+) operating at 500MHz a scratchpad memory (SRAM) of 64KB and a main memory (SDRAM) of 128MB. We profile the application for execution times using SimItARM and TI-CCStudio v3.3 simulators for the StrongARM and TI-C64X+ processors respectively. For the energy consumption, we use JouleTrack [22] and functional level power analysis model of TI-C6X [16] for the StrongARM and TIC64X processors. Whereas the dataflow between the actors was measured using PinComm [12] and the memory accesses by Cachegrind. For this experiment we profiled 15 different sample videos of resolution 352x480 for every frame, and took the worst case values for each actor. An unroll factor of two was used for the ASDFG. A higher unroll factor might improve the quality of our results, but the amount of time required for the exploration would significantly increase.

A. Effects of varying Scratchpad sizes

Fig. 9 shows that varying the amount of scratchpad memory for the buffers significantly effects the energy-execution time trade-off. We can observe that in order to meet the same timing deadline with a smaller scratchpad memory available, a higher energy schedule is often required. The scratchpad sizes affect several aspects of scheduling. Smaller scratchpads can cause the buffer sizes to shrink resulting in more context switches and less parallelism. Also, they cause more DAG edges to be mapped onto the main memory thus increasing the execution time and energy consumption of the tasks connected to those edges. Furthermore, in order to meet the deadlines some tasks may have to be moved onto a faster but more energy hungry processor (in this case the TI-C64X+) thus causing higher energy schedules for the same timing performance.

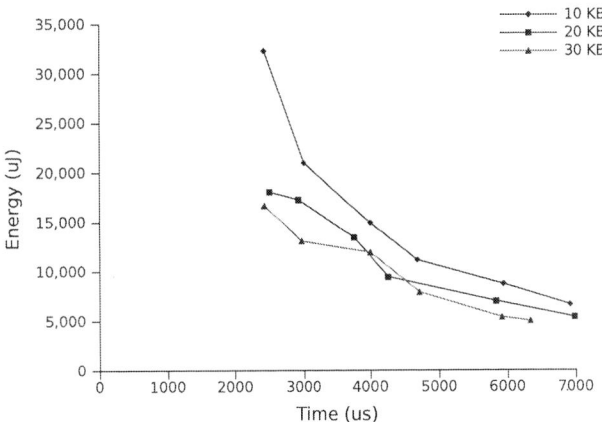

Fig. 9. Energy throughput trade-off under varying scratchpad sizes

B. Memory sharing buffers vs baseline allocation

Efficient utilization of memory space is very important for a scratchpad based system. We study the effects of *memory sharing* optimization [8], [20] on the trade-off between throughput and execution time. In the memory sharing approach buffer sizes are allowed to varry over time. Therefore, the same memory space can be used by different buffers. In the baseline allocation approach buffer sizes remain static. We simulated the baseline appraoch by using a *StepAtStart* function instead of a *Pulse* function in the *MemUsage* function in Section V-A3. Fig. 10 shows that significantly better schedules on the energy-execution time trade-off can be constructed for a memory sharing buffer allocation when compared to a baseline allocation. It is important to note that a memory sharing implementation is expected to require more runtime defragmentation overhead compared to a baseline allocation. However, defragmentation effects are not modeled in this simulation.

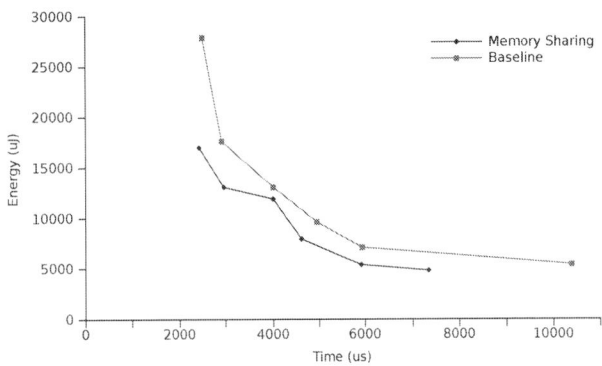

Fig. 10. Memory sharing vs baseline allocation

C. Illustrating the exploration phase

During the exploration phase, the solver attempts to find better alternatives given its best known solution thus far. Fig.

11 shows a screenshot from the ILOG profiler, illustrating how the solution converges in a feasible amount of time.

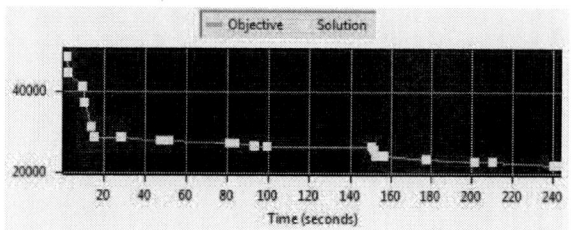

Fig. 11. IBM ILOG screenshot showing convergence of the solution.

VII. RELATED WORK

The problem addressed in this paper crosscuts two concerns in the domain of memory management for embedded systems, and we therefore subdivide the related work in two sections accordingly.

A. Buffer dimensioning

Several techniques are available for calculating the minimum buffer requirements such that there are no deadlocks [9], [1], while other calculate buffer requirements for *rate-optimal* schedules [11], [19]. Between these two extremes, other techniques explore trading of throughput and buffer size [23], [27], [28]. However, none of these techniques takes platforms with complex memory hierarchies into account, such as today's embedded devices. For such platforms the actual execution times of actors might depend on whether data is mapped onto a fast scratchpad memory or a slower flash memory. Therefore, the buffer size throughput trade-off also depends on the sizes of scratchpads.

The concept of letting buffers share the same memory space has been studied in [20] on a coarse grain level, where buffers with non-overlaying lifetimes can share the same memory space. In [8] on a fine grain level where even buffers with overlapping lifetimes are allowed to share the same memory space. These techniques have been shown to significantly reduce the overall memory requirements of streaming applications. However, our goal is to provide users with better energy efficiency and quality of service. Therefore we study the impact of such an optimization in the trade-off between execution time and energy.

B. Scratchpad allocation techniques

The problem of content selection for scratchpads has been extensively studied for C-like application models. [13] presents a survey of scratchpad allocation techniques. Most of these techniques suffer from the lack of information about the program structure [3] and are sometimes not applicable to applications with non-affine accesses, pointers, passing by reference, dynamic assignments etc. Therefore the source code usually needs to be 'cleaned' before these techniques could be applied [18]. In contrast to all these approaches we propose a scratchpad allocation technique at the level of dataflow graphs.

The literature on scratchpad allocation for dataflow graphs is quite limited [6], [4], [3]. In [6] a scratchpad aware scheduling technique is presented that maps both code and data segments of an application described as a dataflow diagram. The trade-off between context switches and buffer sizes is modeled but parallelism and energy costs are not considered. [4] and [3] present methodologies to dynamically allocate code and data of a Hetrochronous Dataflow Graph to scratchpads in a Harvard Architecture, but buffer sizes are not explored.

VIII. CONCLUSION & FUTURE WORK

In this paper, our energy aware technique for mapping SDFGs to scratchpad based MPSoCs called SAMOSA is presented. Tasks are mapped onto different processors whereas data buffers are mapped onto various types of memories. The buffer dimensioning is scratchpad aware and tradeoffs with context switches and parallelism are explored. The content selection criteria for scratchpads in a multiprocessor system implicitly considers data access densities of buffers as well as the lengths of different execution paths.

We used SAMOSA to explore the trade-off between energy and execution time for mapping the H.264 decoder to a scratchpad based MPSoC. Our results show that scratchpad sizes significantly affect the trade-off between energy and execution time Fig. 9. Therefore, we conclude that scratchpad awareness in energy-aware mapping of streaming applications to MPSoCs is important. Our results also show that letting buffers share the same memory space allows significantly better mappings on the energy-time trade-off for a scratchpad based system Fig. 10.

In the future we would like to use bio-inspired multi-objective evolutionary techniques and heuristics for a faster exploration phase and apply SAMOSA to more complex applications and platforms. We aim to extend this technique for platforms with multi hop communication, such as Network-on-Chip based platforms. We also intend to apply the concept of system-scenarios based design [10] to deal with dynamism in the application and use Polyhedral Process Networks [25] as application model for more accurate information data access patterns.

IX. ACKNOWLEDGEMENTS

This research is partially funded by the Flemish Agency for Innovation by Science and Technology (IWT) through a Strategic Basic Research project OptiMMA (Contract No IWT-060831) and by the research fund K.U.Leuven.

REFERENCES

[1] M. Ade, R. Lauwereins, and J. Peperstraete, "Data Memory Minimisation For Synchronous Data Flow Graphs Emulated On DSP-FPGA Targets," in *Design Automation Conference, 1997. Proceedings of the 34th*, jun 1997, pp. 64 –69.

[2] R. Banakar, S. Steinke, B.-S. Lee, M. Balakrishnan, and P. Marwedel, "Scratchpad memory: design alternative for cache on-chip memory in embedded systems," in *Proceedings of the tenth international symposium on Hardware/software codesign*, ser. CODES '02. New York, NY, USA: ACM, 2002, pp. 73–78. [Online]. Available: http://doi.acm.org/10.1145/774789.774805

[3] S. Bandyopadhyay, "Automated memory allocation of actor code and data buffer in heterochronous dataflow models to scratchpad memory," Master's thesis, EECS Department, University of California, Berkeley, Aug 2006. [Online]. Available: http://www.eecs.berkeley.edu/Pubs/TechRpts/2006/EECS-2006-105.html

[4] S. Bandyopadhyay, T. H. Feng, H. D. Patel, and E. A. Lee, "A scratchpad memory allocation scheme for dataflow models," EECS Department, University of California, Berkeley, Tech. Rep. UCB/EECS-2008-104, Aug 2008. [Online]. Available: http://www.eecs.berkeley.edu/Pubs/TechRpts/2008/EECS-2008-104.html

[5] S. S. Bhattacharyya, G. Brebner, J. W. Janneck, J. Eker, C. von Platen, M. Mattavelli, and M. Raulet, "OpenDF: a dataflow toolset for reconfigurable hardware and multicore systems," SIGARCH Comput. Archit. News, vol. 36, pp. 29–35, June 2009. [Online]. Available: http://doi.acm.org/10.1145/1556444.1556449

[6] W. Che and K. Chatha, "Scheduling of synchronous data flow models on scratchpad memory based embedded processors," in Computer-Aided Design (ICCAD), 2010 IEEE/ACM International Conference on, nov. 2010, pp. 205 –212.

[7] A. Darte, Yves, and F. Vivien, Scheduling and Automatic Parallelization, 1st ed. Birkhäuser Boston, Mar. 2000.

[8] M. H. Foroozannejad, M. Hashemi, T. L. Hodges, and S. Ghiasi, "Look into details: the benefits of fine-grain streaming buffer analysis," in Proceedings of the ACM SIGPLAN/SIGBED 2010 conference on Languages, compilers, and tools for embedded systems, ser. LCTES '10. New York, NY, USA: ACM, 2010, pp. 27–36. [Online]. Available: http://doi.acm.org/10.1145/1755888.1755894

[9] M. Geilen, T. Basten, and S. Stuijk, "Minimising buffer requirements of synchronous dataflow graphs with model checking," in Proceedings of the 42nd annual Design Automation Conference, ser. DAC '05. New York, NY, USA: ACM, 2005, pp. 819–824. [Online]. Available: http://doi.acm.org/10.1145/1065579.1065796

[10] S. V. Gheorghita, M. Palkovic, J. Hamers, A. Vandecappelle, S. Mamagkakis, T. Basten, L. Eeckhout, H. Corporaal, F. Catthoor, F. Vandeputte, and K. D. Bosschere, "System-scenario-based design of dynamic embedded systems," ACM Trans. Des. Autom. Electron. Syst., vol. 14, pp. 3:1–3:45, January 2009. [Online]. Available: http://doi.acm.org/10.1145/1455229.1455232

[11] R. Govindarajan, G. R. Gao, and P. Desai, "Minimizing buffer requirements under rate-optimal schedule in regular dataflow networks," J. VLSI Signal Process. Syst., vol. 31, pp. 207–229, July 2002. [Online]. Available: http://portal.acm.org/citation.cfm?id=598549.598651

[12] W. Heirman, D. Stroobandt, N. R. Miniskar, R. Wuyts, and F. Catthoor, "Pincomm: Characterizing intra-application communication for the many-core era," Parallel and Distributed Systems, International Conference on, vol. 0, pp. 500–507, 2010.

[13] M. Idrissi Aouad and O. Zendra, "A Survey of Scratch-Pad Memory Management Techniques for low-power and -energy," in 2nd ECOOP Workshop on Implementation, Compilation, Optimization of Object-Oriented Languages, Programs and Systems (ICOOOLPS'2007). Berlin Allemagne: ECOOP, 2007, pp. 31–38.

[14] A. Jantsch, Modeling Embedded Systems and SoC's: Concurrency and Time in Models of Computation (Systems on Silicon), 1st ed. Morgan Kaufmann, Jun. 2003.

[15] P. Laborie, "IBM ILOG CP Optimizer for Detailed Scheduling Illustrated on Three Problems," in Integration of AI and OR Techniques in Constraint Programming for Combinatorial Optimization Problems, ser. Lecture Notes in Computer Science, W.-J. van Hoeve and J. Hooker, Eds. Springer Berlin / Heidelberg, 2009, vol. 5547, pp. 148–162. [Online]. Available: http://dx.doi.org/10.1007/978-3-642-01929

[16] J. Laurent, N. Julien, E. Senn, and E. Martin, "Functional level power analysis: An efficient approach for modeling the power consumption of complex processors," in Proceedings of the conference on Design, automation and test in Europe - Volume 1, ser. DATE '04. Washington, DC, USA: IEEE Computer Society, 2004, pp. 10 666–. [Online]. Available: http://portal.acm.org/citation.cfm?id=968878.968987

[17] E. A. Lee and D. G. Messerschmitt, "Synchronous data flow," Proceedings of the IEEE, vol. 75, no. 9, pp. 1235 – 1245, sept. 1987.

[18] J.-Y. Mignolet and R. Wuyts, "Embedded multiprocessor systems-on-chip programming," Software, IEEE, vol. 26, no. 3, pp. 34 –41, may-june 2009.

[19] O. Moreira, T. Basten, M. Geilen, and S. Stuijk, "Buffer sizing for rate-optimal single-rate data-flow scheduling revisited," Computers, IEEE Transactions on, vol. 59, no. 2, pp. 188 –201, feb. 2010.

[20] P. Murthy and S. Bhattacharyya, "Shared buffer implementations of signal processing systems using lifetime analysis techniques," Computer-Aided Design of Integrated Circuits and Systems, IEEE Transactions on, vol. 20, no. 2, pp. 177 –198, feb 2001.

[21] J. Pino and E. Lee, "Hierarchical static scheduling of dataflow graphs onto multiple processors," Acoustics, Speech, and Signal Processing, IEEE International Conference on, vol. 4, pp. 2643–2646, 1995.

[22] A. Sinha and A. P. Chandrakasan, "Jouletrack: a web based tool for software energy profiling," in Proceedings of the 38th annual Design Automation Conference, ser. DAC '01. New York, NY, USA: ACM, 2001, pp. 220–225. [Online]. Available: http://doi.acm.org/10.1145/378239.378467

[23] S. Stuijk, M. Geilen, and T. Basten, "Throughput-buffering trade-off exploration for cyclo-static and synchronous dataflow graphs," Computers, IEEE Transactions on, vol. 57, no. 10, pp. 1331 –1345, oct. 2008.

[24] W. Thies, M. Karczmarek, and S. Amarasinghe, "Streamit: A language for streaming applications," in International Conference on Compiler Construction, Grenoble, France, Apr 2002. [Online]. Available: http://groups.csail.mit.edu/commit/papers/02/streamit-cc.pdf

[25] S. Verdoolaege, "Polyhedral process networks," in Handbook of Signal Processing Systems, S. S. Bhattacharyya, E. F. Deprettere, R. Leupers, and J. Takala, Eds. Springer US, 2010, pp. 931–965.

[26] M. A. Viredaz and D. A. Wallach, "Power evaluation of a handheld computer," Micro, IEEE, vol. 23, no. 1, pp. 66 – 74, jan/feb 2003.

[27] Y. Yang, M. Geilen, T. Basten, S. Stuijk, and H. Corporaal, "Exploring trade-offs between performance and resource requirements for synchronous dataflow graphs," in Embedded Systems for Real-Time Multimedia, 2009. ESTIMedia 2009. IEEE/ACM/IFIP 7th Workshop on, oct. 2009, pp. 96 –105.

[28] E. Zitzler, J. Teich, and S. S. Bhattacharyya, "Evolutionary algorithms for the synthesis of embedded software," IEEE Trans. Very Large Scale Integr. Syst., vol. 8, pp. 452–456, August 2000. [Online]. Available: http://portal.acm.org/citation.cfm?id=349683.358382

978-1-4577-0671-4/11 $26.00 © 2011 IEEE

A System Level Power Consumption Estimation for MPSoC

Santhosh Kumar RETHINAGIRI*, Rabie Ben ATITALLAH† and Jean-Luc DEKEYSER*

*IINRIA Lille Nord Europe, Université de Lille1 , France

Email: santhosh-kumar.rethinagiri@inria.fr and jean-luc.dekeyser@inria.fr

†LAMIH, Université de Valenciennes et du Hainaut Cambrésis, Valenciennes, France

Email: rabie.benatitallah@univ-valenciennes.fr

Abstract—This paper proposes an efficient Hybrid System Level (HSL) power estimation methodology for MPSoC. Within this methodology, the Functional Level Power Analysis (FLPA) is extended to set up generic power models for the different parts of the system. Then, a simulation framework is developed at the transactional level to evaluate accurately the activities used in the related power models. The combination of the above two parts lead to a hybrid power estimation that gives a better trade-off between accuracy and speed. The proposed methodology has several benefits: it considers the power consumption of the embedded system in its entirety and leads to accurate estimates without a costly and complex material. The proposed methodology is also scalable for exploring complex embedded architectures. The usefulness and effectiveness of our HSL methodology is validated through a typical mono-processor and multiprocessor embedded system designed around the Xilinx Virtex II Pro FPGA board.

I. INTRODUCTION

Due to the ongoing nano-miniaturisation in chip production, estimation of power consumption is becoming a critical pre-design metric in complex embedded systems such as Multi-Processor System-on-Chip (MPSoC). In current industrial and academic practices, power estimation using low level CAD tools is still widely adopted, which is clearly not suited to manage the complexity of modern embedded systems. Recently, the ITRS [7] and HiPEAC [1] roadmaps promote *power defines performance* and *power is the wall*. Facing this issue, designers should calculate the power consumption as early as possible in the design flow to reduce the time-to-market and the development cost. Today, system level power estimation is considered a vital premise to cope with the critical design constraints. However, the development of tools for power estimation at the system level is in the face of extremely challenging requirements such as the efficient power modeling methodology, the rapid system prototyping, and the accurate power estimates.

At the system level, the power estimation process is centered around two correlated aspects: *the power model granularity* and *the system abstraction level*. The first aspect concerns the granularity of the relevant activities on which the power model relies. It covers a large spectrum that starts from the fine-grain level such as the logic gate switching and stretches out to the coarse-grain level like the hardware component events. In general, fine-grain power estimation yields to a more correlated model with data and to handle technological parameters, which is tedious for system level designers. On

the other hand, coarse-grain power models depend on micro-architectural activities that cannot be determined easily. The second aspect involves the abstraction level on which the system is described. It starts from the usual Register Transfer Level (RTL) and extends up to the algorithmic level. In general, going from low to high design level corresponds to more abstract description and then coarser activity granularity. The power evaluation time increases as we go down through the design flow and the accuracy depends on the extraction of each relevant activity and the characterization methodology to evaluate the related power cost. In order to have an efficient power estimation methodology, we should find a better trade-off between these two aspects.

To answer the above challenges, we propose a new Hybrid System Level (HSL) power consumption estimation methodology for complex embedded systems. A key word in our contribution is *hybridization* between abstraction levels. Almost all the previous studies focus on power estimation for a given abstraction level without overcoming the wall of *speed/accuracy trade-off*. The idea here is to build up a hybrid power estimation tool that combines Functional Level Power Analysis (FLPA) for hardware power modeling and Transactional Level Modeling (TLM) simulation technique for rapid system prototyping and fast power estimation. Basically, the FLPA is used for processor power modeling. In the frame of this work, it will be extended to cover the other hardware components used in the MPSoC. After that, we go further in terms of scalability to target homogeneous multiprocessor architectures. The functional power estimation part is coupled with a fast SystemC [11] simulator in order to obtain the needed micro-architectural activities for power models, which allows us to reach a superior bargain between accuracy and speed.

This paper is organized as follows. Section II presents the related works, section III exposes the proposed HSL power estimation methodology. In Section IV, the power modeling methodology is applied to a typical MPSoC designed around Virtex II Pro FPGA board. To evaluate our approach in terms of accuracy, speed and scalability, experimental results are presented in Section V.

II. RELATED WORKS

Among the developed tools for power consumption estimation at the system level, we quote tools based on micro-architectural cycle-level simulation such as Wattch [4] and Simplepower [14]. They define fine-grain power models by

[1] http://www.hipeac.net/system/files/hipeacvision.pdf

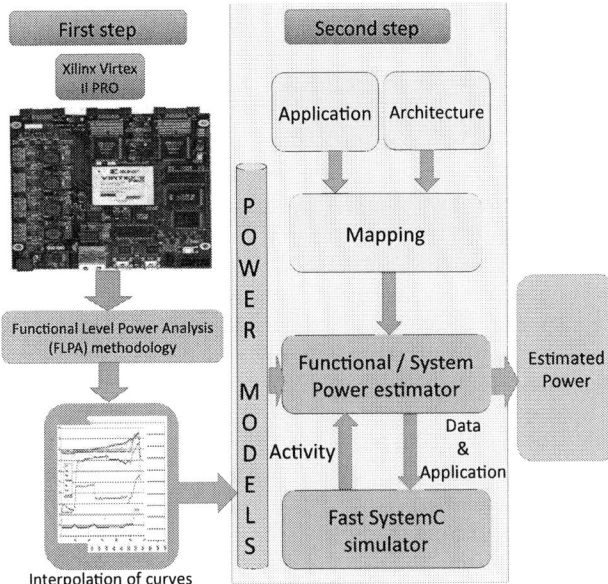

Fig. 1. Hybrid System Level (HSL) power estimation methodology

characterizing component features such as a set of instructions or functional blocks using analytic power laws. The contributions of the internal unit activities are calculated and added together during the execution of the program on the micro-architectural simulator. This approach needs low-level description of the architecture that are often difficult to obtain and with a least significant amount of simulation time.

In an attempt to reduce simulation time, recent efforts have been done to build up fast simulators using *Transaction Level Modeling* (TLM) [3]. SystemC [11] and its TLM 2.0 kit have become a de facto standard for the system-level description of Systems-on-Chip (SoC) by the means of offering different coding styles. Nevertheless, power estimation at the TLM level is still under research and is not well established. In [9] and [10], a methodology is presented to generate consumption models for peripheral devices at the TLM level. Relevant activities are identified at different levels and granularities. The characterization phase is however done at the gate level: from where they deduce the activity and power consumption for the higher level. Using this approach for recent processors and systems is not realistic. Dhawada et al. [5] proposed a power estimation methodology for a monoprocessor PowerPC and CoreConnect-based system at the TLM level. Their power modeling methodology is based on a fine-gsrain activity characterization at the gate level, which needs a huge amount of development time. Due to a high correlation with data, a power estimation inaccuracy of 11% is achieved. Compared to the previous works, our proposed methodology for power estimation also partially uses SystemC/TLM simulation with coarse grain power models. In addition, our methodology is applied for heterogeneous multiprocessor architectures.

For the functional level, Tiwari et al. [13] have introduced the concept of Instruction Level Power Analysis (ILPA). They associate a power consumption model with instructions or instruction pairs. The power consumed by a program running on the processor can be estimated using an Instruction Set Simulator (ISS) to extract instruction traces, and then adding up the total cost of the instructions. This approach suffers from the high number of experiments required to obtain the model. In addition, it can be applicable only for processors.

To overcome this drawback the *Functional Level Power Analysis* (FLPA) was proposed [8], which relies on the identification of a set of functional blocks that influence the power consumption of the target component. The model is represented by a set of analytical functions or a table of consumption values which depend on functional and architectural parameters.

Once the model is build, the estimation process consists of extracting the appropriate parameter values from the design, which will be injected into the model to compute the power consumption. Based on this methodology, the tool SoftExplorer [6] has been developed and included in the recent toolbox CAT [12]. It includes a library of power models for simple to complex processors. Only a static analysis of the code, or a rapid profiling is necessary to determine the input parameters for the power models. However, when complex hardware or software is involved, some parameters may be difficult to determine with precision. This lack of precision may have a non-negligible impact on the final estimation accuracy. In order to refine the value of sensible parameters with a reasonable delay, we propose to couple SystemC/TLM simulation with functional power modeling technique. Thus, by this way a reasonable trade-off between estimation speed and accuracy will be reached.

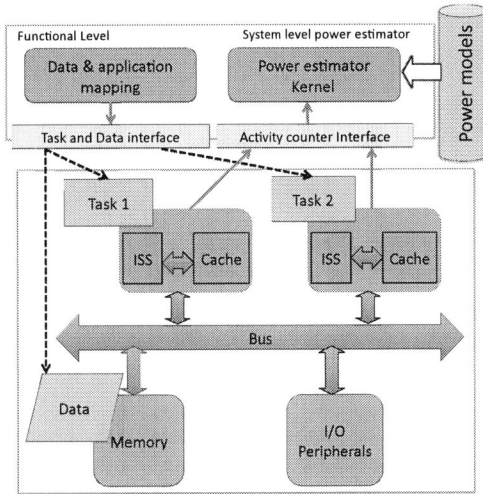

Fig. 2. HSL power estimator tool functioning

III. HYBRID SYSTEM LEVEL POWER ESTIMATION METHODOLOGY

This section exposes our proposed HSL power estimation methodology that is divided into two parts as shown in

TABLE I
GENERIC POWER MODEL PARAMETERS

Algorithmic	Name	Description
	τ	External memory access rate
	γ	Cache miss rate for a processor
	β	Instruction per cycle rate
Architectural	$F_{processor}$	Frequency of the processor
	F_{bus}	Frequency of the bus
	N	Number of processors

Fig. 1. The **first part** concerns the power model elaboration for the system hardware components. In our framework, the FLPA methodology is used to develop generic power models for different target platforms. The main advantage of this methodology is to obtain power models, which rely on the functional parameters of the system with a reduced number of experiments. As explained in the previous section, FLPA comes with few consumption laws, which are associated to the consumption activity values of the main functional blocks of the system. The generated power models have been adapted to system level design, as the required activities can be obtained from a system level environment. For a given platform, the generation of power model is done at once. To estimate the power consumption of an MPSoC system, the first step is to divide the architecture into different functional blocks and then to cluster the components that are concurrently activated when the code is running.

There are two types of parameters: *algorithmic parameters* that depend on the executed algorithm (typically the cache miss or instruction per cycle rates for a processor and area utilization for a hardware accelerator) and *architectural parameters* that depend on the component configuration set by the designer (typically the clock frequency). For instance, Table I presents the common set of parameters of our generic power model. These sets of parameters are defined for a general class of MPSoC. Additional parameters can be identified for complex processors based-architecture such as Superscaler or VLIW (Very Long Instruction Word).

The second step is the characterization of the embedded system power consumption when the parameters vary. These variations are obtained by using some elementary assembly programs (called scenario) or built in test vectors elaborated to stimulate each block separately. Characterization can be performed by measurements on real boards. Finally, a curve fitting of the graphical representation will allow us to determine the power consumption models by regression. The analytical form or a table of values expresses the obtained power models. This power modeling approach was proven to be fast and precise. In our work, this approach has been applied to model power consumption for processor, memory, reconfigurable hardware, and I/O peripherals.

The **second part** of the methodology defines the architecture of our HSL power estimator that includes the *functional power estimator* and *fast SystemC simulator* as shown in Fig.1. The functional power estimator evaluates the consumption of the target system with the help of the elaborated power models

from the first part. It takes into account the architectural parameters (e.g. the frequency, the number of processors, the processor cache configuration, etc.) and the application mapping. It also requires the different activity values on which the power models rely. In order to collect accurately the needed activity values, the functional power estimator communicates with a fast SystemC simulator at a TLM level. The combination of the above two components described at different abstraction levels (functional and TLM) leads to a hybrid power estimation that gives a better trade-off between accuracy and speed. The vital function of system level power estimation is to offer a detailed power analysis by the means of a complete simulation of the application. This process is initiated by the functional power estimator through *the data and task interface* (Fig. 2). In this way, the mapping information is transmitted to the fast TLM SystemC simulator. Our simulator consists of several hardware components which are instantiated from the SoCLib [1] library in order to build a virtual prototype of the target system. We highlight that processors are described using Instruction Set Simulator (ISS) that sequentially executes the instructions and has no notion of concurrency of micro-architecture. In our previous framework [2], we presented an accurate TLM simulation technique that allows to evaluate the MPSoC performances. In the power estimation step, the simulator collects the activities that are influenced by the application and the input data. At the end of the simulation, the values of the activities are transmitted to the power consumption models or power estimator kernal using *the activity counter interface* in order to calculate the global power consumption as illustrated in Fig. 2. As we have stated before, the following section will discuss the **first part** i.e., the elaboration of the power model for the Xilinx Virtex II Pro FPGA platform by using FLPA methodology.

IV. POWER MODEL ELABORATION

In order to prove the usefulness and the effectiveness of the proposed power estimation methodology, we used a PowerPC 405-based architectures implemented into the Xilinx Virtex II Pro FPGA (XupV2Pro) platform. The Virtex II Pro FPGA contains two PowerPC 405 processors that have a 16KB, 2-way set associative instruction and data caches. In addition, this FPGA has a large number of configurable logic blocks (CLB) for implementing hardware accelerators. Each processor has access to the on-chip memory (BRAM) and the off-chip memory (SDRAM) via the Processor Local Bus (PLB). We used the JPEG (Joint Photographic Experts Group) application as a benchmark. The JPEG application consists of 4 main tasks: conversion RGB (Red, Green and Blue) to YUV (luminance, blue chrominance, and red chrominance components), Discrete Cosine Transform (DCT), Quantization, and Huffman coding. We also used the H.264/AVC baseline profile decoder that supports intra and inter-coding, and entropy coding with Context-Adaptive Variable-Length Coding (CAVLC). This section presents the power model elaboration which is the **first part** of our methodology. As explained above, we used the FLPA methodology to generate generic power models for the target system. As a first step, we divided the architecture into different functional blocks such as the processor, the memory system, the reconfigurable logic, etc. Then, we started the characterization of each component in order to extract the related power consumption models.

Processor power model: Table II shows the power consumption models for the PowerPC405 processor and its memory system. These models predict consumption of the processor kernel and the I/Os parts separately, since distinct supplier devices power them with constant voltage: 1.5V for the processor and 2.5V for the SDRAM and I/Os respectively. The obtained power models shown in the Table II depend also on the memory mapping. For this reason, there are different power models for on-chip memory (BRAM) and for external memory (SDRAM). The input parameters on which the power models rely are the frequency of the processor ($F_{processor}$(MHz)), bus frequency (F_{bus} (MHz)), and the cache miss rate ($0 < \gamma < 100$ (%)). The system designer chooses the frequency of the processor and bus while cache miss rate is considered as an activity of the processor, which could be extracted from the simulation environment. According to these power models, the static consumption is dominant which is a drawback of the FPGA technology. For this reason, the latest FPGA circuits come with an optimized static power factory setting.

TABLE II
CONSUMPTION MODELS FOR THE POWERPC 405 PLATFORM

Mapping	Voltage	Power laws
BRAM	1.5V	P(mW) = 0.40 $F_{processor}$ + 3.24 F_{bus} + 74
	2.5V	P(mW) = 5.37 F_{bus} + 1588
SDRAM	1.5V	P(mW) = 0.38 $F_{processor}$ + 3.45 F_{bus} + 79
	2.5V	P(mW) = 4.1γ + 6.3F_{bus} + 1599

Extrapolation for homogeneous multiprocessor architectures: The above developed power models will be used in the framework of system level estimation of homogeneous multiprocessor architectures that may contain several processors. This approach is mandatory in the design flow for two reasons. First, system level estimation can be achieved with acceptable accuracy of 10-1000x faster than the physical level. Second, it allows exploring architectures that cannot be implemented due to the hardware resource limitation or the unavailability of the target component. For instance, we cannot exceed two PowerPC based architecture using our XupV2Pro platform. Thus, it is important to have a scalable approach to address the complex system power/energy estimation issue. The equation 1 will be considered for the total system power (P_{total}) estimation. In addition to the processor (P_{p_i}), the equation involves the power consumption of the synchronization part (P_{sync}) required to access the shared memory (P_{mem}) and the shared I/O resources ($P_{I/O}$).

$$P_{total} = \sum_i P_{p_i} + P_{mem} + P_{sync} + P_{I/O} \qquad (1)$$

In our XupV2Pro platform, synchronization between parallel tasks running on different processors or hardware accel-

erators is performed by a call to a hardware mutex. Several experiments have been conducted to evaluate the additional power cost of this hardware component. This study includes three parameters which are the number of masters, and the processor & bus frequencies. Experimental results show that the mutex power consumption depends mainly on the PLB frequency.

V. SYSTEM LEVEL POWER ESTIMATION RESULTS

A. Monoprocessor architecture

For the **second part** of our HSL power estimation methodology, a system level prototype of a PowerPC based architecture has been developed. This prototype uses different SystemC models especially the ISS for the target processor. Furthermore, the cache parameters and the bus latencies are set to emulate the real platform behaviour. A set of counters are injected into the simulator to determine the occurrences of the main activities. For the PowerPC processor, the following counters are used for different cache miss rates: read data miss, write data miss and read instruction miss. For the example of ARM cortex A8 processor, additional counters should be defined for IPC and Level-2 cache miss rates computing.

Fig. 3 shows the detailed results of the activities fetched by the fast SystemC simulator for each task of the JPEG application. From these results several remarks can be drawn. First, we can notice that instruction cache miss rates and read data miss rates are very low when compared with write data miss rates. This is due to the reduced task kernel and data pattern sizes that are very low compared to the cache size (16 KB), which decreases the access to the external memory and thus having a minimal effect on the dynamic power consumption. However, with the new submicron technologies the effect of the static power consumption cannot be neglected. For this reason, a softcore processor such as the Microblaze comes with reconfigurable cache size to fit with the application requirements. Second, the data write miss rates have a high impact on the total power consumption of the system. This is because of the algorithm structure, which does not favour the reuse of data output arrays and the usage of write-through cache policy. Therefore, the statistics collected in Fig. 3 could help in tuning the application structure for a better optimization of the system power consumption.

In the next step, we estimated the total power consumption of each task using the power models shown in Table II (SDRAM mapping). Fig. 4 illustrates the results and shows the comparison between the proposed HSL methodology, SoftExplorer tool introduced in Section II, and the real board measurements. First, our power estimator has a negligible average error equals to 1.59%, which offers better accuracy than SoftExplorer which presents an average error of 4.32%. The average error obtained here is negligible due to the dominance of static power. For this reason, we calculate again the average error without taking into account the static power. Our methodology produces an average error of 4.3% and the SoftExplorer gives around 9.4% in comparison with real board measurement. This study offers a detailed power

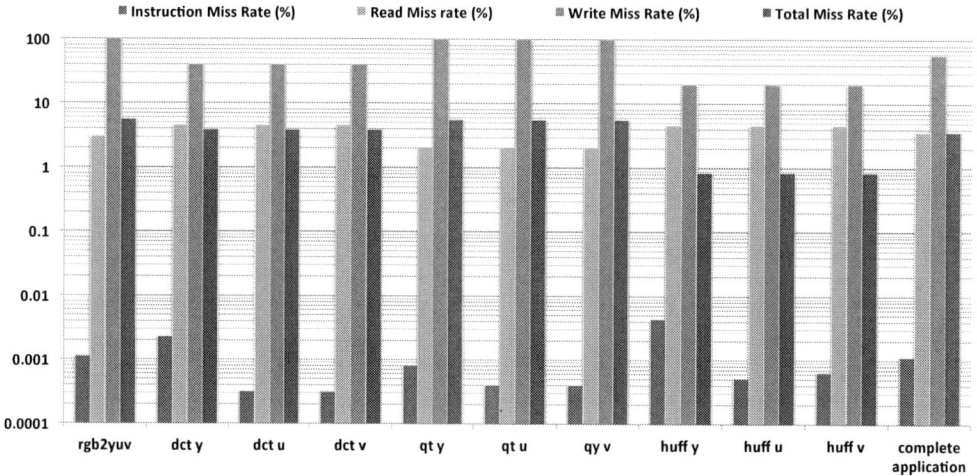

Fig. 3. JPEG application cache miss rates

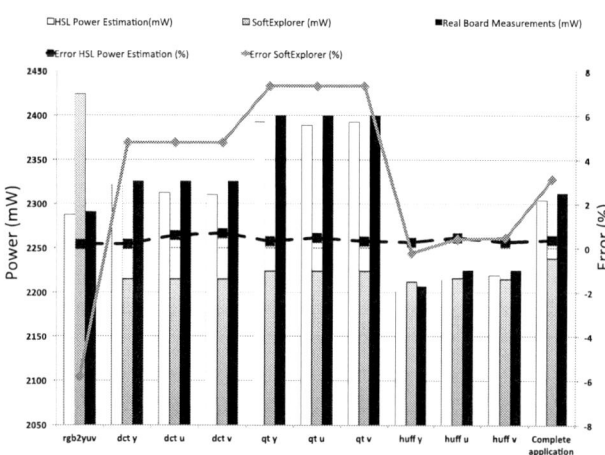

Fig. 4. Power estimation accuracy for the JPEG application

B. Homogeneous multiprocessor architecture

The second study involves an homogeneous architecture with identical processors to run the JPEG application. To evaluate the impact of the number of processors on the execution time and total energy/power consumption, we executed the JPEG on systems with 1 to 8 processors. The PowerPC frequency was set to 300MHz and the PLB frequency to 100MHz. All the processors execute the same workload but on different image macroblocs. Fig. 6 reports the execution time in *ms* and the total energy consumption in *mJ* and Fig. **??** shows the power estimation of multimedia benchmarks for homogeneous multiprocessor (two PowerPC) architecture.

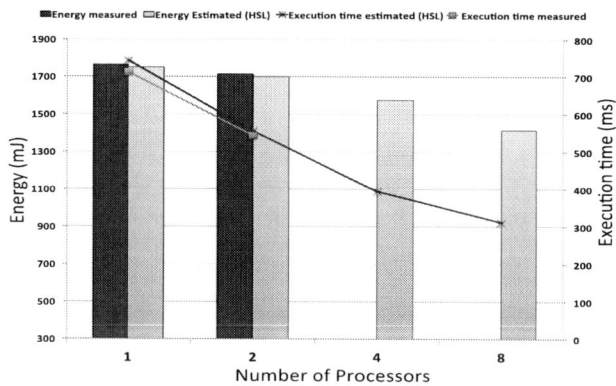

Fig. 6. Execution time and energy variation according to the number of processors

Compared to real board energy measurements, our HSL estimator achieved an error of 0.79% and 3.49% for one and two processors respectively. This accuracy is obtained because of three main reasons. First, power models are extracted from real board measurements. Second, our methodology considers the synchronization part while using multiprocessor system.

analysis for each task in order to help designers to detect peaks of consumption and thus to propose efficient mapping or optimization techniques. In order to evaluate the accuracy of our tool, we carried out power estimation on several image & signal processing benchmarks. Fig. 5 illustrates the power estimation results by showing the comparison between the proposed HSL power estimation methodology, SoftExplorer and the real board measurements. Our proposed methodology has a negligible average error of 1.24%, which offers better accuracy than SoftExplorer with an average error of 6.34% when compared to the real board measurements. This is due to better accuracy of the captured activities in the simulator rather than the static analysis or rapid profiling of the C or assembly code in SoftExplorer.

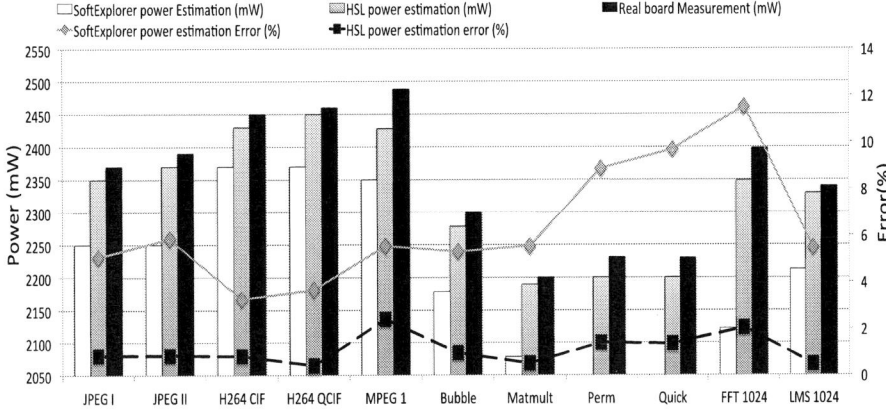

Fig. 5. Comparison of power estimation accuracy of HSL tool vs SoftExplorer (monoprocessor architecture)

Finally, additional activities that are intrinsic in parallel processing such as shared data communication overheads are accurately evaluated by using our SystemC simulator. The above mentioned reasons encourage us to consider architectures with a higher number of processors in the context of exploring new complex MPSoC. Fig. 6 shows that, for the implemented JPEG parallel application, adding processors to the system decreases the execution time, which improves the system performance. This variation is not linear because the processors share resources, which generates conflicts at some times, and reduces the speed-up as waiting cycles are added to the processors execution. In terms of energy consumption, we observe that until a certain number of processors, the total system energy consumption decreases as the number of execution cycles is reduced, and then it tends to stabilize as the system performance improves. But increasing the number of processors over a certain limit tends to be ineffective, as it just adds new conflicts at the bus level, leading to more waiting cycles.

VI. CONCLUSION

This paper presents a hybrid system level estimation methodology for MPSoC power-aware design. Indeed, power/energy constraints are considered as a major challenge when the system runs on batteries. Thus, designers must take these constraints into account as early as possible in the design flow. First, a power modeling methodology has been defined to address the global system consumption that includes processors, memory, reconfigurable hardware, and etc. Secondly, the functional power modeling part is coupled with a fast SystemC simulation technique to obtain the needed micro-architectural activities for the power models, which allows us to reach accurate estimates. With such proposed methodology, the designer can explore several implementation choices: monoprocessor and homogeneous multiprocessor. The future works of this project will focus on more complex heterogeneous platforms. Furthermore, in order to obtain more accurate power estimations, some power model refinements

must be realized. This is the case for the data exchanges between hardware and software tasks respectively executed on hardware resource and on processor which are currently estimated at high level of abstraction.

REFERENCES

[1] The Soclib Website. https://www.soclib.fr/.
[2] R. B. Atitallah, S. Niar, , and J.-L. Dekeyser. Mpsoc power estimation framework at transaction level modeling. In *The 19th International Conference on Microelectronics (ICM 2007)*, 2007.
[3] G. Beltrame, L. Fossati, and D. Sciuto. ReSP: A Nonintrusive Transaction-Level Reflective MPSoC Simulation Platform for Design Space Exploration. *Computer-Aided Design of Integrated Circuits and Systems, IEEE Transactions on*, 28(12):1857–1869, Dec. 2009.
[4] D. Brooks, V. Tiwari, and M. Martonosi. Wattch: a framework for architectural-level power analysis and optimizations. In *Proceedings of the 27th annual international symposium on Computer architecture*, pages 83–94, 2000.
[5] N. Dhanwada, R. A. Bergamaschi, W. W. Dungan, I. Nair, P. Gramann, W. E. Dougherty, and I.-C. Lin. Transaction-level modeling for architectural and power analysis of powerpc and coreconnect-based systems.
[6] S. Dhouib, J.-P. Diguet, D. Blouin, and J. Laurent. Energy and power consumption estimation for embedded applications and operating systems. *Journal of Low Power Electronics (JOLPE)*, 5(3), 2009.
[7] ITRS. Design, 2010 edition. http://public.itrs.net/, 2010.
[8] J. Laurent, N. Julien, and E. Martin. Functional level power analysis: An efficient approach for modeling the power consumption of complex processors. In *Proceedings of the Design, Automation and Test in Europe Conference*, Munich, 2004.
[9] I. Lee, H. Kim, P. Yang, S. Yoo, E. Chung, K.Choi, J.Kong, and S.Eo. Powervip: Soc power estimation framework at transaction level. In *Proc. ASP-DAC*, 2006.
[10] N.Dhanwada, I. Lin, and V.Narayanan. A power estimation methodology for systemc transaction level models. In *International conference on Hardware/software codesign and system synthesis*, 2005.
[11] Open SystemC Initiative. Systemc, 2008. World Wide Web document, URL: http://www.systemc.org/.
[12] J. D. S. Douhib. Model driven high-level power estimation of embedded operating systems communication and synchronization services. In *Proceedings of the 6th IEEE International Conference on Embedded Software and Systems*, China, May 25-27 2009.
[13] V. Tiwari, S. Malik, and A. Wolfe. Power analysis of embedded software: A first step towards software power minimization. In *Transactions on VLSI Systems*, 1994.
[14] W. Ye, N. Vijaykrishnan, M. Kandemir, and M. Irwin. The Design and Use of SimplePower: A Cycle Accurate Energy Estimation Tool. In *Design Automation Conf*, June 2000.

OpenCL implementation of Cholesky Matrix Decomposition

Claudio Brunelli, Eero Aho, Heikki Berg
Nokia Research Center
Tampere, Finland
Email: claudio.brunelli@nokia.com

Abstract— This paper presents some OpenCL implementations for Cholesky decomposition, a very popular algorithm used in linear algebra and signal processing applications. The Cholesky algorithm represents a very interesting candidate for OpenCL implementation since it contains sequential parts besides parallel ones. Furthermore, one step involves just a small amount of calculations. These characteristics pose challenges which call for suitable techniques to overcome the limitations of the language. We propose several versions of the implementation of the Cholesky algorithm, then provide an analysis of the trade off between complexity and performance offered by each of them. We also analyze the differences between execution of the program on GPU and on multicore CPU.

I. INTRODUCTION

Cholesky decomposition is a very popular and very efficient method to obtain the factorization of symmetric and positive-definite matrixes into the product of a lower triangular matrix and its conjugate transpose [1]. For this reason the Cholesky operation is also known as *square root of a matrix*.

Cholesky algorithms can be utilized for instance as a numeric means of solving linear systems of equations. Other possible applications are in the Monte Carlo method for simulating systems, Kalman filters, and Digital Signal Processing (DSP) algorithms for radio technologies such as LTE (Long Term Evolution) or other OFDM based protocols [2].

GPUs (Graphics Processing Units) have traditionally been developed and used only for games. However, in the last few years, a new trend has emerged based on the observation that GPUs have a lot of arithmetic computational power which remains unused when the user is running non-gaming applications. By exploiting the aforementioned computational power for other applications, a whole new application domain for graphic processors has been born that nowadays is commonly referred to with the term GPGPU (General Purpose GPU).

It is necessary to use dedicated programming paradigms in order to exploit GPGPU computation. An example of such paradigm which is getting increasingly popular nowadays is OpenCL (Open Computing Language). OpenCL is a framework for writing programs that are executed on heterogeneous platforms consisting of CPUs, GPUs, and other (possibly multicore) processors [3]. OpenCL standard includes a C99 based language called OpenCL C for writing so called kernels (that are functions executed on OpenCL devices, e.g. in GPU) and OpenCL API that is used to interface kernels to the main C program running on the host (usually CPU). OpenCL

provides parallel computing using both task-level and data-level parallelism.

Not all general-purpose applications are suitable for implementation on GPGPU architectures. Due to the overhead of transferring data across the system bus and the internal architecture of the graphic processors, it is necessary that the applications exhibit a convenient ratio between memory and arithmetic operations, so that the data transfer overhead can be compensated by the faster execution on GPU of the arithmetic operations. More importantly, such operations should be as independent as possible from each other, in order to fully exploit the abundance of computation engines of the GPUs.

The Cholesky decomposition represents a very interesting and challenging candidate for GPGPU implementations because it contains some steps where parallelism is abundant, but others which are made of only one (or a few) operations. On top of this, the Cholesky algorithm steps must be executed sequentially, posing extra challenges.

So far, there are only a few previous works which implement Cholesky decomposition on GPU, and only one of them seems to be done using OpenCL [4]. In [4] the authors describe the Relevance Vector Machine algorithm accelerated by utilizing GPUs. The algorithm needs recursive Cholesky decomposition, which is implemented on GPUs. Compared to [4], we use another version of the Cholesky algorithm and we provide an analysis of the bottlenecks and issues in the algorithm together with a set of different solutions to overcome them.

In [5], a set of algorithms is presented (such as matrix multiplication, and different algorithms for matrix decomposition: LU, QR and Cholesky). The authors present a CUDA implementation of those algorithms running on multiple GPUs. To achieve high performance, they use a set of advanced techniques such as look-ahead, overlapping CPU and GPU computation, auto tuning, etc. They are able to achieve a best speedups equal to 4x when compared to a quad core CPU.

In [6], the authors present solvers for linear systems of equations based on Cholesky and other matrix factorization techniques (LU and QR). The authors implement these techniques on GPU in CUDA language (also exploiting existing libraries such as LAPACK and BLAS). Their results are available in their MAGMA library.

In [7], a novel way of programming GPUs is presented: the authors propose an approach based on the combination of the FLAME library for dense linear algebra and a run-time system. Cholesky decomposition is used as a test case, showing how the performance achieved using the proposed library is competitive with the state of the art.

In [8], the authors push the previous approach to the extreme: proposing a compiler that is able to convert to CUDA language some code written in an high-level programming language such as the C programming language.

In [9], a high-performance implementation of QR decomposition on GPU is presented. The authors analyze performance for real-valued matrices and propose general guidelines to achieve high performance in dense linear algebra procedures for GPUs.

This paper is organized as follows: Section II briefly illustrates the Cholesky algorithm. Section III explains some keywords needed to understand the details of the implementation section. Section IV illustrates different versions of OpenCL implementations of the Cholesky decomposition. Section V presents the results of different OpenCL implementations of the Cholesky algorithms, discusses their characteristics and compares them. Finally, in Section VI we draw the conclusions.

II. Cholesky algorithm for matrix factorization

Cholesky decomposition is made up of basically three steps:

- Square rooting
- Scaling
- Updating submatrix

These three steps are performed iteratively over the k-th SxS submatrix of a given NxN matrix A. In other words, given the square matrix A the Cholesky decomposition returns the triangular matrix L such that:

$$A = L^H * L \qquad (1)$$

Where L^H denotes the Hermitian transpose of matrix L. Let us now consider a 4x4 matrix as an example. The square rooting step (step 1) replaces the k-th element on the main diagonal of matrix A with its square root, as depicted in figure 1. The scaling step (step 2) divides the elements of the k-th column by the result of step 1, as depicted in figure 2. Note that this step cannot begin before step 1 has finished. The updating submatrix step (step 3) basically performs multiply-accumulate operations on the elements of the lower-triangular matrix sized (k-1)*(k-1) using the results of step 2, as depicted in figure 3. Note that this step cannot begin before step 2 has finished.

III. Exploration topics

In this section, some terms are described to give understanding for the evaluations.

Global work size. Global work size describes the total executed work-item count.

Local work size or *work-group size*. The work-items are collected to work-groups. The work-items in a work-group are executed on the same compute unit, share local memory and work group barriers [11]. For good performance with NVidia

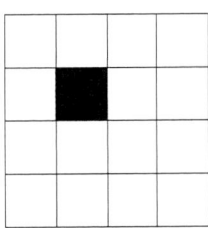

Fig. 1. Step 1: square rooting for iteration k=1, element A(k,k) is replaced by its square root.

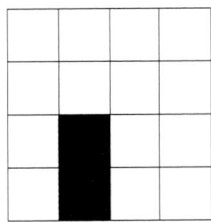

Fig. 2. Step 2: scaling for iteration k=1, elements in the column under element A(1,1) are divided by the result of step 1.

GPU, local work group size should be multiple of 32 (NVidia warp size) [12]. The respective size with contemporary AMD GPUs is 64 (AMD wavefront size) [13]. When using work group size 1, NVidia and AMD GPUs use respectively only 1/32 or 1/64 of the ALU performance, thus in principle it is good to avoid such a solution. Nonetheless, when targeting OpenCL execution on the CPU (using Mac OS X 10.6.4 and XCode 3.2.3) a work group size equal to 1 is the only choice which is allowed by the OpenCL implementation. This limits the achievable performance and/or the portability of the same OpenCL code; on the other hand when targeting the CPU the penalty due to the data transfer overhead is smaller, leaving some room for speed-up.

Barrier or *synchronization.* In OpenCL, there are two types of barriers: a command-queue barrier and a work-group barrier [11]. In host side, the OpenCL API provides a function to enqueue a command-queue barrier. This barrier command ensures that all the previously enqueued commands have been finished before the commands after the barrier command. In device side, the OpenCL C language provides work-group barrier function to be used in kernel code. All the work-items of a work-group must execute the barrier before any are allowed to continue going beyond the barrier.

The used synchronization method naturally affects the performance. Synchronization in host side means that the execution needs to returned back to host side before executing another kernel with clEnqueueNDRangeKernel -function. According to our experiments, one clEnqueueN-DRangeKernel -OpenCL API function call takes at minimum about 200-300us in our GPU environment (HP Elite-Book 8530w laptop computer, Windows 7 operating system and NVidia SDK for OpenCL). Kernel side barrier delays depend a lot about work-group size and if local memory accesses or global memory accesses are synchro-

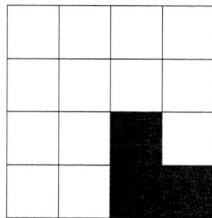

Fig. 3. Step 3: submatrix updating for iteration k=1, elements in the lower triangular submatrix A(r,c) such that r and c are greater than k are processed based on the result of step 2.

nized by using `barrier(CLK_LOCAL_MEM_FENCE)` or `barrier(CLK_GLOBAL_MEM_FENCE)` -function, respectively. According to our experiments, global memory barrier takes noticeable more time (10-20x) than local memory barrier.

Coalesced memory accesses. In NVidia GPU, the high priority optimization is coalescing global memory accesses [12]. Sixteen (NVidia half warp) global memory loads and stores can be coalesced as one 64B (16x32bit) transaction. Coalescing the memory accesses largely improves the achieved bandwidth compared to several short memory accesses. However, to allow coalescing, the program code needs to be formulated so that adjacent memory addresses are accessed [12].

IV. IMPLEMENTATION OF THE CHOLESKY DECOMPOSITION

Cholesky decomposition is not an easily parallelizable algorithm; it has sequential parts besides parallel parts. In other words, each parallel part needs to be completely over before it is possible to start executing the following step, thus it is necessary to utilize proper synchronization mechanisms to achieve this goal. In particular, such mechanisms can be either consist of splitting the algorithm into several OpenCL kernels (synchronize in host side), or rely on *barrier* function calls (synchronize in device side). The first solution is intrinsically safe but implies a lot of overhead which deteriorates the overall performance achieved, while the second is faster but works only when applied to threads which belong to the same work group. This chapter describes different OpenCL implementations of the Cholesky algorithm. As a reference we compare the performance of the OpenCL implementations versus the C implementation of the algorithm as presented in [10].

A. Cholesky reference C implementation

The pseudo-code reported below is an example of straightforward Cholesky decomposition C implementation. In the code, N is the input matrix width and height (NxN) for a complex valued input matrix A. As mentioned in Section II, the main loop contains three steps named STEP 1, STEP 2, and STEP 3. The fourth step (STEP 4) is outside the main loop. STEP 4 fills the upper triangle of the matrix A with zeros since the result is only the lower triangular of the matrix.

```
for (k=0;k<N;k++)
```

```
{
  //STEP 1
  DiagonalElement = complexSqrt(A[k][k]);
  A[k][k] = DiagonalElement;

  //STEP 2
  for (i=k+1;i<N;i++)
  {
    A[i][k] =
    complexDiv( A[i][k], DiagonalElement );
  }

  //STEP 3
  for (j=k+1;j<N;j++)
  {
    for (i=j;i<N;i++)
    {
      tmpConj = conjugate(A[j][k]);
      tmpMul = complexMul( A[i][k], tmpConj );
      tmp = complexSub( A[i][j], tmpMul );
      A[i][j] = tmp;
    }
  }
}

//STEP 4
// The result matrix is lower triangular.
// Filling the upper triangle with zeroes
for (k=0;k<N;k++)
{
  for (j=k+1;j<N;j++)
  {
    A[k][j].re = (float) 0.0;
    A[k][j].im = (float) 0.0;
  }
}
```

B. Cholesky decomposition OpenCL implementation

Table I shows the details of our OpenCL implementations. There are seven different versions. Four of them are synchronized (barrier) in OpenCL host side (`HostSync_v1`, ..., `HostSync_v4`) and three are synchronized in OpenCL device size (`DevSync_v1`, ..., `DevSync_v3`). In this section the Host is considered to be the CPU whereas the Device is the graphics card (in our case an NVidia GPU).

In the different versions, there are different memory access patterns. For instance, column-wise memory accesses are used in the straightforward implementation according to what is specified in the reference C-implementation. However, row-wise accesses are better from a GPU perspective since they allow coalescing of the global memory accesses [12]. This is due to the fact that in the C host code the matrix is stored in row-major order, thus elements belonging to the same row are stored in adjacent memory locations. When the OpenCL kernel accesses the matrix in row-wise manner it can then access adjacent memory locations, allowing coalesced memory

978-1-4577-0671-4/11 $26.00 © 2011 IEEE

TABLE I

OpenCL implementation details. N = matrix width and height (NxN), k = loop index (See pseudo-code).

	Sync.	Access order	Kernel count	Global work size	Local work size	Note
HostSync v1	Host	Column-wise	3	1,N,NxN	1,1,1x1	
HostSync v2	Host	Column-wise	3	1,N,NxN	1,M,Mx1	M=min(N,512)
HostSync v3	Host	Row-wise	3	1,N,NxN	1,M,Mx1	M=min(N,512)
HostSync v4	Host	Row-wise	3	1,N-k,(N-k)x(N-k)	(N-k)x1	N max 512
DevSync v1	Device	Column-wise	1	NxN	Nx1	N max 512
DevSync v2	Device	Row-wise	1	NxN	Nx1	N max 512
DevSync v3	Device	Row-wise	1	(N-k)x(N-k)	(N-k)x1	N max 512

transfers.

The kernel count is also different in different versions. Host side synchronized versions have three distinct OpenCL kernels, whereas device side synchronized versions have only one OpenCL kernel. Therefore, global work size (as well as local work size) is represented by three comma-separated values in host synchronized versions in Table I. Due to NVidia specific restrictions the local work size can be at maximum 512.

The following pseudo-code represents host side synchronized versions (OpenCL host code). At first the matrix A is written to device global memory. Then, three kernels are executed within a loop of N iterations. Each of the kernels has a different global work size and local work size. After the loop is completed the result matrix (which is still contained in A due to in-place execution) is read back from the device. Specifically, the following pseudo code depicts the `HostSync_v2` and `HostSync_v3` versions, even though `HostSync_v1` and `HostSync_v4` are very similar. Note that N = matrix width and height (NxN), and A = matrix.

```
WriteBufferToDevice(A);
M=min(N,512);
for (k=0;k<N;k++)
{
  //STEP 1
  gws = 1; //Global work size
  lws = 1; //Local work size
  ExecuteKernel(STEP1, gws, lws, A, N, k);

  //STEP 2
  gws = N; //Global work size
  lws = M; //Local work size
  ExecuteKernel(STEP2, gws, lws, A, N, k);

  //STEP 3 and STEP 4
  gws = {N, N}; //Global work size
  lws = {M, 1}; //Local work size
  ExecuteKernel(STEP3&STEP4, gws, lws, A,
                N, k);
}
ReadBufferFromDevice(A);
```

The following pseudo-code is the OpenCL host code for device side synchronized versions (`DevSync_v1`, `DevSync_v2`, and `DevSync_v3`). The functionality of all the three kernels is combined into a single OpenCL kernel. The synchronization between the steps is done inside the kernel code via barriers. The kernel code utilizes also the local memory present in the graphics card: each work group has local memory of size (N x 2 x sizeof(float)) bytes that is used as internal storage for the computation in the GPU.

```
WriteBufferToDevice(A);
for (k=0;k<N;k++)
{
  //STEP 1, STEP 2, STEP 3, and STEP 4
  gws = {N, N}; //Global work size
  lws = {N, 1}; //Local work size, N = 512
  lms = N*2*sizeof(float); //Local memory
                           //size
  ExecuteKernel(ALL_STEPS, gws, lws, A, N,
                k, lms);
}
ReadBufferFromDevice(A);
```

V. RESULTS

The described implementations were evaluated with varying matrix sizes. As mentioned in [5] the overhead related to kernel launch deteriorates the efficiency for fine-grain computations, thus when the matrix size is not large enough. On the other hand, we considered also fairly small matrixes in our analysis because they are commonly found in some DSP algorithms for radio applications.

A. OpenCL on GPU

The measurements have been done with HP EliteBook 8530w laptop computer and Windows 7 operating system. The CPU is an Intel Core 2 Duo T9600 @ 2.8 GHz with 4GB RAM. The GPU is an NVidia Quadro FX 770M equipped with 32 CUDA cores and 512MB RAM. The used GPU is passive cooled and is a low/mid-range device when compared to the most powerful NVidia GPUs (which feature several hundred CUDA cores).

Table II shows the results in terms of the time [ms] taken to execute the implementations described. The C-version results are shown as a reference. Note that some versions of the implementations are restricted to N less or equal to 512. Version `HostSync_v1` utilizes GPU resources poorly due to small local work size. Just by increasing the local work size (like in `HostSync_v2`) the performance is increased up to 2.8x compared to `HostSync_v1` with N = 512. Version `HostSync_v3` accesses and computes the result row by row instead of the normal column by column order. For small N values this actually causes a small decrease in performance when compared to `HostSync_v2` (up to 7% with N = 8). However, with bigger matrix sizes the performance improvement is up to 5.8x with N = 7168. This is mainly due to the usage of coalesced global memory accesses. Version `HostSync_v4` decreases the executed work item count as the current loop index k increases. However, this gives only small improvement compared to `HostSync_v3`.

With small N values, device synchronized versions (`DevSync_vx`) are noticeably faster than host synchronized versions (`HostSync_vx`). With N = 128, device synchronized versions are at least two times faster than host synchronized versions. This is due to the fact that host synchronized versions run three times more kernels (three times more clEnqueueNDRangeKernel -function calls). As mentioned, according to our evaluation one clEnqueueNDRangeKernel -function call takes at least 200-300us even though the GPU is actually using about 15us to compute the respective result for STEP 1 and STEP 2 of Cholesky composition. With bigger matrix size N, the relative difference decreases since the actual computation is taking more time.

Table III shows the relative performance when comparing OpenCL coded GPU performance to the reference C-version (running on the CPU). With small matrix sizes, all the OpenCL implementations have clearly smaller performance than the reference C-version. However, the relative difference decreases quickly when N increases. The device synchronized `DevSync_v3` version shows the best OpenCL performance (even though it is restricted to a maximum of N = 512) obtaining 0.74x performance when compared to the C-version. With bigger matrix sizes (N = 7168), the maximum performance improvement achieved is 5.96x (with `HostSync_v3`) compared to reference C-version.

For reference, image scaling with cubic 6x6 interpolation utilizes 6x6 original image pixels to generate a single scaled image pixel [14]. It is reasonably complex but straightforward to make parallel implementation with OpenCL. With the same HP EliteBook 8530w laptop computer, we achieved 16x improvement with OpenCL on GPU compared to respective reference C-implementation. This gives us a reference of upper performance limit with the used computer. Cholesky decomposition has serial and parallel parts, thus is not a natural candidate for OpenCL implementation, but in spite of that we still achieved a speed-up close to the half of the aforementioned 16x speed-up.

B. OpenCL on multicore CPU

As a further study we run the OpenCL implementations of Cholesky on a multicore CPU. We use an Intel Nehalem processor @2.8 GHz, that is a powerful processor featuring 8 physical cores which can be utilized by the software as 16 logical cores. The host machine is an Apple workstation running Mac OS 10.6 Snow Leopard, which supports OpenCL natively.

As explained before, one drawback of OpenCL kernels running on CPU (at least with the used OpenCL driver implementation) is that the maximum work group size allowed is one. This implies that usually the kernel code needs to be at least partially rewritten, and in the worst case the CPU implementation must be entirely recoded. We run on the Nehalem CPU the versions `HostSync_v1`, `HostSync_v3` (with work group size equal to 1 instead of N or NxN), and `DevSync_v2` (with work group size equal to 1).

The results of the OpenCL kernels on the multicore CPU are reported in table IV. Table V shows the corresponding speed-up when compared to the reference C implementation. A first interesting experiment we made was to check the allocation of the computation across the cores using the Shark profiler present in Snow Leopard. We noticed that while the C implementation was running on only one core, the OpenCL kernel was distributed across all the cores.

Another interesting remark is done about how the speed-up of the different versions changes differently than when targeting the GPU. In particular, it can be seen that by using a single kernel rather than three pays off only for matrix sizes smaller than 300: in those cases the computation is not enough to compensate the kernel calling overhead and having a single kernel is beneficial. Beyond that threshold, though, the computation is significant enough to compensate the kernel call overhead: on the GPU implementation this still leads to an advantage of DevSync versions, while on the CPU it does not. This is probably due to the fact that `DevSync_v2'` version utilizes only slow global memory instead of fast local memory (due to local work group size restrictions when targeting the CPU).

With OpenCL on multicore CPU, the maximum achieved performance improvement is 5.15x compared to reference C-version (with N = 7168 and `HostSync_v3'`).

VI. CONCLUSION

This paper presented several versions of OpenCL implementation for Cholesky matrix decomposition. The usage of OpenCL allowed us to run different versions of the Cholesky algorithm on GPU and also on multicore CPU. Transferring the computation to the OpenCL device (GPU or CPU) introduces some overhead (kernel invocation, data transfer, etc.) which impacts significantly the overall performance. The effects of this can especially be seen in our experiments when the input matrix has a small size: in those cases the C program executing on the CPU gives noticeably better performance than the OpenCL implementation on GPU or CPU.

The partially serial nature of the Cholesky algorithm posed interesting and challenging problems that we solved in different ways using different implementations. We measured the performance of each implementation and examined the impact of the choices operated for each version. In the best case, our OpenCL implementation guarantees a 6x speedup when compared to the C reference implementation of the algorithm.

TABLE II

TIME [MS] TAKEN TO EXECUTE OPENCL KERNELS ON NVIDIA GPU. NOTE THAT SOME VERSIONS ARE RESTRICTED TO N SMALLER OR EQUAL TO 512.

N	8	40	128	300	512	1024	2048	4096	7168
Reference C version	0,007	0,20	5,3	55	365	3290	45200	746000	4716000
HostSync v1 (col, lws=1,1,1x1)	10.250	36.40	148.0	831	3220	23970	180400	1437000	7807000
HostSync v2 (col, lws=1,M,Mx1)	10.350	33.80	109.0	324	1130	12550	104700	959000	4568000
HostSync v3 (row, lws=1,M,Mx1)	11.140	34.20	111.0	291	670	3315	22400	155000	789000
HostSync v4 (row, lws=(N-k)x1)	10.300	34.10	110.0	287	660	n.a.	n.a.	n.a.	n.a.
DevSync v1 (col, lws=Nx1)	6.390	15.30	48.2	251	1060	n.a.	n.a.	n.a.	n.a.
DevSync v2 (row, lws=Nx1)	6.410	15.20	46.3	230	695	n.a.	n.a.	n.a.	n.a.
DevSync v3 (row, lws=(N-k)x1)	6.500	15.00	47.9	156	495	n.a.	n.a.	n.a.	n.a.

TABLE III

SPEED-UP IN EXECUTING OPENCL KERNELS ON THE GPU VS. C IMPLEMENTATION ON THE CPU. NOTE THAT SOME VERSIONS ARE RESTRICTED TO N SMALLER OR EQUAL TO 512.

N	8	40	128	300	512	1024	2048	4096	7168
Reference C version	1.000	1.00	1.00	1.00	1.00	1.00	1.00	1.00	1.00
HostSync v1 (col, lws=1,1,1x1)	0,001	0,005	0,04	0,07	0,11	0,14	0,25	0,52	0,60
HostSync v2 (col, lws=1,M,Mx1)	0,001	0,006	0,05	0,17	0,32	0,26	0,43	0,78	1,03
HostSync v3 (row, lws=1,M,Mx1)	0,001	0,006	0,05	0,19	0,55	0,99	2,01	4,80	5,96
HostSync v4 (row, lws=(N-k)x1)	0,001	0,006	0,05	0,19	0,55	n.a.	n.a.	n.a.	n.a.
DevSync v1 (col, lws=Nx1)	0,001	0,012	0,11	0,22	0,34	n.a.	n.a.	n.a.	n.a.
DevSync v2 (row, lws=Nx1)	0,001	0,013	0,11	0,24	0,53	n.a.	n.a.	n.a.	n.a.
DevSync v3 (row, lws=(N-k)x1)	0,001	0,013	0,11	0,35	0,74	n.a.	n.a.	n.a.	n.a.

TABLE IV

TIME [MS] TAKEN TO EXECUTE OPENCL KERNELS ON INTEL NEHALEM MULTICORE CPU.

N	8	40	128	300	512	1024	2048	4096	7168
Reference C version	0.001	0.057	1.93	21.86	207	2097	23162	361271	2517300
HostSync v1 (col, lws=1,1,1x1)	0.700	4.220	22.62	98.97	328	1880	14273	143986	917630
HostSync v3' (row, lws=1,1,1x1)	0.500	4.068	18.43	78.26	244	1563	11998	92614	488878
DevSync v2' (row, lws=1x1)	0.296	2.330	16.54	86.51	323	2211	21248	197316	1121610

TABLE V

SPEED-UP IN EXECUTING OPENCL KERNELS ON INTEL NEHALEM MULTICORE CPU VS. C IMPLEMENTATION ON THE SAME PROCESSOR.

N	8	40	128	300	512	1024	2048	4096	7168
Reference C version	1.000	1.000	1.000	1.00	1.00	1.00	1.00	1.00	1.00
HostSync v1 (col, lws=1,1,1x1)	0.001	0.013	0.085	0.22	0.63	1.11	1.62	2.51	2.74
HostSync v3' (row, lws=1,1,1x1)	0.002	0.014	0.104	0.28	0.85	1.34	1.93	3.90	5.15
DevSync v2' (row, lws=1x1)	0.003	0.024	0.116	0.25	0.64	0.95	1.09	1.83	2.24

REFERENCES

[1] G.H. Golub and C.F.V. Loan: Matrix Computations (3rd ed.), chapter 4, pp.133-205. John Hopkins Univeristy Press, Baltimore, MD, USA, 1996.

[2] L.M. Davis: Scaled and Decoupled Cholesky and QR Decompositions with Application to Spherical MIMO Detection. In: Proc. of Wireless Communications and Networking, New Orleans, LA, USA, 2003, Vol.1, pp. 326 – 331.

[3] http://en.wikipedia.org/wiki/Opencl.

[4] D. Yang, G. Liang, D. D. Jenkins, G. D. Peterson, and H. Li: High Performance Relevance Vector Machine on GPUs. In: Proc. of Symposium on Application Accelerators in High Performance Computing (SAAHPC), Knoxville, TN, 2010.

[5] V. Volkov and J.W. Demmel: Benchmarking GPUs to Tune Dense Linear Algebra. In: Proc. of ACM/IEEE Conference on Supercomputing, Austin, Texas, 2008.

[6] S. Tomov, R. Nath, H. Ltaief and J. Dongarra: Dense Linear Algebra Solvers for Multicore with GPU Accelerators. In: Proc. of IEEE International Symposium on Parallel and Distributed Processing, Workshops and Phd Forum, 2010, pp.1-8.

[7] M. Marques, G. Quintana-Orti, E.S. Quintana-Orti and R. van de Geijn: Using Graphics Processors to Accelerate the Solution of Out-of-Core

Linear Systems. In: Proc. of International Symposium on Parallel and Distributed Computing, Lisbon, Portugal, 2009, pp. 169–176.

[8] A. Leung, N. Vasilache, B. Meister, M. Baskaran, D. Wohlford, C. Bastoul and R. Lethin: A mapping path for multi-GPGPU accelerated computers from a portable high level programming abstraction. In: Proc. of Workshop on General-Purpose Computation on Graphics Processing Units, Pittsburgh, Pennsylvania, 2010, pp. 51–61.

[9] A. Kerr, D. Campbell and M. Richards: QR Decomposition on GPUs In: Proceedings of Workshop on General Purpose Processing on Graphics Processing, Washington, D.C., 2009, pp. 71–78.

[10] G.H. Golub and C.F. Van Loan: Matrix Computations (Johns Hopkins Studies in Mathematical Sciences)(3rd Edition).

[11] Khronos OpenCL Working Group, The OpenCL Specification, Vers. 1.0, Rev. 48, Nov. 2009.

[12] NVIDIA OpenCL Best Practices Guide, Vers. 2.3, Aug. 2009.

[13] AMD, Heterogeneous Computing OpenCL and the ATI Radeon HD 5870 (Evergreen) Architecture, slide set, Mar. 2010, available: http://developer.amd.com.

[14] T.M. Lehmann et al. Survey: interpolation Methods in Medical Image Processing In: IEEE Trans. Medical Imaging, Vol. 18, No. 11, Nov. 1999.

Low-power Arithmetic Unit for DSP Applications

Mehdi Modarressi[1,2], Seyyed Hossein Nikounia[1], Amir-Hossein Jahangir[1]

[1]Department of Computer Engineering, Sharif University of Technology, Tehran Iran

[2]IPM School of Computer Science, Tehran, Iran

modarressi@ce.sharif.edu, nikoonia@ce.sharif.edu, jahangir@sharif.edu

Abstract—**DSP algorithms are one of the most important components of modern embedded computer systems. These applications generally include fixed point and floating-point arithmetic operations and trigonometric functions which have long latencies and high power consumption. Nonetheless, DSP applications enjoy from some interesting characteristics such as tolerating slight loss of accuracy and high degree of value locality which can be exploited to improve their power consumption and performance. In this paper, we present an application-specific result-cache that aims to reduce the power consumption and latency of DSP algorithms by reusing the results of the arithmetic operations executed on the same (or approximately the same) inputs. Our proposal improves the area overhead and hit ratio of previously proposed result-caches by figuring out the frequent operands of the arithmetic operations and allocating the cache entries to them. This mechanism exploits the fact that in most DSP applications, one of the operands of the arithmetic operations is input-data independent and takes its value from a finite set of coefficients. The experimental results show that the proposed approach can significantly reduce the power consumption of arithmetic operations by providing a high hit ratio using a small cache.**

Keywords-component; Arithmetic; Result-cache; Low-power.

I. INTRODUCTION

DSP (Digital Signal Processing) algorithms are the major components of multimedia and telecommunication applications, hence play an important role in embedded computer systems. With the increasing presence of power-hungry signal processing functions such as audio and video encoding and decoding in such (usually battery-powered) devices, low-power design of such algorithms is of primary importance, especially in hand-held and portable devices. Not only the power reduction is a critical issue in handheld devices, but also it is important in all VLSI systems, due to the complex temperature management issues and related cooling and reliability challenges.

Almost all DSP algorithms consist of floating-point multiplication, division and trigonometric function evaluation operations which have usually long latencies. Besides, arithmetic operations are a major source of power consumption in typical DSP applications. Some characteristics of DSP applications, however, have a high potential to mitigate the power and latency problems of arithmetic operations. The most important properties of DSP applications which have been exploited in some previous work are inaccuracy-tolerance and locality of values [1]. Inaccuracy-tolerance states that slight quality degradation may be unperceivable to the humans which are the final target of most multimedia devices. In addition, in some other

applications, including radar processing and CORDIC calculation, by relaxing the constraint of producing exact results, arithmetic operations can be made faster and more energy efficient [2], while the results are still acceptable.

Locality of values characterizes the input workload of DSP algorithms and says that at each time slot, the arithmetic operations are executed on the operands with the values close to each other. This locality, for example, comes from the similarity of the pixels in a region of a picture and the graceful changes of a typical speech signal.

Several existing methods exploit the above-mentioned properties of DSP applications to reduce the power consumption of arithmetic operations. Taking advantage of locality of values was first addressed in [3]. Methods like multi-precision multiplier [4], truncated multiplier [5], and adaptive precision multipliers [6] exploit the inaccuracy tolerance property, while some other ones such as region-level approximate computation buffer [7] and fuzzy memoization [1] take advantage of the both characteristics in order to decrease the arithmetic unit power consumption.

In this paper, we propose an application-specific cache for caching and reusing the results of recently executed arithmetic operations which gives a high hit ratio with a small cache size.

Most existing arithmetic result-caches use conventional cache architectures to capture the value locality of the computations. They store the operands and the result of an arithmetic operation in the cache and reuse the saved results when the current operands of the operation match (either partially or completely) the corresponding saved ones. However, in most of the frequently-used DSP applications such as DCT, FFT, and filtering, one of the operands comes from the input data and the other is a coefficient specified by the algorithm (for example trigonometric coefficients in FFT and DCT). Our proposed method benefits from the fact that the coefficients form a finite set of values and can be determined based on the target application independently from the input data. Our caching method extracts the set of data-independent operands for a given application and allocates one or multiple cache blocks to each coefficient. Each block stores the recent operands multiplied to the corresponding coefficient along with the calculated result. As a result, the data-independent operand of each arithmetic operation is used as the index and the other one acts as a tag. This is in contrary to the existing result caches where both operands are considered as tag.

We also use the inaccuracy-tolerance of DSP applications to improve the cache hit-ratio and minimize the conflict among the values stored for different coefficients. This is done by manipulating the least significant bits of the coefficients (by which the cache is indexed) in order to distribute the cache space among the coefficients (allocate non-overlapping cache blocks to each coefficient). We also present a pre-lookup procedure to reduce the power consumption of set-associative caches and benefit from both worlds: the high hit ratio of set-associative caches and the energy-efficiency of direct-mapped caches.

This small cache can be either attached to the IP cores of the target DSP application if the application is encapsulated in an IP core or to the corresponding functional unit if the application is integrated into a processor data-path as a functional unit.

In this paper, like other related work, we focus on multiplication operations, but this method can be applied to other arithmetic operations with different granularities (from simple arithmetic units to complex functions).

The rest of the paper is organized as follows. In Section II, we present the details of the proposed caching method. Section III evaluates the proposed method and finally, Section IV concludes the paper.

II. APPLICATION-SPECIFIC RESULT CACHE

Result-cache stores the results of arithmetic operations to reuse them upon detecting the same operands. If the cache-access triggered by an arithmetic operation results in a hit, the answer appears quickly and the arithmetic operation can be cancelled. On a miss, the arithmetic unit writes the result into the cache at the same time as it sends the result to the output.

In the existing result caches, the operands play a role similar to the memory addresses in a conventional cache. In other words, both operands contribute in indexing the table (using a hashing function) and also act as a tag (by matching the two passed operands to the stored ones) to select the correct data. However, in common DSP applications such as DCT, FFT, and filters, one of the multiplication operands comes from the input data while the other one is an algorithm-specific coefficient from a finite set of values which can be either a constant vector (in Filters) or the result of a trigonometric function (in DCT and FFT). Theses coefficients can be determined for every application independently of the input values. This characteristic can be considered as the third important property of DSP applications along with the aforementioned inaccuracy-tolerance and locality of value properties.

In order to exploit this characteristic, we propose an application-specific caching mechanism for DSP applications to simplify the structure, reduce the size, and increase the hit ratio of the cache. In our approach, the designer specifies the set of input-independent arithmetic operands of the algorithm and allocates one or more cache blocks to every coefficient to store the recent values multiplied to it. In this way, the coefficients are used for indexing the cache while the input operands act as a tag during the lookup procedure.

Like conventional caches, this cache could be either direct-mapped or set-associative.

Fig. 1 shows a 2-way set-associative result cache for an FIR filter. In this figure, some parts of the cache, like the LRU bits and valid flags are not shown. When an input data is multiplied by a coefficient, it is compared to the saved operands at the cache entries indexed by the coefficient. In case of a hit, the result of multiplication which is also saved at the same entry is forwarded to the processor as the result. Otherwise, the multiplication is performed and the operand and the result are replaced with one of the ways of the related set by the LRU algorithm.

In order to increase the hit ratio of this cache, at the cost of losing some degree of accuracy, we ignore some of the least significant bits of the mantissa of the operands. These bits are ignored during the lookup procedure and also are not stored in cache. Let us refer to the number of ignored LSBs as *level of fuzziness*. When there is a hit, to make up the bit-width difference between the expected and the stored result, the not-stored bits of the mantissa are all filled with zeros except the first one which is set to one.

In order to index the cache entries by coefficients, we again take advantage of the inaccuracy-tolerance of multimedia and DSP applications and change the low-order bits of the coefficients in such a way that each of them indexes (i.e. points to) a unique cache set. For example, assume a DSP application with 8 coefficients and a result cache with 8 sets. We consider the three LSBs of the mantissa of the coefficients as index and if two or more coefficients have the same LSB, keep the original LSB value of one of these coefficients unchanged, but replace the LSB of the others with some values from 000 to 111 that is not repeated in any other coefficients. Using these three bits as the cache index, each coefficient points to a unique and separate entry, hence there is no conflict among the data saved for different coefficients. This approach guarantees that the result-cache can be accessed by the coefficients in the same way as a traditional cache. Please note that we can alternatively use a table to keep the cache entries assigned to each coefficient. However, extracting the cache addresses directly from the numerical value of the coefficients will lead to faster and more power- and cost-efficient implementation of the caching mechanism for small caches.

Compared to a conventional cache, the proposed cache structure significantly reduces the cache size, as only one operand of each multiplication is stored.

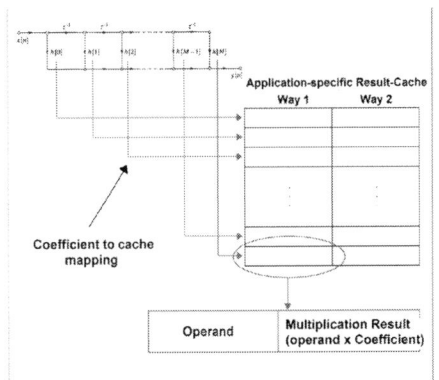

Figure 1. The structure of the proposed result-cache for the FIR filter algorithm

To power consumption of a set-associative cache increases with the associativity degree. The main reason for this higher power consumption is that when the associativity degree of a cache increases, the number of tag comparisons per access increases proportionally, leading to more power consumption. To benefit from the higher hit-ratio of set-associative caches while avoiding their side-effect on power consumption, we use a two-step lookup procedure consisting of an initial tag filtering followed by the main tag comparison. Several previous work have used a two-phase cache access to reduce the power consumption of the cache [8][9].

Tag filtering of the first step is a light-weight lookup which performs the comparisons using a few bits (1 or 2 bits) of the tags. The ways that do not match in this initial comparison are eliminated at this step. There may be one or more than one matches in the first steps on which the full comparison is performed at the second step. Our simulation results show that in an 8-way set-associative cache, the 2-bit comparison of the first step filters out 5 ways, on average. Thus, the comparison of the remaining *N-2* bits (where *N* is bit-width of the tag) is done on 3 tags.

Like other two-step look-up caches [8][9], this approach leads to a two phase access and can potentially increase the cache access time by one clock cycle to perform the filtering comparison. However, as we use very small-size caches, the cache can operate very fast. An analysis by CACTI shows that a cache access (light-weight and full tag comparisons) can be completed within the single clock cycle of a 1GHz processor, hence it does not violate the system timing.

III. EVALUATION

In this section, the proposed cache architecture is implemented on three key DSP applications, namely FIR filter, DCT, and FFT [10]. The applications and proposed caching mechanism is implemented in MATLAB and the circuit-level parameters, like the energy per cache access, are extracted from the CACTI cache simulator [11] in 45nm technology. The cache size is determined based on the application and an LRU replacement policy is used. The input data and coefficients are single-precision floating point numbers with 23-bit mantissa. Each cache way (block) holds a 32-bit operand (used as tag) and a 32-bit multiplication result.

The coefficient values are initialized based on the application and remain fixed during the application execution time. To evenly allocate the cache sets to the coefficients according to the policy of Section II, some least significant bits of the mantissa of the coefficients are considered and modified in order to guarantee that each coefficient points to a unique cache set. Our simulation results show that these modifications do not significantly affect the relative error of the results.

We first use a low-pass FIR filter with the order of 42 (with 42 coefficients). The input of the algorithm is a speech signal in the WAV format.

Figure 2 shows the hit ratio of the proposed 42-set cache for different levels of fuzziness and various cache sizes. This figure shows that an 8-way set associative cache, which keeps the 8 recent operands for each coefficient, gives a hit ratio as high as 89% when the fuzziness level is set to 5, i.e. when 5 least significant bits of the operands' mantissa do not participate in the lookup procedure.

Figure 3 compares the power consumption of a conventional FIR filter with an FIR filter that uses the proposed caching mechanism in its multiplication unit.

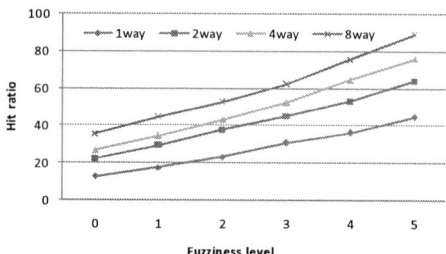

Figure 2. The hit ratio of the proposed result-cache for the FIR filter

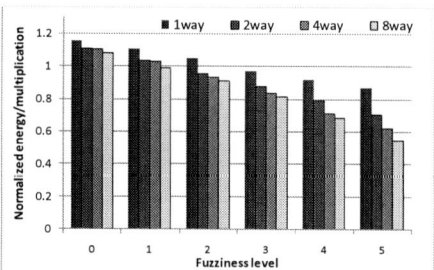

Figure 3. The power consumption of the conventional FIR filter and the FIR filter that uses the proposed caching mechanism. The results are normalized wit respect to the power consumption of the conventional FIR filter.

The results are normalized to the power consumption of the conventional FIR filter to give a better understanding of the obtained improvements. We have calculated the cache energy per access (read and write) using the CACTI tool. The power consumption of the single-precision multiplication is also extracted from a state-of-the art multiplier in [12]. We have also calculated the cache energy consumed when writing a new data to the cache and included it in the reported energy results. The static power consumption of the cache is also included in the results.

Our mechanism adds a cache lookup (hence its energy consumption and latency) to every arithmetic operation. As Figure 3 indicates, when the hit ratio of the cache is low, the energy per multiplication of the proposed mechanism is worse than the energy of a conventional system. However, for higher hit ratios, the power saving obtained by reusing the results not only completely compensates this extra energy overhead, but also offers up to 45% reduction in the energy consumption of the multiplication unit.

Fuzziness level introduces a trade-off between the cache hit ratio and result accuracy. Higher fuzziness levels means higher cache hit ratio which in turn, reduces the arithmetic unit power consumption and latency, but at the cost of more precision loss of the computations. The proper fuzziness level may be different for different applications and is determined by the degree of the inaccuracy tolerance of the application, i.e. the degree to which the application can tolerate the inaccuracy of the arithmetic unit results without degrading the final output quality. To investigate this trade-off, Figure 4 shows percentage of average relative error of the results of the FIR filter for different levels of fuzziness. The average relative error is calculated as

$$100 \times \sum_{\text{for all calcualted results}} 1/_n \times \frac{|R_{\text{exact}} - R_{\text{approx}}|}{R_{\text{exact}}}$$

978-1-4577-0671-4/11 $26.00 © 2011 IEEE

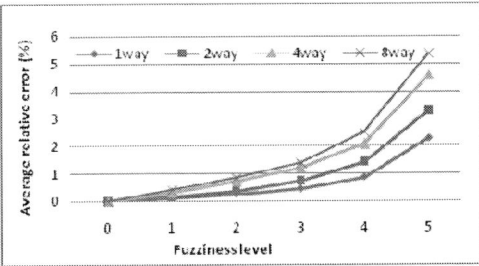

Figure 4. Relative error for different levels of fuzziness

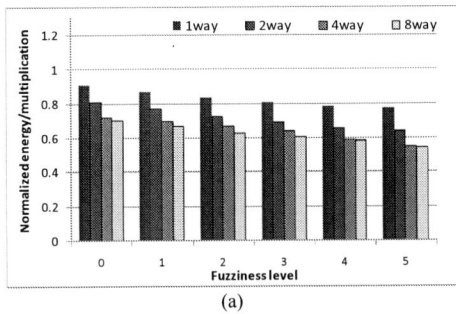

(a)

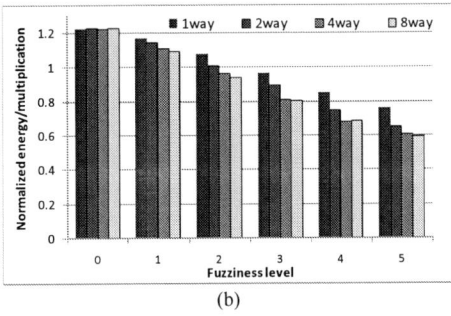

(b)

Figure 5. The normalized power consumption of (a) FFT algorithm and (b) DCT algorithm with using the proposed caching mechanism

Where n is the number of generated results of the algorithm (e.g. FIR filter) and R_{exact} and R_{approx} denote the exact value of the algorithm and the approximate value generated by the algorithm enriched with our caching mechanism, respectively.

Figure 4 shows that the maximum hit ratio (obtained by the fuzziness level of 5) is gotten at the cost of 5.5% relative error at the algorithm output that is still acceptable for many applications.

An 8-input one-dimensional Fast Fourier Transform (FFT) and the parallel version of 3-dimentional Discrete Cosine Transform (DCT) presented in [13] are the other important DSP algorithms we selected as benchmark.

The 1-D FFT and the 3-D DCT algorithms have 12 and 32 coefficients, respectively.

Figure 5 shows the power saving obtained by the proposed caching mechanism for different fuzziness levels and cache associativity degrees.

Following the same trend as for the previous benchmark, up to 40% power reduction for 3D-DCT and 44% power

reduction for 1-D FFT can be obtained when utilizing an 8-way set-associate result cache with the fuzziness level of 5.

IV. CONCLUSION

In this paper, we presented an application-specific result-cache for DSP applications which improves the area overhead and the hit ratio of previous arithmetic result-cache proposals. This method exploits the fact that in many DSP algorithms, one of the operands of arithmetic operations are the algorithm coefficients which are input-data independent and can be selected from a finite set of values. By using the coefficients as the address to index the cache, instead of caching it along with the other operand, we can significantly reduce the cache size. We also presented some techniques in order to simplify the cache indexing and reduce the cache energy per access. The simulation results using some important and frequently used DSP algorithms showed that the proposed caching mechanism can offer a considerable cache hit ratio by a small cache size, which leads to up to 45% reduction in the energy consumption of arithmetic units.

REFERENCES

[1] C. Alvarez, J. Corbal, and M. Valero, "Fuzzy memoization for floating-point multimedia applications" in *IEEE Transactions on Computers*, Vol. 54, No. 7, pp. 922-927, July 2005.

[2] D. Kelly, B. Phillips, S. Al-Sarawi, "Approximate signed binary integer multipliers for arithmetic data value speculation", in *Proc. of the conference on design and architectures for signal and image processing*, 2009.

[3] E. Richardson, "Caching function results; faster arithmetic by avoiding unnecessary computation", Technical report, SUN Microsystems, 1992.

[4] J. Y. F. Tong, D. Nagle, and R. A. Rutenbar, "Reducing power by optimizing the necessary precision/range of floating-point arithmetic," in *IEEE Trans. VLSI Syst.*, vol. 8, pp. 273.285, June 2000.

[5] M.J. Schulte, J.E. Stine, and J.G. Jansen "Reduced Power Dissipation through Truncated Multiplication," in *IEEE Alessandro Volta Memorial Workshop on Low-Power Design*, 1999.

[6] A. Sinha, A. Chandrakasan, "Energy efficient filtering using adaptive precision and variable voltage", in *IEEE international ASIC/SOC conference*, 1999.

[7] X. Cheng, et al., "Region-Level Approximate Computation Reuse for Power Reduction in Multimedia Applications", in *Proc. of ISLPED*, pp.119-125, 2005.

[8] S. Hessabi, M. Modarressi, M. Goudarzi, H. Javan-Hemmat, "A Table-Based Application-Specific Prefetch Engine for Object-Oriented Embedded Systems", in *International Conference on Embedded Computing Systems: Architectures, Modeling, and Simulation*, 2006.

[9] Yen-Jen Chang, Shanq-Jang Ruan, and Feipei Lai, "Sentry tag: an efficient filter scheme for low power cache", in *Proc. of ACSAC*, pp. 135-140, 2002.

[10] A. Oppenheim, et al., *Discrete-time Signal Processing*, Prentice Hall Pubs., 1999.

[11] S. Thoziyoor, N. Muralimanohar, J. H. Ahn and N. P. Jouppi, "CACTI 5.1", *Technical Report HPL-2008-20*, HP Laboratories, 2008.

[12] S. Galal, and M. Horowitz, "Energy-efficient floating point unit design", in *IEEE Transactions on Computers*, to be published, DOI: 10.1109/TC.2010.121.

[13] M Modarressi, H. Sarbazi-Azad, "Parallel 3-Dimensional DCT Computation on k-Ary n-Cubes", in *Proc. of the International Conference on High Performance Computing in Asia Pacific Region*, pp.91-97, 2005.

978-1-4577-0671-4/11 $26.00 © 2011 IEEE

Co-designs of Parallel Rijndael

Issam W. Damaj
Computer Engineering Program
Division of Sciences and Engineering
American University of Kuwait
P.O.Box 3323, Safat, Kuwait 13034
Email: idamaj@auk.edu.kw

Abstract— State-of-the-art Field Programmable Gate Arrays (FPGAs) have inspired the innovation of hardware/software co-design methodologies that provide a high-level of abstraction in the design process. In this paper, we explore the effectiveness of a formal methodology in the co-design of parallel versions of the Rijndael cryptographic algorithm. The investigated methodology employs the functional paradigm for specifications, derived concurrency, and hardware mapping. Several implementations are developed with different performance characteristics. The refined designs are tested under *RC-1000* reconfigurable computer with its two million gates *FPGA*.

I. INTRODUCTION

The continuous need for secrecy and the increase in the use of computer networks in all areas of human activity had drawn researchers' attention for designing the most secured, efficient and fast cryptographic algorithm. For many years, the adopted encryption standard was the Data Encryption Standard (*DES*). The *DES* was widely used by a large number of financial services and industries to protect sensitive applications. The advances in cryptanalysis techniques and the emergence of specialized hardware had broken the *DES* and proved its inadequacy for use in many applications. This raised the need for a more secured and efficient standard: the Advanced Encryption Standard (*AES*).

The *Rijndael* [1] is the winner of the five finalist algorithms chosen for the *AES* competition held by the National Institute of Standards and Technology (*NIST*). Currently, the *Rijndael* is known as the *AES* and available as a Federal Information Processing Standard [2].

The emergence of the new computing paradigm, Reconfigurable Computing (*RC*), introduces novel techniques for accelerating many areas of application, such as, information coding, digital signal processing, space and solar applications, biomedical engineering, networking, and security [3].

RC-systems combine the flexibility offered by software and the performance offered by hardware [4]. It requires a reconfigurable hardware, and a software design environment that aids in the creation of configurations for the reconfigurable hardware [5]. *RC*-systems can build reconfigurable specific purpose processors by programming the fine components of *FPGAs*. *FPGAs* have shown a dramatic increase in their density and a hybridization of their organization over the last few years. For example, companies such as *Xilinx* and *Altera* have enabled the production of *FPGAs* with several millions of gates like *Virtex-7* family and *Stratix-V* [6], [7].

FPGAs have inspired and assisted the development of high-level co-design methodologies to facilitate hardware design. *FPGAs* have made prototyping easy, enabled on-the-fly testing, and the instant realization of parallel designs [8], [9], [10], [3], [11], [12].

This paper investigates a methodology that enables high-level of abstraction in the process of hardware design. The practical realization of this methodology, as evidenced by a case study from cryptography, is the main motivation behind the presented findings.

The proposed methodology is a step-wise refinement approach for developing parallel algorithms from functional specification [13], [14]. The refinement then captures the concurrency of the desired design using Communicating Sequential Processes (*CSP*) models [15]. Finally, the method compiles the described concurrent behavior to hardware using *Handel-C* [3], [12]. This paper will focus on the development of the *Rijndael* cipher without detailing the key-scheduler refinements.

The *Rijndael* in hardware have been presented in [16] where the authors proposed a reconfigurable system that uses different algorithms to enhance the randomization of the ciphertext and the security levels of the *Rijndael*. Rodriguez et al presented in [17] a high performance encryption and decryption core for the *Rijndael*. The proposed architecture implemented on a single-chip *FPGA* using a fully pipelined approach. Ashruf et at in [18] investigated several *Rijndael* implementations based on a suggested processor employing two types of *FPGAs* from *Xilinx* and *Altera*. In [19], the authors presented the design and hardware implementation of a *Rijndael* processor for multimedia applications. A tightly-coupled hardware accelerator using FPGAs for the *Rijndael* is proposed in [20].

The remaining sections of the paper are organized so that Sections II, III and IV introduces the development model by developing the *Rijndael*. The method and the chosen algorithm are analyzed and evaluated in Section V. Section VI concludes the paper.

II. THE FUNCTIONAL SPECIFICATION OF THE RIJNDAEL

The proposed methodology is captured in Figure 1. The functional notation is used for specifying algorithms. The proposed specification style will focus on the use of higher-order functions (such as *map*, *filter*, and *fold*). The functional

978-1-4577-0671-4/11 $26.00 © 2011 IEEE

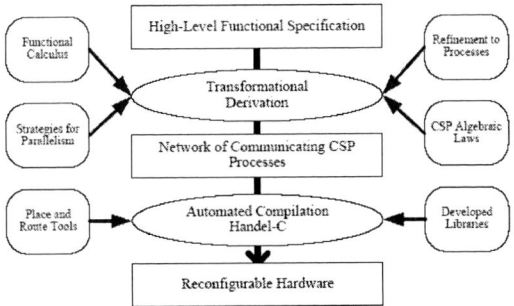

Fig. 1. An overview of the transformational derivation and the hardware realization processes.

specifications are tested on the component, integration and system levels using *Hugs98 Haskell* compiler.

The *Rijndael* algorithm is a 9, 11, or 13-round symmetric block cipher operating on 128, 192, or 256-bit input data and a key of 128, 192, or 256 bits in length. A *Rijndael* round includes byte substitution step (where each byte of the input data block is replaced by its substitute in an S-box), rows shifting step, column mixing step (employing multiplication over GF(2^8)), and adding a round key step (XORing a subkey with an input data block). In addition to the rounds, *Rijndael* algorithm has a starting add round key step, and a final stage consisting of byte substitution and rows shifting steps.

The *Rijndael* cipher requires the generation of 11 key expansion states from the original 128-bit private key. Each state is a (4×4) matrix of 8-bit elements. The cipher in our case inputs 128-bit (1 state) of plaintext and outputs the corresponding ciphered data.

For readability we introduce the following type definitions:

```
type PrivateKey = [Bool]
type State = [[[Bool]]]
```

A functional specification formulates the *rijndael* block cipher as a function *rijndael*. This function inputs a list of states and a private key and outputs a ciphered list of states. The function works by mapping a function *rijndaelState* to all input states. The function *rijndaelState* is responsible for a single state encryption. The functions *rijndael* and *rijndaelState* are specified as follows:

```
rijndael :: [State] -> PrivateKey -> [State]
rijndael statesIn key = map (rijndaelState key) statesIn

rijndaelState :: State -> PrivateKey -> State
rijndaelState stateIn key = finalR
  where
  initialR = addRoundKey stateIn (sKeysStates!!0)

  midRounds = foldl singleRound initialR
                    (take 9 (drop 1 sKeysStates))

  finalR = addRoundKey ((shiftRows.subBytes) midRounds)
                       (last sKeysStates)

  sKeysStates = keySchedule key
```

The function *rijndaelState* performs the ciphering as shown in the Concurrent Process Model (CPM) of Figure 2. The functions used to specify the *Rijndael* building blocks are as follows:

- The function *singleRound* formulating a *Rijndael* round.
- The function *addRoundKey* responsible for XORing a data state with a key state.
- The function *subBytes* responsible for substitution using the S-Box.
- The function *mixColumns* responsible for data states column mixing using multiplication in GF(2^8).

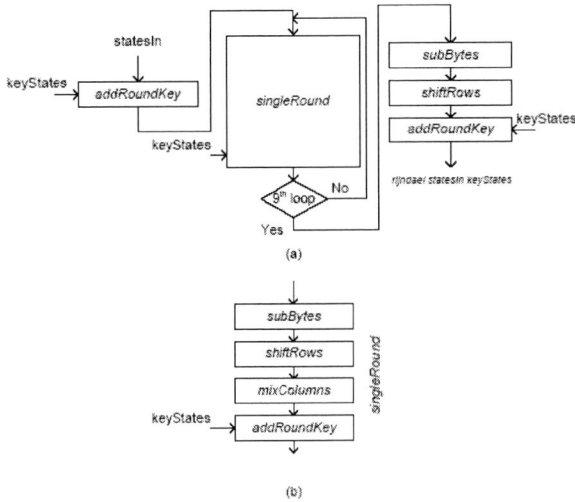

Fig. 2. (a) The function *rijndael* specifying the *Rijndael* block cipher. (b) A single round

The rounds computations are specified using the higher-order function *foldl* of the function *singleRound* over the expanded key states from a function *keySchedule*. The function *keySchedule* is responsible for generating the subkey states. The function *singleRound* could be specified as the functional composition of three functions over an input state and a set of pre-generated key states. The specification of *singleRound* is as follows:

```
singleRound :: State -> State -> State
singleRound stateIn sKeys =
    addRoundKey ((mixColumns.shiftRows.subBytes) stateIn)
    sKeys
```

The function *addRoundKey* zips two states with the function *exor*. As the states are lists of lists (matrices), the zipping is on two levels. The two-level zipping specifies a zip with a zipper on all lists of *statein* and *keystate*. The specification of *addRoundKey* is as follows:

```
addRoundKey :: State -> State -> State
addRoundKey statein keystate =
```

```
zipWith (zipWith exor) statein keystate
```

The function *subBytes* maps another mapping with a function *sBox* over all elements of the state *statein*. The specification of the function *subBytes* is as follows:

```
subBytes :: State -> State
subBytes statein = map (map sBox) statein
```

The function *shiftRows* maps all state rows with the function *shift* for different shift values as follows:

```
shiftRows :: State -> State
shiftRows statein =
    transpose (mapWith [id, (shift 1),
                        (shift 2), (shift 3)]

                    (transpose statein))
```

The function *mixColumns* maps all state columns with the function *singleMix* for mixing. Mixing is done with the aid of multiplication over $GF(2^8)$ formulated in the function *gfMul2*. The formulation of *gfMul2* follows the efficient specification presented in [1]. The functions *mixColumns* and *singleMix* are specified as follows:

```
mixColumns :: State -> State
mixColumns statein = map singleMix statein

singleMix :: Column -> Column
singleMix [c1, c2, c3, c4] = [s1, s2, s3, s4]
  where
    abcd = foldr1 exor [c1, c2, c3, c4]
    a = map gfMul2 (zipWith exor [c1,c2,c3,c1]
                                 [c2,c3,c4,c4])

    [s1, s2, s3, s4] = map (foldr1 exor)
        [[c1, abcd, (a!!0)],
        [c2, abcd, (a!!1)], [c3, abcd, (a!!2)],
                        [c4, abcd, (a!!3)]]
```

III. THE BEHAVIORS OF PARALLEL RIJNDAEL

Based on the functional specification, the behaviors of the parallel *Rijndael* are refined. The refinement between functional specification and networks of *CSP* processes is based on a well establishment pool of provably correct transformation rules [3], [21], [11]. This refinement step is called the transformational derivation, as shown in Figure 1, and supported by strategies for parallelism, *CSP* laws, and the rules of refinement.

The main communication entities used for data refinement are *Item*, *Stream*, *Vector*, and some of their combined forms. The suggested data refinement is based on the message passing technique. The *Item* corresponds to a basic type, such as an *Integer*. The *Stream* is a purely sequential method of communicating a list of values on a channel and an end of transmission (*EOT*) on a different channel. $\langle A \rangle$ is a *Stream* of some type A. $\lfloor A \rfloor_n$ is a *Vector* of type A and of length n. In a *Vector* all items are communicated in parallel. Whenever dealing with multi-dimensional data structures, combinations of vectors and/or streams are used.

As an example, consider the following functional specification for a function *double*:

```
double :: Int -> Int
double x = 2*x
```

The refinement is straight forward and doesn't include a choice for parallelism. The following data refinement considers the input and the output to/from the function *double* as a 32-bit *Integer* items:

$$double :: Int32 \rightarrow Int32$$

After finalizing the decision on the desired level of parallelism by choosing the appropriate data refinement, the refinement is continued by looking into functions that are to be refined to *CSP* processes.

The refinement of the function *double* is the process *DOUBLE*. The refinement is denoted by the following:

$$double \sqsubseteq DOUBLE$$
$$DOUBLE = (in?x \rightarrow SKIP); (out!(x \times 2) \rightarrow SKIP)$$

The process *DOUBLE* describes the behavior of the function *double* by specifying the steps of execution. *DOUBLE* will input x and then outputs its double.

The data and process refinements to *CSP* are provably correct. The main constructs that will produce possibly a parallel behavior are high-order functions; a big refined set of these functions is presented in [11], [12].

At this stage, the refinements of the input and output of the function implementing the *Rijndael* determines the desired degree of parallelism. The data could be refined to streams, vectors, or combination of both as in the following refinements:

- Stream-based:
  $$rijndael :: \langle\langle\langle Int8 \rangle\rangle\rangle \rightarrow Int128 \rightarrow \langle\langle\langle Int8 \rangle\rangle\rangle$$
- Fully Pipelined:
  $$rijndael :: \lfloor\lfloor Int8 \rfloor_4\rfloor_4 \rightarrow Int128 \rightarrow \lfloor\lfloor Int8 \rfloor_4\rfloor_4$$

Different refinements of the function *rijndael* and its subfunctions enable for implementations with different degrees of parallelism. In this paper, we will only detail the refinements of a stream-based and a fully-pipelined implementations. We will provide a bird's view of other possible refinements as an evidence of the high degree of development flexibility the proposed method enjoy.

In a first fully-pipelined design, the *Rijndael* rounds are refined as a pipeline ($\gg$) performing the encryption in parallel ($\|$) with a process representing the key-scheduling (*KEYSCHED*). The rounds are replicated with synchronization (pipelined) using the parallel process *VVFOLD*. *VVFOLD* is the pipeline refinement of the high-order function *fold*. *RIJNDAEL* is described in *CSP* as follows:

$$RIJNDAEL =$$
$$KEYSCHED \| ((ADDROUNDKEY$$
$$\gg_4 VVFOLDL(SINGLEROUND) \gg_4$$
$$SUBBYTES \gg SHIFTROWS \gg STORE_{sov}(temp));$$
$$(PRD_{vov}(temp) \gg_4 ADDROUNDKEY))$$

where,
the function *singleRound* is refined to ($\sqsubseteq$) the *CSP* process *SINGLEROUND*
addRoundKey $\sqsubseteq$ *ADDROUNDKEY*

$subBytes \sqsubseteq SUBBYTES$
$shiftRows \sqsubseteq SHIFTROWS$

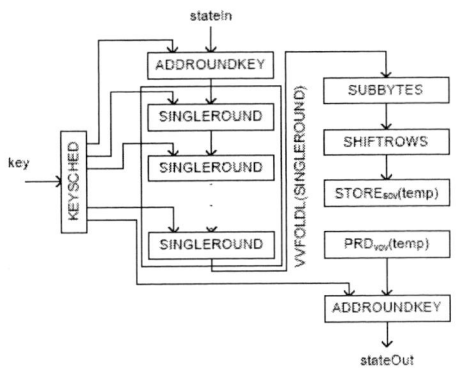

Fig. 3.   The process *RIJNDAELSTATE*

The functional composition in the function *singleRound* is refined to processes piping in the process *SINGLEROUND*. Again for this function, the input and output states are refined to vectors.

$singleRound :: \lfloor\lfloor Int8\rfloor_4\rfloor_4 \rightarrow \lfloor\lfloor Int8\rfloor_4\rfloor_4 \rightarrow \lfloor\lfloor Int8\rfloor_4\rfloor_4$
$SINGLEROUND = (SUBBYTES \gg SHIFTROWS \gg MIXCOLUMNS)$
$\qquad\qquad\qquad\qquad\qquad \parallel ADDROUNDKEY$

The process *ADDROUNDKEY* implements a two-level parallel vector zipping with the XORing process *EXOR*. The *CSP* process *ADDROUNDKEY* is suggested as follows:

$addRoundKey :: \lfloor\lfloor Int8\rfloor_4\rfloor_4 \rightarrow \lfloor\lfloor Int8\rfloor_4\rfloor_4 \rightarrow \lfloor\lfloor Int8\rfloor_4\rfloor_4$
$ADDROUNDKEY = VZIPWITH(VZIPWITH(EXOR))$

VZIPWITH is a parallel version of the high-order function *zipwith*.

The function *subBytes* will be intently refined to a sequential process to avoid the replication of the expensive hardware use of the process *SBOX*. This will, expectedly, introduce a bottle neck in the design as a quid pro quo for less silicon area. The input of *subBytes* will be refined to a vector of vectors to match the piping suggested in the process *SINGLEROUND*, while the output will be refined to a stream of streams. Again, the refinement will employ a two-level application of a higher-order process (*SMAP*). *SMAP* is a sequential version of the high-order function *map*. The process *SUBBYTES* is implemented as follows:

$subBytes :: \lfloor\lfloor Int8\rfloor_4\rfloor_4 \rightarrow \langle\langle Int8\rangle\rangle$
$SUBBYTES = STORE_{vov}(temp);$
$(PRD_{sos}(temp) \triangleright SMAP(SMAP(SBOX)))$

The processes *SHIFTROWS* and *MIXCOLUMNS* refinements are straightforward. Data refinement has been done to enable matching with other processes in a pipe refinement. The descriptions are as follows:

$shiftRows :: \langle\langle Int8\rangle\rangle \rightarrow \langle\lfloor Int8\rfloor_4\rangle$
$SHIFTROWS = STORE_{sos}(temp); TRANSPOSE(temp);$
$(PRD_s([0,1,2,3]) \parallel PRD_{sov}(temp)) \triangleright SMAPWITH(VSHIFT)$
$\gg STORE_{sov}(temps)); TRANSPOSE(temps); PRD_{sov}(temps)$

Then,

$mixColumns :: \langle\lfloor Int8\rfloor_4\rangle \rightarrow \lfloor\lfloor Int8\rfloor_4\rfloor_4$
$MIXCOLUMNS = SMAP(SINGLEMIX) \gg STORE_{sos}(temp); PRD_{vov}$

In the presented refinement, *STORE* and *PRD* processes are used to enable piping of processes having different refined datatypes from the input side, output side, or both. The storing and producing of middle results through all the studied cases introduce a major effect on the expected speeds of designs' implementations. Including or avoiding such processes is left to the designer according to the suggested designs' refinements and the capabilities of the targeted hardware.

In a second stream-based design, the previous fully-pipelined design is replaced by a single-round, sequential, low speed design having less computational resources. The key states are passed as a stream, while the states are refined to vectors of vectors.

The data refinement for *rijndael* is kept the same as in the first design, in other words, the only change is in the communication between *KEYSCHED* and the encryption rounds. The higher-order process *SVFOLDL* is used to construct a sequential implementation with the process *SINGLEROUND* over the stream of the generated key states. *SVFOLDL* is the sequential refinement of the high-order function *fold*. This design is pictured in Figure 4.

$RIJNDAELSTATE =$
$KEYSCHED \parallel ((ADDROUNDKEY \gg_4$
$SVFOLDL(SINGLEROUND) \gg_4$
$SUBBYTES \gg SHIFTROWS \gg STORE_{sov}(temp));$
$(PRD_{vov}(temp) \gg_4 ADDROUNDKEY))$

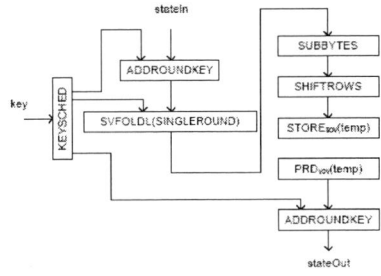

Fig. 4.   The process *RIJNDAELSTATE* using *SVFOLDL*

A third partially-pipelined design allows for the control over the number of rounds to be replicated in parallel. Accordingly, the third design enables the control over the hardware resources needed to implement the *Rijndael*. The more the replicated rounds the larger the silicon area needed for implementation.

This design is depicted in Figure 5 and described as follows:

$RIJNDAEL = KEYSCHED \parallel$
$((ADDROUNDKEY \ggg_4$
$((PRD(n) \rhd VVFOLDL(SINGLEROUND))$
$\parallel SVFOLDL(SINGLEROUND)) \ggg_4$
$SUBBYTES \ggg SHIFTROWS \ggg STORE_{sov}(temp));$

$(PRD_{vov}(temp) \ggg_4 ADDROUNDKEY))$

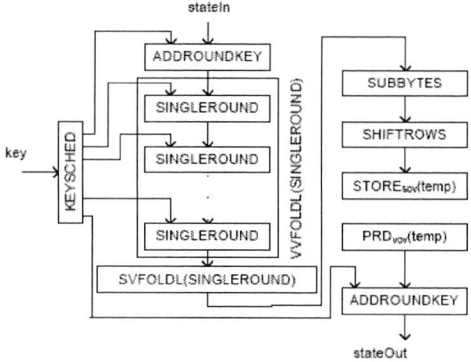

Fig. 5. The process *RIJNDAELSTATE*, third partially-pipelined design

At this point, we would like to stress the flexibility in developing straightforward designs with different degrees of parallelism. More designs could be suggested by using off-the-shelf refinements.

The bottleneck introduced by suggesting a sequential refinement for the function *subBytes* could be removed by refining this function with a higher degree of parallelism. In a fourth stream-based design, we modify the second stream-based design by changing the refinement of *subBytes*. The process *SUBBYTES* is to have a stream of vectors as output instead the previous stream of streams.

The effect of this modification is apparent from the following *CSP* descriptions:

$subBytes :: \lfloor \lfloor Int8 \rfloor_4 \rfloor_4 \rightarrow \langle \lfloor Int8 \rfloor_4 \rangle$
$SUBBYTES = STORE_{vov}(temp); (PRD_{sov}(temp) \rhd SMAP(VMAP(SBOX)))$

The process *SHIFTROWS* refinement is then modified to the following:

$shiftRows :: \langle \lfloor Int8 \rfloor_4 \rangle \rightarrow \langle \lfloor Int8 \rfloor_4 \rangle$
$SHIFTROWS = STORE_{sov}(temp); TRANSPOSE(temp);$
$(PRD_s([0, 1, 2, 3]) \parallel PRD_{sov}(temp)) \rhd SMAPWITH(VSHIFT))$
$\ggg STORE_{sov}(temps)); TRANSPOSE(temps); PRD_{sov}(temps)$

This design is included for the sake of later measuring the effect of widening the bottleneck on the performance of the fully-sequential design.

IV. HARDWARE IMPLEMENTATIONS OF THE RIJNDAEL

Based on the refined networks of *CSP* processes, we include samples of the *Handel-C* code used in the realization of the hardware circuit.

The function *singleRound* formulates a single encrypting round using functional decomposition of three functions. This

formalism has been refined to piping in *CSP*, thereby *SINGLEROUND* will be implemented as follows in *Handel-C*:

```
macro proc SingleRound
        (vovStateIn, vovKStateIn,vovStateOut) {
StreamOfStreamsOfItems(sosOut1, Int8);
StreamOfVectorsOfItems(sovOut2, 4, Int8);
VectorOfVectorsOfItems(vovOut3, 4, 4, Int8);

par{
    SubBytes(vovStateIn, sosOut1);
    ShiftRows(sosOut1, sovOut2);
    MixColumns(sovOut2, vovOut3);
    AddRoundKey(vovOut3, vovKStateIn, vovStateOut);}}
```

The macro *VZipWith* implementing the process *VZIPWITH* is coded as follows:

```
macro proc VZipWith(n, v1,  v2, vOut, F){
    par(i = 0; i < n; i++){
        F(v1[i], v2[i], vOut[i]);}}
```

Where *F* is a macro with two inputs and one output.

V. PERFORMANCE ANALYSIS AND EVALUATION

The development is originated from a specification with higher-level of abstraction. The functional specifications are clear and concise. At this level of development, the correctness of the specification is insured by construction. The correctness is carried forward to the next stage of development by applying provably correct rules of refinement. The rules of refinement are highly flexible and can produce various designs with different performance characteristics. The refinement steps are done by combining off-the-shelf reusable instances of basic building blocks.

In Table I, we present the findings for the encryption designs implementation. Nevertheless the implemented designs run with low speeds, the different degrees of designs' parallelism is reflected through their throughput. The second stream based design runs with a throughput of 769 Kbps occupying an area of 8713 Slices. The third partially-pipelined design is a 3.9% faster (800.4 Kbps) than the second design with a 109% larger area (18244 Slices). The bottleneck reduction done through the fourth stream-based design increased the speed wrt the second design by 22% (891.9 Kbps). The area occupied by the fourth design is 10655 Slices. A fifth purely sequential design achieved a speed of 392.6 Kbps occupying an area of 4346 Slices.

Manually designed high-speed implementations for the *Rijndael* (1.9379 Gbps) have been reported by Elbirt et al [22]. Gaj et al in [23] presented another high-speed implementation for the *Rijndael* (414.2 Gbps). Rodriguez et al in [17] investigated an optimised implementation of the *Rijndael* achieving a throughput of 4.121 Gbps.

VI. CONCLUSION

The main goal of the paper is to explore and extend a systematic approach for high-level hardware development. The development starts by specifying algorithms using the functional notation. Parallel implementations are then derived in the form of a *CSP* network of processes. The designs are then

978-1-4577-0671-4/11 $26.00 © 2011 IEEE

TABLE I

TESTING RESULTS OF *Rijndael* ENCRYPTION

Designs Metrics	1st Fully-pipelined	2nd Stream-based	3rd Partially-Pipelined (5 Parallel and 4 Sequential)	4th Stream-Based Subbytes a Stream of Vectors	5th Sequential
Number of Gates		114264 NANDs	225431 NANDs	127960 NANDs	52840 NANDs
Number of Occupied Slices		8713 Slices (45%)	18244 Slices (95%)	10655 Slices (55%)	4346 Slices (22%)
Total Equivalent Gate Count		158559 Gates	296893 Gates	154477 Gates	97159 Gates
Number of Cycles	NA	NA	NA	NA	9442 Cycles
Maximum Frequency of a Design		26 MHz	54.2 MHz	26.07 MHz	61.8 MHz
Measured Throughput		NA	NA	NA	837.8 Mbps
Measured Execution Time		166.4 Micro Sec.	159.9 Micro Sec.	143.6 Micro Sec.	458 Micro Sec.
Throughput		769 Kbps	800.4 Kbps	891.9 Kbps	392.6 Kbps

implemented under *Handel-C* and tested under *FPGAs*. The paper have presented several co-designs of parallel *Rijndael*. The development method is proven to be reliable and flexible. The produced specification is concise and correct by construction. The development have produced compiled functional specifications that could be reused with other cryptographic algorithms. The parallel behavior is correct by refinement. Further reasoning of the parallel behavior is possible with the available *CSP* descriptions. The *CSP* networks were easily captured using *CPMS*. The matching between *CSP* and *Handel-C* has provided rich softcore libraries that could be reused with other cryptographic algorithms. The methodology has provided a high-level framework. The performance characteristics of the derived implementations were more expensive than manual implementations as a quid pro quo for flexibility and systematic development. Future work includes the automation of the development processes and the optimization of the realization for more economical implementations with higher throughput.

REFERENCES

[1] J. Daemen and V. Rijmen, "AES proposal: Rijndael," 1998.

[2] N. I. of Standards and Technology, "Announcing the advanced encryption standard AES," Federal Information Processing Standards Publication 197, Tech. Rep., November 2001.

[3] I. Damaj, "Parallel algorithms development for programmable logic devices," *Advances in Engineering Software*, vol. 37, no. 9, pp. 561–582, 2006.

[4] K. Compton and S. Hauck, "Reconfigurable computing: a survey of systems and software," *ACM Comput. Surv.*, vol. 34, no. 2, pp. 171–210, 2002.

[5] Y. Li, T. Callahan, E. Darnell, R. Harr, U. Kurkure, and J. Stockwood, "Hardware-software co-design of embedded reconfigurable architectures," in *Proceedings of the 37th Annual Design Automation Conference*, ser. DAC '00. New York, NY, USA: ACM, 2000, pp. 507–512. [Online]. Available: http://doi.acm.org/10.1145/337292.337559

[6] Altera, "Information available from," http://www.altera.com, 2011.

[7] Xilinx, "Information available from," http://www.xilinx.com, 2011.

[8] J. Bowen, M. Fränzle, E. Olderog, and A. Ravn, "Developing correct systems," in *Proc. 5th Euromicro Workshop on Real-Time Systems*. IEEE Computer Society Press, 1993, pp. 176–187.

[9] S. Abdi and D. Gajski, "Provably correct architecture refinement," Center for Embedded Computer Systems at University of California Irvine, Irvine-USA, Technical Report CECS0329, September 2003.

[10] J. Peng, S. Abdi, and D. Gajski, "Automatic model refinement for fast architecture exploration," in *the Asia-Pacific Design Automation Conference*, January 2002, p. 332337.

[11] A. E. Abdallah, "Derivation of Parallel Algorithms: From Functional Specifications to csp Processes," in *Proceedings of Mathematics of Program Construction*, ser. Lecture Notes in Computer Science, B. Moller, Ed., vol. 947. Springer-Verlag, August 1994, pp. 67–96.

[12] A. E. Abdallah and J. Hawkins, "Formal Behavioural Synthesis of Handel-C Parallel Hardware Implementation for Functional Specifications," in *Proceedings of the 36th Annual Hawaii International Conference on System Sciences*. IEEE Computer Society Press, January 2003, pp. 278–288.

[13] R. Bird and P. Wadler, *Introduction to Functional Programming*. Prentice-Hall, 1988.

[14] R. Bird, *Introduction to Functional Programming Using Haskell*. Addison Wesley, 1999.

[15] C. A. R. Hoare, *Communicating Sequential Processes*. Prentice-Hall, 1985.

[16] M. Jing, C. Hsu, T. Truong, Y. Chen, and Y. Chang, "The diversity study of AES on FPGA application," in *Proceedings of the IEEE International Conference on Field-Programmable Technology*. IEEE, December 2002, pp. 390–393.

[17] F. Rodriguez-Henriquez, N. Saqib, and A. Diaz-Perez, "4.2 Gbit/s single-chip FPGA implementation of AES algorithm," *ELECTRONICS LETTERS*, vol. 39, no. 15, July 2003.

[18] R. Ashruf, G. Gaydadjiev, and S. Vassiliadis, "Reconfigurable implementation for the AES algorithm," in *Proceedings of ProRISC 2002*, November 2002, pp. 169–172.

[19] Z. Medien, M. Machhout, B. Bouallegue, L. Khriji, A. Baganne, and R. Tourki, "Design and hardware implementation of qoss-aes processor for multimedia applications," *Trans. Data Privacy*, vol. 3, pp. 43–64, April 2010. [Online]. Available: http://portal.acm.org/citation.cfm?id=1747335.1747338

[20] A. Irwansyah, V. P. Nambiar, and M. Khalil-Hani, "An aes tightly coupled hardware accelerator in an fpga-based embedded processor core," in *Proceedings of the 2009 International Conference on Computer Engineering and Technology - Volume 02*. Washington, DC, USA: IEEE Computer Society, 2009, pp. 521–525. [Online]. Available: http://portal.acm.org/citation.cfm?id=1510526.1511109

[21] A. E. Abdallah, "Synthesis of massively pipelined algorithms for list manipulation," in *Proceedings of the European Conference on Parallel Processing, EuroPar'96, LNCS 1024, (Springer Verlag, 1996)*, L. Bouge, P. Fraigniaud, A. Mignotte, and Y. Robert, Eds. Springer Verlag, 1996, pp. 911–920.

[22] A. Elbirt and C. Paar, "An FPGA implementation and performance evaluation of the Serpent block cipher," in *Proceedings of the 2000 ACM/SIGDA eighth international symposium on Field programmable gate arrays*. New York - USA: ACM Press, 2000, pp. 33 – 40.

[23] K. Gaj and P. Chodowiec, "Fast implementation and fair comparison of the final candidates for advanced encryption standard using field programmable gate arrays," *Lecture Notes in Computer Science*, vol. 2020, pp. 84–100, 2001.

978-1-4577-0671-4/11 $26.00 © 2011 IEEE

A Set of Traffic Models for Network-on-Chip Benchmarking

Esko Pekkarinen, Lasse Lehtonen, Erno Salminen, Timo D. Hämäläinen
Tampere University of Technology, Department of Computer Systems

Abstract—This paper presents a set of **9** application traffic models for benchmarking Networks-on-Chip designs. Common benchmarks allow fair comparison, reproduction of research results, and accelerate NoC development. The set is based on real applications found in literature and executable on freely available benchmarking tool called Transaction Generator (TG). It was found that traffic target distribution is far from uniform and bandwidth requirements vary very much between tasks and between applications. TG and the model source codes are freely available. Models are stored in XML format and are based on task graphs with on average 15 processing tasks. The average communication load is about 2 GByte/s.

Index Terms—Multiprocessor System-on-Chip, Network-on-Chip, benchmarking

I. Introduction

A modern Multiprocessor System-on-Chip (MP-SoC) integrates many Processing Elements (PEs) into a single chip to execute complex applications, such as multimedia players or wireless communication. System design has shifted towards combining existing Intellectual Properties (IPs) from different vendors to fulfill the desired functionality. However, parallel processing increases the requirements for the Network-on-Chip (NoC) [1], [2] which provides interconnections between PEs, memories and off-chip interfaces.

A lot of research and design effort has been invested in NoCs in order to fulfill the stringent performance requirements. Optimizing the resources and application broadens the already large design space and the need for efficient tools is obvious. This applies to NoCs as well, yet they still lack common benchmarks for fair comparison of designs [3]. The problem is currently addressed by the OCP-IP Network-on-chip benchmarking workgroup which seeks to define common benchmarks and benchmarking methods for NoCs [4].

This work presents a set of traffic models and is part of the OCP-IP standardization. The overall flow is depicted in Figure 1. We analyze the traffic patterns of parallel applications based on literature and create executable models for them. The models can be used in a SystemC tool called Transaction Generator (TG) that creates traffic load to the benchmarked NoC, and for collecting statistics [5], [6].

This paper concentrates on model creation. The main contribution is a set of 9 application models for TG tool. This paper is organized as follows. In Section II, related work is presented and Section III introduces the Transaction Generator tool. Details regarding the proposed benchmarking set are given in Section V. Finally, conclusions are drawn.

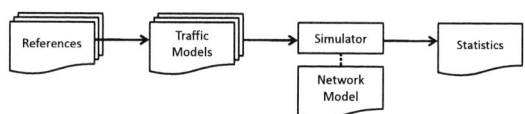

Fig. 1.   Overview of traffic model flow.

II. Related work

NoC benchmarks has been divided into four categories. A *synthetic* benchmark creates traffic to NoC attempting to mimic the behavior of target application. *Algorithm-based kernels* focus on the key algorithm isolated from the application. Running the *actual application* is naturally the best possible benchmark, but most often it is not an option. Varying *combinations* of all three are also possible. So far, most NoC research has been based on proprietary test cases which complicates direct comparison and reproduction of findings [1].

Two aspects of network traffic must be considered: *spatial* (where to send) and *temporal* (when to send). Spatial target distribution can be purely synthetic, such as random, bit rotation or tornado traffic; or based on statistics from real application. Temporal properties include data rate, burstiness, and dependencies. Another approach is presented in [7] where three statistical components are used to cover the spatio-temporal traffic properties of 25 applications: *Hurst exponent* H for burstiness, p for hop distance and σ for distribution ratio of injected traffic.

Three applications from E3S benchmark suite [8] have been used earlier by Hu *et al.* for NoC study [9]. Noxim [10] is a simulator tools that allows evaluation of throughput, delay and power consumption for customised NoCs. SDF^3 [11] generates Synchronous Dataflow Graphs (SDFGs) which can be used to mimic e.g. multimedia applications with cyclic behavior. This work is based on [9], [12], [13], [14], [15], [16], [17] and more details about them are given in section V.

All works cited above are synthetic and based on real applications (except the generic cases). We use the same approach but intercommunicating tasks are used to create traffic to NoC instead of independent statistical components. Furthermore, all presented tools and traffic models are freely available for download.

III. Transaction Generator

We utilize Transaction Generator which is a freely available MP-SoC evaluation tool written in SystemC [14]. Abstract

978-1-4577-0671-4/11 $26.00 © 2011 IEEE

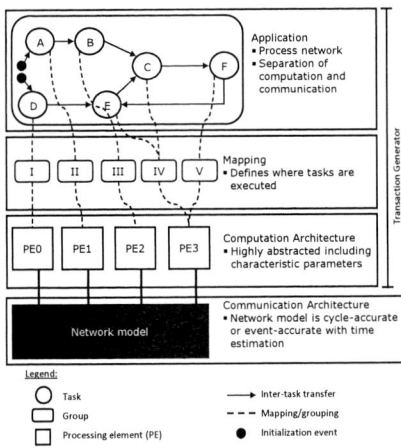

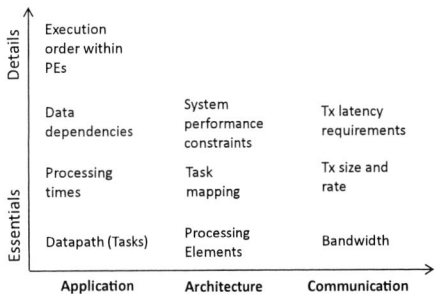

Fig. 2. Transaction Generator overview. Task network models application mapped on multiple PEs. An abstract network model provides the interconnection.

Fig. 3. Modeling parameters addressed by TG.

high level application workload model is based on task graphs and described in eXtensible Markup Language (XML) format [5]. TG creates traffic to NoC according to the description file and measures various performance metrics during simulation. The concept is shown in Figure 2.

A task is activated when it receives a data token. Then it "executes" a certain amount of computation (practically waits), and initializes data transfers to other tasks. Most transfers are simple write operations and reads are supported with memories. Tasks can perform processing and transfers simultaneously. In addition to tasks, there are timers to create stimulus to model the environment. TG automatically collects statistics regarding data amounts, latencies, PE utilization and task execution counts.

Designer can easily modify the mapping of the application tasks to the PEs which are modeled in abstract manner. Note that TG kernel is decoupled from the NoC. NoCs are interfaced with Open Core Protocol (OCP) and they can be descibed at Register Transfer Level (RTL) and Transaction Level, the latter obtaining up to 80x speedup in simulation [6]. Current implementation takes about 5000 lines of SystemC for TG and about 15000 lines of Java for optional graphical user interface.

IV. TRAFFIC MODELING

Models always impose certain inaccuracy that must be accepted,since otherwise model creation would be too cumbersome. Our earlier experiments with TG suggest that error remains mostly below 10% compared to the FPGA prototype which is an acceptable level in NoC benchmarking. The modeling parameters are depicted in Figure 3 and explained next.

A. Parameters

Based on our experience companies are not willing to share details regarding their applications or details are not gathered. Hence one must rely on literature. However, many details of test application are often omitted because they are

not necessarily the main focus of the paper, for example when discussing design methodology. Modeling is possible with rather little information, at minimum only the data rates i.e. bandwidth between PEs is sufficient. However, minimal models do not allow other evaluation than basic NoC benchmarking, for example CPU utilization cannot be measured.

Computation can be refined into a set of tasks, resulting in at least one task per PE and perhaps much more. The processing time of a task on the given PE can be measured either in clock cycles or seconds. Dependencies force tasks to wait until the operands are present which makes the model more accurate. For example, two transmissions from separate tasks are required before computation can be started. TG initializes transfers at the end of the computation. In real applications this may not always be the case and data can be sent at several stages during computation. At the most accurate level the execution order of tasks within one PE is also known.

Architecture defines the set of resources, for example PEs and memories, that is absolutely mandatory information in case that task-level information is not available. Task mapping defines on which PE each task is executed. Optimal mapping is rarely intuitive and it is one of the main problems of design space exploration. Although mapping isn't an essential part of a traffic model, it contributes to the measured performance. Some systems impose constraints to PEs, such as maximum frequency cannot exceed 200 MHz.

Bandwidth is a minimum requiement and it can be derived from other parameters if available. Exact transfer (Tx) sizes refine the communication and affect the packetization at network interfaces. Transfer sizes can be derived from bandwidth by dividing the overall bandwidth by transfer rate. The rate is based on an estimate or the requirements of the target application. Transfers may have constraints, such as maximum latency, similarly to system level constaints.

B. Model capture for TG

Figure 4 shows an example of a task graph for audio video (AV) benchmark. It has 40 tasks (nodes) and 57 unidirectional connections (edges). Bandwidth in MB/s is marked next to each edge. In [9] the application was mapped to 16 PEs and hence not all bandwidths are known. The processing times and transfer sizes are also unknown.

The XML model description is rather easily generated. The previous example is less than 2000 lines of XML and

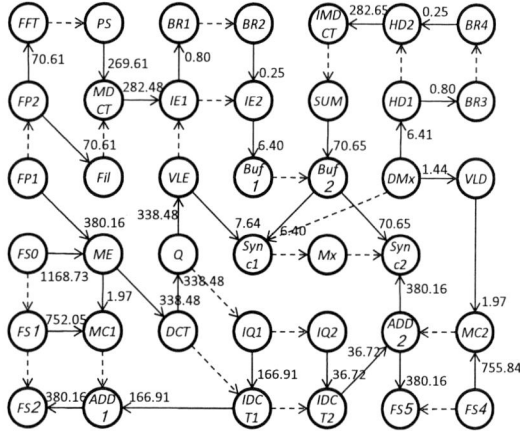

Fig. 4. Example of system [9] traffic graph. Nodes present tasks and edges unidirectional data channels. Dashed edges present transfers within a PE and therefore their bandwidths are omitted. Channel bandwidths are given in MB/s.

```
<task name="MDCT" id="39" class="general">
  <in_port id="3936"/>
  <in_port id="3938"/>
  <out_port id="3913"/>
  <trigger dependence_type="and">
    <in_port id="3936"/>
    <in_port id="3938"/>
    <exec_count>
      <op_count>
        <int_ops>
          <polynomial>
            <param value="0" exp="0"/>
          </polynomial>
        </int_ops>
      </op_count>
      <send out_id="3913" prob="1">
        <byte_amount>
          <polynomial>
            <param value="8828" exp="0"/>
          </polynomial>
        </byte_amount>
      </send>
      <next_state value="READY"/>
    </exec_count>
  </trigger>
</task>
```

Fig. 5. XML description of task MDCT. Data transfer to task IE1 (port 3913) is initialized only after data from tasks PS and Fil has been received. The task executes no computation.

creation took less than 4 hours once we had got familiar with the used notations in the original article. Description of task MDCT near top-left corner is given in Figure 5. It demonstrates the XML usage and how the total number of lines is somewhat expanded by repetitive structures. The task has two input ports and is triggered when data arrives in both of them (dependence_type=and). In this case there is no computation (int_ops value=$0 \cdot x^0 = 0$) and the task only sends 8828 bytes when activated. This model assumes that each task is executed 32000 times per second. Hence $32\ kHz \cdot 8828\ B = 282\ MB/s$.

V. PROPOSED BENCHMARKING SET

Our proposed set contains 9 traffic models that focus on multimedia and telecommunication applications derived from

literature. SoC is a natural platform in both application domains. Many telecommunication applications process several parallel signal streams and both require heavy computation capacity. Both domains have real-time requirements regarding minimum bandwidth or maximum latency.

This set serves as a starting point for standardized evaluation and will undergo updating and refinement by the OCP-IP workgroup. However, it is already ready-to-use with TG as it is. Evaluator may choose to modify the platform (e.g. NoC type or frequency) but the applications should remain unchanged for benchmarking purposes.

A. Telecommunication benchmarks

We have selected four applications having different communication loads. Channel Equalizer presents the smallest communication of 19 MB/s in total [13]. UMTS and OFDM receivers both have modest total bandwidths, 94 and 196 MB/s respectively [15], [16]. The largest by far is an application by Ericsson Radio Systems with 4488 MB/s [12]. It is heavy in communication but exact functionality of the application is unknown.

Channel Equalizers are needed in radio receivers to correct the distortion, caused e.g. by signal reflection from buildings and vehicles. In portable systems adaptive equalization which immediately reacts to changes in channel characteristics is one of the key features.

UMTS receivers are used mostly in mobile phones. Incoming serial data is branced into parallel streams. In our models one task processes all parallel streams. Orthogonal Frequency Division Modulation (OFDM) is used for radio and television transmissions. OFDM data is transferred over several channels, each carrying different data, and receiver selects the desired channel by filtering it. Both methods enable the efficient use of the overall wireless bandwith for many simultaneous transfers.

B. Multimedia

Data-parallel MPEG-4 encoder is based on our earlier work [14] and the only one whose computation is known in detail and addressed by the model. One master processor assigns the computation to several identical slave processors and merges the results. The number of slaves is configurable and XML file is automatically generated. Here, values are for QCIF and 9 slave CPUs whereas having 16CIF and with 36 slaves has a total bandwidth requirement of about 170 MB/s. It is modest due to local data memories at PEs.

In other cases PEs and data rates have been reported, but their exact functionality is unknown. The MPEG-4 decoder, Multiwindow Display (MWD) and Video Object Plane Decoder (VOPD) have been used earlier for benchmarking e.g. in [17]. In the first one, communication is concentrated to/from memories whereas the two others have more pipeline style structure.

Audio video benchmark model is derived from the E3S suite [8]. The presented values in this paper are based on earlier NoC study by Hu et al. [9]. The structure of the task graph is one of the most complicated. Note that tasks have at most two targets which favors hierachical NoCs that can exploit locality.

978-1-4577-0671-4/11 $26.00 © 2011 IEEE

TABLE I

TRAFFIC CHARACTERISTICS

#	Application	Structure				PE data rate [MB/s]				XML lines
		#tasks	#edges	avg #dst	max #dst	max eject	max emit	avg emit	Total emit	
1	Ericsson Radio System [12]	16	26	1.6	6	576	3 072	280	4 488	1 108
2	Channel Equalizer [13]	12	20	1.7	4	10	13	5	19	818
3	UMTS receiver [15], [16]	7	11	1.6	6	61	61	24	94	592
4	OFDM receiver [15], [16]	7	12	1.7	4	64	64	49	196	466
5	MPEG-4 encoder [14]	229	505	2.2	19	<1	1	<1	4	13 063
6	MPEG-4 decoder [17]	12	26	2.2	7	897	897	206	2 466	943
7	MWD [17]	12	22	2.4	4	480	480	132	1 184	682
8	VOPD [17]	12	14	1.2	2	800	500	282	3 378	712
9	AV benchmark [9]	40	57	1.4	2	1549	1169	423	6 772	1 635
	min	7	11	1.2	2	10	13	5	19	466
	avg	39	77	1.7	6	555	782	178	2 325	2 094
	max	229	505	2.2	19	1 549	3 072	482	6 772	13 063

With almost 7000 MB/s it's highly demanding application for the NoC.

C. Summary

Traffic characteristics of the released models are shown in table I. We noticed that average size of a task graph is around 15 tasks, excluding the MPEG-4 encoder that is generated with scripts. On average each task has only two targets and in 6 graphs there are tasks that have at least 4 targets. In 4 cases there is one very active source, i.e. a task sending large fraction of the whole traffic, for example 68% for the first graph. Similarly, sinks that eject large fraction of traffic, can be identified in 4 cases. This shows that traffic is far from uniform and great variation can be seen within a single application and between applications.

For complete application descriptions some general assumptions were made: Internal transfers are either minimal messages or copies of outgoing transfers. If separate tasks aren't reported, we assume one task per PE. Tasks with more than one incoming channels are triggered when data is received in all inputs. Bidirectional edges (in original article) are split into two unidirectional channels with half the bandwidth. If no apparent system input exists, tasks with no incoming channels are triggered by timers simultanuosly at short interval to run the application.

TG allows combining multiple task graphs to model heterogenous behavior. For example, one part of large MP-SoC computes telecommunication function and the other multimedia processing. Another example is that telecommunication stack changes dynamically between UMTS and WLAN which can be achieved with timers and few additional scheduling tasks.

VI. CONCLUSION

We have presented a set of traffic models for NoC benchmarking purposes. It is freely available along with the necessary tools and serves as a starting point for the OCP-IP standardization. Transaction Generator and the presented traffic models are available under Lesser General Purpose License (LGPL) at http://www.tkt.cs.tut.fi/research/nocbench/

ACKNOWLEDGEMENT

This work has been supported by the Academy of Finland.

REFERENCES

[1] E. Salminen, On Design and Comparison of On-Chip Networks, PhD Thesis, Tampere University of Technology, Publication 872, 2010, 230 pages.

[2] T. Bjerregaard and S. Mahadevan. A survey of research and practices of Network-on-chip. ACM Computing Surveys. 38 (1). pp. 1-51. 2006.

[3] R. Marculescu. U. Ogras. P. Li-Shiuan. N.E. Jerger. Y. Hoskote. Outstanding Research Problems in NoC Design: System, Microarchitecture, and Circuit Perspectives. IEEE Trans. Computer-Aided Design of Integrated Circuits and Systems, vol. 28. no. 1. pp. 3-21. Jan. 2009

[4] E. Salminen. K. Srinivasan and Z. Lu. OCP-IP Network-on-chip benchmarking workgroup. White paper, OCP-IP, 2010.

[5] E. Salminen, C. Grecu, T. D. Hämäläinen, A. Ivanov, Application modeling and hardware description for Network-on-chip benchmarking, IET Computers & Digital Techniques, Sep. 2009, vol. 3, no. 5, Special issue on Network-on-chip, pp. 539-550.

[6] L. Lehtonen, E. Salminen, T. D. Hämäläinen, Analysis of Modeling Styles on Network-on-Chip Simulation, in Norchip, Nov. 2010, 4 pages

[7] V. Soteriou. H. Wang. L.-S. Peh. A statistical traffic model for on-chip interconnection network. in MASCOTS. Sep. 2006. pp. 104-166.

[8] R. Dick. Embedded System Synthesis Benchmark Suites (e3s). [Online]. Available: http://ziyang.eecs.umich.edu/ dickrp/e3s/

[9] J. Hu. U. Ogras. and R. Marculescu. System-level buffer allocation for application-specific networks-on-chip router design. IEEE Trans. Comput.-Aided Design Integr. Circuits Syst.. vol. 25. no. 12. Dec. 2006. pp. 2919-2933

[10] M. Palesi. D. Patti. F. Fazzino. Noxim the NoC simulator. [Online]. Available: http://noxim.sourceforge.net/

[11] S. Stuijk. M. Geilen. T. Basten. SDF3: SDF for Free, in ACSD, June 2006, pp. 276-278.

[12] Z. Lu. A. Jantsch. TDM virtual-circuit configuration for NoC. IEEE Trans. VLSI systems. vol. 16. no. 8. Aug. 2008. pp. 1021-1034

[13] A. Moonen. M. Bekooij. R. van den Berg. J. van Meerbergen. Evaluation of the throughput computed with a dataflow model - A case study. ES Reports. ESR-2007-01. Mar. 2007

[14] T. Kangas. K. Kuusilinna. T. Hämäläinen. Scalable Architecture for SoC Video Encoders. J. VLSI Signal Processing-Systems for Signal, Image, and Video Technology. Mar. 2006. Vol.44. pp. 79-95.

[15] P. Wolkotte. Exploration withing the Network-On-Chip Paradigm. PhD Thesis. University of Twente. CTIT Ph.D.-thesis series No. 09-133. 2008.

[16] G. Rauwerda. Multi-standard adaptive wireless communication receivers. PhD Thesis. University of Twente. CTIT Ph.D.-thesis series No. 08-109. 2007. 241 p.

[17] D. Bertozzi, A. Jalabert, S. Murali, R. Tamhankar, L. Benini and G. de Micheli. Noc synthesis flow for customized domain spesific multiprocessor systems-on-chip. IEEE Trans. Parallel Distrib. Syst. vol. 16. no 2. pp. 113-129. Feb. 2005.

Effects of Loop Unrolling and Use of Instruction Buffer on Processor Energy Consumption

Vladimír Guzma, Teemu Pitkänen, Jarmo Takala
Department of Computer Systems
Tampere University of Technology
Tampere, Finland
Email: name.surname@tut.fi

Abstract—In the area of Embedded Systems, instruction memories are one of the critical components consuming significant amounts of energy. Existence of a relation between size of the compiled program, and consequently required size of the instruction memory, and the compiler optimization flags is well-known. In particular, loop transformations such as loop unrolling, while having potential to increase performance dramatically, often cause unreasonable growth in the size of the required instruction memory, causing loss of benefit of lower cycle count from overall system energy point of view.

One method how to decrease energy consumption of the memories is use of instruction buffers. Often executed loops are stored in the buffer and executed from there, while main memory is not read.

In this paper, we show how the compiler flag, controlling loop unrolling, influences the structure of the loops in the program. While unrolling improves performance, unrolled loops can disappear from the program completely, or grow to unreasonable size where use of instruction buffer brings no benefits from the energy point of view.

I. Introduction

Loop transformations are extensively studied methods how to improve performance of the code from several different points of view. In this paper we concentrate on *loop unrolling*, loop transformation that replaces the body of the loop with several copies of the body of the loop. Loop control is adjusted as well. This transformation reduces overhead of loop execution such as test testing for loop termination condition and can improve instruction scheduling and other optimization. Impact of loop unrolling on Instruction Buffer size was studied in [16] [7]. The obvious impact of use of the loop unrolling is the increase of the size of the resulting code. However, when used as a tool to improve instruction level parallelism, partial loop unrolling could improve performance significantly without extensively extending code size. On the other hand, the well understood methods how to improve performance of loop in a processors with complex memory hierarchies are use of repeat buffers or instruction caches. Additional appreciated effect of use of the instruction buffers is reduction in the energy consumed by the instruction memories. While the instructions are executed form the buffer, the memory is not read and only idle power is used.

Since many DSP applications have large amounts of instruction level parallelism to explore with relatively simple structure, they are good candidates for implementation of custom sized instruction buffer, tailored for specific loop size. On the other hand, the simplicity of the loop structures is also often exploitable by loop unrolling, improving performance significantly.

Multiple methods for reduction in the energy requirements of memory hierarchies were surveyed in [4]. Natural extension of this mechanism is storing already decoded instructions in the buffer [2]. This saves cost of repeated decoding of the instructions in loop body.

In [6], the energy consumption of processor is analyzed, when using addressed, compiler controlled, fixed size instruction register file.

Our previous work [9] describes actual methods how to find out suitable size of the instruction buffer for specific application by using after code generation analysis and trace results from the simulation.

In this paper, we use same analysis mechanism, however we apply it in combination with compiler optimization designed to control loop unrolling. We vary loop unrolling threshold and analyze its effect on structure of the loops present in the scheduled code. We show that the use of unrolling can have unexpected effects on overall energy consumption of the memories, as well as usability of the instruction buffer.

II. Experimental Setup and Measurement Methodology

In order to determine possible effects of loop unrolling optimization on the memory size, memory energy consumption, instruction buffer size and energy consumption of memories when the instruction buffer is used, we selected subset of CHStone [10] benchmark. Algorithms tested are presented in Table I. We designed sample test architecture using TTA-based Co-design Environment (TCE) [11] [1], which gives us flexibility to modify the global control unit with addition of the instruction buffer and buffer control state machine. TCE toolset uses the LLVM Compiler Infrastructure [13] [15] as a frontend and custom code generation backend. Since loop unrolling is one of the optimizations performed in frontend, we use *loop unrolling threshold* as varying command line flag. The number passed as a parameter with this flag indicates the size of the loop in the LLVM machine format, which when reached will stop LLVM from unrolling more. Therefore, the higher threshold, the larger the loop can become - more copies

978-1-4577-0671-4/11 $26.00 © 2011 IEEE

of the loop body can be unrolled. For each of the selected threshold sizes, we scheduled benchmark and collected the cycle counts and the number of instructions that the program binary has.

Once the binaries for different sizes of loop unrolling threshold were generated, we analyzed them to find simple loops suitable for storing in the instruction buffer, as well as the number of iterations those loops perform for our sample input using instruction trace [9]. We also find the most beneficial loop; one where the number of iterations multiplied by number of instruction is greater than for any other candidate loop.

Based on data above, we selected for each of the benchmark cases the size and the number of the memories required to store the program binary. Since only single line is read from the memory at a time, in the presence of several memory blocks other blocks remain idle. In this particular experiment we did not try to change layout of the binary in order to store the functions with the loops into the single largest memory block, and therefore we computed average memory power. For each of the memory blocks SX defines number of lines the memory block has, IX defines the idle power consumed when block is not read and AX defines the active – read power of the memory block. For example, for three memory blocks, we compute weighted average power as follows:

$$AveragePower = ((A1 + I2 + I3) * S1$$
$$+ (I1 + A2 + I3) * S2 + (I1 + I2 + A3) * S3)$$
$$/(S1 + S2 + S3)$$

For the instruction buffer power consumption, we used linear approximation based on results of our previous gate level simulations of the buffer [9]. Our data show that the power of buffer for the instruction width of 270 bits grows by 0.2mW with each added line of the buffer. This power is used when the buffer is accessed. Buffer control state machine consumes 2.24mW and is running all the time, whether or not buffer is used. Therefore, in case architecture has added buffer and that is not used at all, the power consumption will increase by mentioned 2.24mW nevertheless .

For estimation of the impact of use of the buffer, we used above-mentioned power estimates, the preferred buffer size and optimistic prediction on how much of total cycles could be executed from the buffer.

The results of our energy estimates are in Section III.

III. RESULTS

Table I show results we collected and computed. For each of the benchmarks we selected several different loop unroll thresholds, essentially starting from 0 - no unrolling, and going through thresholds 4, 8, 16, 32, 64, 128, 256, 512, 1024. Choice of powers of two was random. In cases where different consecutive threshold values did not change cycle count or instruction memory size, we omitted duplicate records.

We list, as mentioned in Section II, the threshold value, the achieved cycle count with such limited loop unrolling, and the number of instructions in the instruction memory. In

addition, we also describe selected choice of memories. We chose from the 1024, 512, 256 and 128 instruction memories. For each of them we collected active and idle power and computed weighted average power. We used this power to compute energy required by the memories for given cycle count, assuming our processors runs at 200MHz.

We also list out results of loop coverage analysis. We show how much of the all cycle counts application requires can be executed from the buffer, as well as preferred size of the buffer. These values, together with knowledge of weighted average and idle memory power, and the power consumption of the buffer allowed us to compute estimate of the processor energy requirements when the instruction buffer is used.

Our five benchmark cases show different effects the choice of unroll threshold has on the application memory energy requirements as well as energy requirements when the buffer is used.

In the case of *adpcm* benchmark, listed in Table I(a), we can see that allow loop unrolling leads to significant reduction in the cycle count (drop of about 25%). However, the memory size doubles at the same time. Energy the memory consumes still drops, however, as the number of memory elements remains in most cases 2. Therefore, the drop in cycle count influence drops in memory energy most. From a point of view of instruction buffer, we can see that use of it is profitable only for smaller thresholds. Threshold of 128 is limit where the loops unroll too much and are not suitable for storage in simple buffer, or are fully unrolled. Together with the increased size of the remaining loops, use of buffer becomes expensive. As a result of our analysis, we can say that for the *adpcm* benchmark, the most energy efficient solution would be to allow larger loop threshold of 512 and take advantage of cycle count drop to save energy, without using the instruction buffer. However, in case the memory area is also the issue, threshold of 64 and use of instruction buffer will provide most suitable solution.

The case of the *aes* benchmark, listed in Table I(b) shows different situation. Once again, increase of loop threshold causes around 30% drop in cycle count at expense of the 30% increase in the memory size. In this case, however, loops keep their coverage of 47% to 53% of all executed instruction through all the tests. The significant factor here is the size of the instruction buffer. Even for the threshold of 0, the buffer size required is already 98 instructions. This number grows into the 201 with threshold of 512. Overall therefore, thanks to great number of instructions executed from the buffer, the energy can be saved compared to execution from the memory only, except last two cases, where the increase of buffer size to around 200 from around 100 causes buffer to consume more energy then is saved by keeping all memories idle. We can therefore say that in this case again, the use of the buffer does not bring energy savings overall, and the least energy is required in case unroll threshold is 512, thanks to the lowest cycle count.

Effect of presence of loop nests can be seen with threshold of 64 and 512. In case of loop nest only the most nested

TABLE I

COLLECTED RESULTS FOR BENCHMARK SET

(a) ADPCM

Unroll threshold	Cycle count	Number of Instructions	Memory configuration (1024, 512, 256, 128)	Memory energy (mJ)	Instructions inside loops	Buffer size	Memory and uffer energy (mJ)
0	104210	1212	(1, 0, 1, 0)	24.47	33%	14	19.35
32	95722	1198	(1, 0, 1, 0)	22.47	28%	22	18.91
64	87268	1196	(1, 0, 1, 0)	20.49	20%	14	18.17
128	76712	1280	(1, 0, 1, 1)	19.41	6%	45	19.53
256	75086	1447	(1, 1, 0, 0)	17.58	0%	81	18.39
512	74832	1520	(1, 1, 0, 0)	17.52	0%	154	18.36
1024	74887	2465	(2, 1, 0, 0)	19.20	0%	0	20.04

(b) AES

Unroll threshold	Cycle count	Number of Instructions	Memory configuration (1024, 512, 256, 128)	Memory energy (mJ)	Instructions inside loops	Buffer size	Memory and uffer energy (mJ)
0	30582	1856	(2, 0, 0, 0)	7.30	52%	98	6.05
32	29701	1838	(2, 0, 0, 0)	7.09	51%	101	5.96
64	28716	1889	(2, 0, 0, 0)	6.85	53%	97	5.63
128	27905	2150	(2, 0, 0, 1)	7.17	52%	100	6.09
256	24074	2272	(2, 0, 1, 0)	6.18	47%	161	6.05
512	19222	2496	(2, 1, 0, 0)	4.92	55%	201	5.20
1024	19299	2814	(3, 0, 0, 0)	5.01	47%	198	5.23

(c) Blowfish

Unroll threshold	Cycle count	Number of Instructions	Memory configuration (1024, 512, 256, 128)	Memory energy (mJ)	Instructions inside loops	Buffer size	Memory and uffer energy (mJ)
0	765984	957	(1, 0, 0, 0)	166.93	18%	13	149.17
16	760480	956	(1, 0, 0, 0)	165.73	18%	13	148.94
32	756409	974	(1, 0, 0, 0)	164.84	17%	13	148.29
64	754496	989	(1, 0, 0, 0)	164.42	17%	13	148.09
128	753727	1058	(1, 0, 0, 1)	177.55	17%	13	161.52
256	742339	1193	(1, 0, 1, 0)	174.31	16%	13	160.14
512	742338	1462	(1, 1, 0, 0)	173.88	16%	13	159.83
1024	742477	2093	(2, 0, 0, 1)	191.02	16%	13	176.61

(d) GSM

Unroll threshold	Cycle count	Number of Instructions	Memory configuration (1024, 512, 256, 128)	Memory energy (mJ)	Instructions inside loops	Buffer size	Memory and uffer energy (mJ)
0	24094	1697	(2, 0, 0, 0)	5.75	34%	16	4.51
32	23965	1738	(2, 0, 0, 0)	5.72	31%	14	4.59
64	18153	1972	(2, 0, 0, 0)	4.33	22%	36	3.87
128	17906	2358	(2, 1, 0, 0)	4.59	12%	56	4.47
256	17087	2980	(3, 0, 0, 0)	4.43	12%	91	4.39
512	17224	3665	(4, 0, 0, 0)	4.83	11%	106	4.83
1024	16535	4331	(4, 0, 1, 0)	4.94	11%	115	4.98

(e) SHA

Unroll threshold	Cycle count	Number of Instructions	Memory configuration (1024, 512, 256, 128)	Memory energy (mJ)	Instructions inside loops	Buffer size	Memory and uffer energy (mJ)
0	870525	515	(1, 0, 0, 0)	189.71	54%	22	117.04
64	760523	544	(1, 0, 0, 0)	165.74	45%	33	117.26
128	708909	714	(1, 0, 0, 0)	154.49	35%	69	130.14
256	698133	1017	(1, 0, 0, 0)	152.14	23%	136	150.01
512	643423	1567	(2, 0, 0, 0)	153.63	0%	13	160.81
1024	642535	1776	(2, 0, 0, 0)	153.42	0%	13	160.59

loop can be stored in buffer, as presented here. However, if this nested loop is fully unrolled, the outer loop may became valid candidate for storing in the buffer, increasing number of instruction executed from the buffer which can higher then in case of no unrolling.

The *blowfish* benchmark, listed in Table I(c) shows again very different behavior. In this case, drop in cycle count associated with the increase of the unroll threshold is minimal, about 4%, and at the same time, the size of the memory doubles. This leads to situation where after initial drop in cycle counts, threshold 64 shows the lowest energy requirements. However, with further increase, additional memory module is required and overall memory energy increases. Situation repeats with addition of third memory block in case of threshold 1024. On the other hand, the loop size does not seem to be influenced by the unroll threshold, we can assume that loop could not be unrolled. Loop coverage remains also similar between cases, small fluctuation of 2% caused by optimization of other loops, not suitable for the instruction buffer. Overall, thanks to the lack of changes in the buffer coverage, most energy efficient solution seems to be use of threshold 64 and instruction buffer.

The case of *gsm* benchmark, listed in Table I(d), shows again steady decrease in the cycle count, as the threshold size grows (31% drop), and increase instruction memory size (2.5 times more memory). This leads to situation where least energy consumed by memory requires threshold of 256 and three memory blocks. At the same time, number of instruction execution from the buffer decreases steadily and the buffer size grows. Largely thanks to drop in cycle count, most energy efficient solution when using instruction buffer is also with loop threshold of 256, even though the buffer size is rather large at 91 instructions.

The last benchmark, *sha*, as listed in Table I(e) is a perfect example of our approach. With growing unroll threshold, the cycle count drops (27%), memory requirement grows (3.5 times more memory). This leads to steady decrease in memory energy, until the point where the second memory block is added and energy rises. At the same time, the percentage of instructions executed form loop drops from 53% with threshold 0 to 23% with threshold 256, and falls to zero for larger values of threshold. The buffer size meanwhile increases from 22 instructions for threshold 0 to 136 instructions in case of threshold 256, and falls to 13 for the last two cases, indicating that the loops which could be stored in buffer beforehand are now fully unrolled. This leads to overall result that without a buffer most efficient solution would be use of unroll threshold 256. However, when the buffer is used, more efficient solution is to use threshold of 0 (152mJ vs. 117mJ).

IV. CONCLUSIONS

In this paper we looked into the effect the compiler optimization, loop unrolling, can have on energy requirements of the instruction memory and the viability of use of an instruction buffer.

We showed that simple analysis of the several different cases can provide insight as to what solution would be most energy efficient, and whether or not the use of instruction buffer will actually contribute to finding lowest energy solution.

We showed that there is no single solution to overall question of energy efficiency, since the single compiler optimization parameter influences cycle count, memory size, presence and structure of a loops, as well as required instruction buffer size.

REFERENCES

[1] "TCE: TTA-based codesign environment." [Online]. Available: http://tce.cs.tut.fi

[2] R. S. Bajwa, M. Hiraki, H. Kojima, D. J. Gorny, K. Nitta, A. Shridhar, K. Seki, and K. Sasaki, "Instruction buffering to reduce power in processors for signal processing," *IEEE Trans. Very Large Scale Integr. Syst.*, vol. 5, no. 4, pp. 417–424, 1997.

[3] R. Banakar, S. Steinke, B.-S. Lee, M. Balakrishnan, and P. Marwedel, "Scratchpad memory: design alternative for cache on-chip memory in embedded systems," in *CODES '02: Proceedings of the tenth international symposium on Hardware/software codesign*. New York, NY, USA: ACM, 2002, pp. 73–78.

[4] L. Benini, A. Macii, and M. Poncino, "Energy-aware design of embedded memories: A survey of technologies, architectures, and optimization techniques," *ACM Trans. Embed. Comput. Syst.*, vol. 2, no. 1, pp. 5–32, 2003.

[5] H. Corporaal, *Microprocessor Architectures: From VLIW to TTA*. John Wiley & Sons, 1997.

[6] W. J. Dally, J. Balfour, D. Black-Shaffer, J. Chen, R. C. Harting, V. Parikh, J. Park, and D. Sheffield, "Efficient embedded computing," *Computer*, vol. 41, pp. 27–32, 2008.

[7] J. W. Davidson and S. Jinturkar, "Improving instruction-level parallelism by loop unrolling and dynamic memory disambiguation," in *Proceedings of the 28th annual international symposium on Microarchitecture*, ser. MICRO 28. Los Alamitos, CA, USA: IEEE Computer Society Press, 1995, pp. 125–132. [Online]. Available: http://portal.acm.org/citation.cfm?id=225160.225184

[8] J. A. Fisher, "Very long instruction word architectures and the ELI-512," in *ISCA '83: Proc. 10th int. symp. on Computer architecture*. Los Alamitos, CA, USA: IEEE Computer Society Press, 1983, pp. 140–150. [Online]. Available: http://portal.acm.org/citation.cfm?id=801649

[9] V. Guzma, T. Pitkänen, and J. Takala, "Reducing instruction memory energy consumption by using instruction buffer and after scheduling analysis," in *Proc. Int. Symp. System-on-Chip*, Tampere, Finland, Sep. 29–30 2010, pp. 99–102.

[10] Y. Hara, H. Tomiyama, S. Honda, and H. Takada, "Proposal and quantitative analysis of the CHStone benchmark program suite for practical C-based high-level synthesis," *Journal of Information Processing*, vol. 17, pp. 242–254, 2009.

[11] P. Jääskeläinen, V. Guzma, A. Cilio, and J. Takala, "Codesign toolset for application-specific instruction-set processors," in *Proc. SPIE Multimedia on Mobile Devices*, San Jose, CA, Jan. 29–30 2007, pp. 65 070X–1 – 65 070X–11.

[12] J. Kin, M. Gupta, and W. H. Mangione-Smith, "The filter cache: an energy efficient memory structure," in *MICRO 30: Proceedings of the 30th annual ACM/IEEE international symposium on Microarchitecture*. Washington, DC, USA: IEEE Computer Society, 1997, pp. 184–193.

[13] C. Lattner and V. Adve, "LLVM: A compilation framework for lifelong program analysis & transformation," in *Proc. Int. Symp. Code Generation Optimization*, Palo Alto, CA, March 20–24 2004, p. 75.

[14] W. Tang, R. Gupta, and A. Nicolau, "Power savings in embedded processors through decode filer cache," in *DATE '02: Proceedings of the conference on Design, automation and test in Europe*. Washington, DC, USA: IEEE Computer Society, 2002, p. 443.

[15] The LLVM Team, "The LLVM Compiler Infrastructure Project," http://llvm.org. [Online]. Available: http://llvm.org

[16] S. Weiss and J. E. Smith, "A study of scalar compilation techniques for pipelined supercomputers," in *Proceedings of the second international conference on Architectual support for programming languages and operating systems*, ser. ASPLOS-II. Los Alamitos, CA, USA: IEEE Computer Society Press, 1987, pp. 105–109. [Online]. Available: http://dx.doi.org/10.1145/36206.36191

Applying IP-XACT in Product Data Management

Erno Salminen, Timo D. Hämäläinen and Marko Hännikäinen
Tampere University of Technology, P.O. Box 553, FIN-33101 Tampere, Finland
email:erno.salminen@tut.fi

Abstract— **Key challenge for industrial embedded system companies is product data management (PMD) of rapidly changing requirements, new platforms and own legacy intellectual property. This can be alleviated with IP component reuse methods, platform based design, and Model Driven Development (MDD) methodologies. We propose an open source product integration environment suitable for small and mid-size enterprises (SME) utilizing FPGAs. We extend the use of IP-XACT standard from HW integration at IP-level to board and chip level designs, describe SW as IP-XACT metadata objects, and add reusability status to objects for PDM. We do not modify or extend IP-XACT metadata format to maintain compatibility with standard compliant tools. Instead, we propose naming conventions and usage profiles. An exemplar FPGA-product shows feasibility in practice in our Kactus2 tool. The benefits are covering complete products including SW objects and HW/SW mappings while strictly keeping the standard XML format.**

Keywords: intellectual property (IP), product data management, meta-data, IP-XACT, electronic design automation, multiprocessor system-on-chip

I. INTRODUCTION

Embedded product development and maintenance requires porting the existing functionality to new platforms as new multicore system-on-chip (SoC) platforms and FPGA (field-programmable gate arrays) generations are introduced. Modern SoCs are very complex and design is heavily based on reusable components [1], [2]. Currently, there are about 50 intellectual property (IP) components per chip [3] and integrating them efficiently calls for specialized design tools [4], [5] and standards, such as IP-XACT [6].

Figure 1 depicts an outline of modern product process, in which new platforms and new products overlap over time. Actions are shown on top and artifacts on bottom. Functionality is captured from the requirements management, and realized as applications executed on platforms composed of HW and SW intellectual property blocks. Many tasks, such as requirement management for the next generation and finalizing the software of the current one, are carried out in parallel and new product releases happen often. A release is, for example, bug fix, new functionality, of change is underlying platform while retaining the functionality. It combines the newest interoperable features from aspects above.

However, keeping track of all different versions both at product and IP level is challenging. Furthermore, in addition to actual design data (e.g. source codes), there is plenty of process-related and other auxiliary information that must be handled. Unfortunately, management of this information is omitted in many design flow proposals.

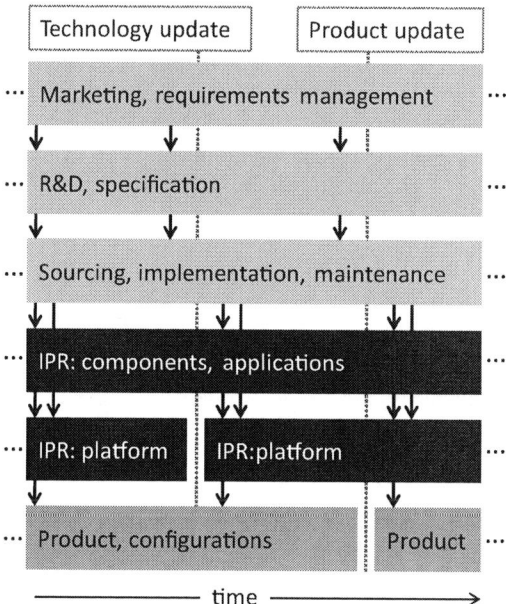

Fig. 1. Continuous process for product creation.

Our goal is a methodology and freely available open source tool, called Kactus2, that allows both design capture [7] and management of essential product information using IP-XACT standard. It is continuation of our previous work in MP-SoC design that utilized model-driven development (MDD) together with reusable library components and standardized interfaces [8]. Since PDM is very wide concept, just the features that easily fit into IP-XACT are considered. The target users are SMEs (small and mid-sized enterprises) that include FPGAs and on- or off-chip processors in their products.

The main contributions of the paper are

- Extending the scope of IEEE1685 [6] (IP-XACT 1.5) above the IP and SoC level for designing products and managing their meta-data
- Categorization of reusable IP components

The rest of the paper is structured as follows. Section II discusses the related works. The basic concept of higher hierarchies and product data management are presented in Sections III, IV and V. Section VI presents taxonomy for reusable IP components and Section VII concludes the paper.

978-1-4577-0671-4/11 $26.00 © 2011 IEEE

II. RELATED WORK

A. Product data management

Product Data Management (PDM) is the discipline of controlling the evolution of a product design and all related product data. It includes managing product structure, its parts, configurations, documents, and workflow. PDM stems from creating part lists (bill of materials) in mechanical engineering but shares many aspects with software configuration and revision management as well [11]. However, they complement each other, as configuration management is used mostly during development phase and PDM at more the early and late stages of the product life cycle (e.g. requirement and maintenance).

Our aim is on embedded system, i.e. to keep track of both HW components and SW artifacts, but also various system configurations and documentation. However, we wish to implement a "light-weight" PDM without separate, often expensive, SW tools. The presented work is a step in right direction: from barebone IP blocks towards products. At the moment, we deliberately leave out the project and workflow related matters.

B. IP formulation for reuse

Management of IP libraries is increasingly difficult due to large number of IPs, dependencies between components, varying source (in-house vs. 3rd party), increasing configuration options, different implementation languages (e.g. VHDL, Verilog, SystemC), different abstraction levels (e.g. RTL vs. TLM), and so on. Especially formulating the components into easily accessible format poses challenges. Some vendor-specific formats exist, but they are not inter-operable.

IEEE1685-2009 standard [6] aids faultless integration of HW IP-blocks in a vendor, IP and tool independent way. The most obvious information are the pin interfaces, list of source codes, and, in case of hierarchy, the sub-blocks and their interconnections. Hence, they form an "electronic data sheet" of an IP or SoC. The standard specifies seven IP-XACT objects, the most important ones being: *component* and *design* which capture the (hierarchical) entities; *bus* and *abstraction* definitions which define the interconnections; and *generator* scripts that are used to process objects or source files. Each object includes a tuple {*VENDOR,LIBRARY,NAME,VERSION*} (VLNV) for identification and references between objects, for example {*TUT, Funbase, DDR_ctrl, 1.0*}.

Figure 2 depicts a simplified IP-XACT design flow. Before integration, all IPs are packetized into IP-XACT components, which means writing the necessary metadata descriptions and storing them into a library. An IP-XACT design includes component instances and interconnections. Hierarchical components have subcomponents whereas so called leaf components can have file sets that list all the necessary files, be it HDL, documentation, or something else. Tools are used to process metadata and automatically generate final executable and configurations. Automation reduces design errors and offers speedup, especially when combined with graphical user interface [7], [9], [10].

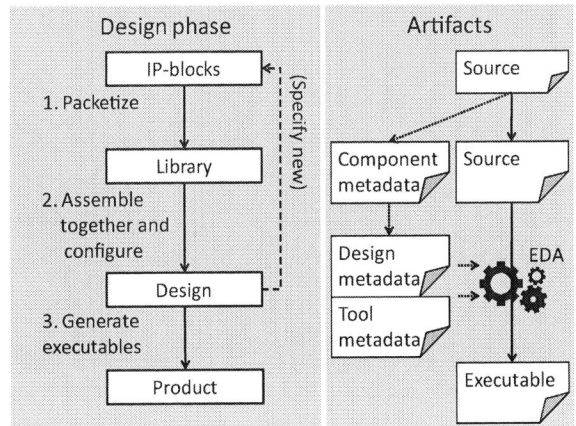

Fig. 2. Simplified IP-XACT design flow.

III. CAPTURING PRODUCT-WIDE HIERARCHY WITH IP-XACT

IP-XACT was developed not only for IP exchange, but also for SoC configuration. We propose to widen its scope to product hierarchies. We keep IP-XACT metadata format untouched, but merely propose usage profiles and naming conventions. This maintains compatibility with standard IP-XACT tools and XML parsers which is of uttermost importance. We should not limit what tools and source formats are used, but define how metadata is defined for each level and how metadata objects are related to each other.

A. Basic concept of product hierarchy

Hierarchical design is fundamental in all engineering and naturally supported by IP-XACT. The most important things are the structure and interfaces. We propose extending the number of hierarchical levels as depicted in Figure 3. The original scope was to integrate *IPs* into a *SoC*, and these two levels are shown on the bottom. However, the same SoC architecture can be implemented using various *chips*, for example ASIC, or FPGA by Altera or Xilinx, each with their own metadata. For example, the pin mapping between logical names and physical pins is chip-specific.

The chips are used in various printed circuit *boards* (PCB), which are used in *products*. For example, the SoC may be first implemented using an off-the-shelf FPGA development board, and then moved to in-house developed board with faster FPGA. Finally, the companys end product consists of one or multiple PCBs. Note that the same board may be used in many products, or that the same product may end up using different board at some stage to obtain cost-savings.

Automated generation of structural HDL is utilized at SoC level and chip-specific pin maps at chip level. The levels above SoC allow keeping track of all product variations, for example if slightly different versions are shipped to different customers. It is notable that all levels include both HW and SW related information but the nature of SW varies. Level

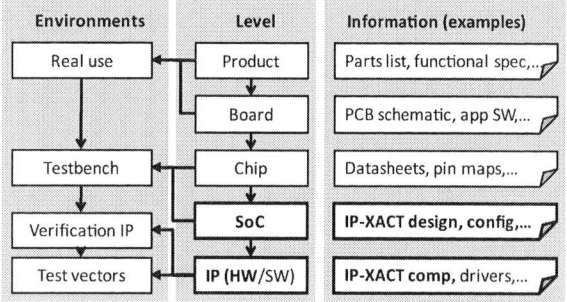

Fig. 3. The concept of utilizing IP-XACT on multiple levels. The original scope of IP-XACT is shown in **boldface** on the bottom (IP+Soc).

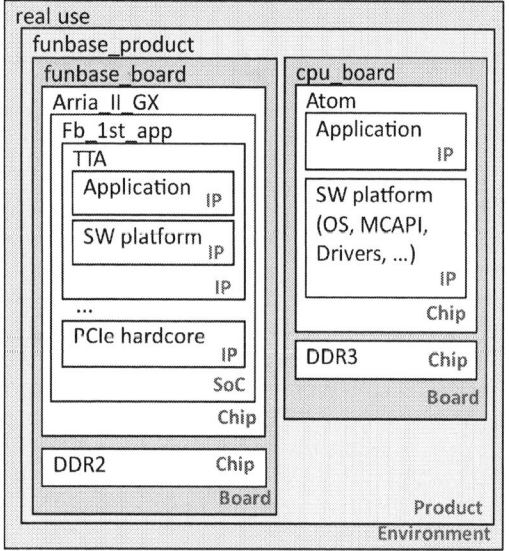

Fig. 4. Example embedded product composition from two boards. For brevity, this figure shows components at all hierarchy levels.

of reusability depends on the structure, configurable elements and dependencies between other IP-XACT objects.

B. SW as component

We wish to describe also SW as IP-XACT components, which gives us method to make VLNV-based references between HW-SW and SW-SW objects. Mapping between SW and HW can be done in two ways. Mapping from HW to SW is implemented by using IP-XACT components hierarchical model view. One view of a CPU refers to RTL implementation and the other references to an executed SW component.

For the mapping from SW to HW, we can use a model parameter in SW component. The value of this parameter is the HW component's instance name in IP-XACT design, or it is already fixed at SW component description. In this way we do not violate IP-XACT metadata format but allow HW-SW mappings from both directions.

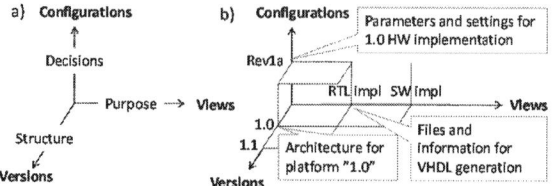

Fig. 7. Three-dimensions of information of an IP-XACT design.

C. Example product

Figure 4 depicts a simplified product example so that the proposed hierarchy is "opened" for brevity. Funbase is abbreviated from Function based design and refers to our current research project. Funbase-product has two boards: one with Altera FPGA-based SoC and an Intel Atom-based processor card. FPGA- board includes DDR2 memory and a SoC containing a TTA processor, accelerators (not shown), and PCIe interface. Processor is actually combination of HW and SW, as it needs application codes and application independent SW like OS and drivers (SW platform). SW is not necessarily "inside" the CPU HW but referred to.

A screen capture from our Kactus2 tool is shown in Figures 5. The left part shows the IP library and the drawing area on the right is used for documenting the structure and connections in Funbase FPGA board. Kactus2 can show library as a flat or using tree view showing the hierarchy. Fig. 6 shows the structure of the SoC inside the Arria II FPGA. It is very convenient to navigate between hierarchy levels and keep them consistent when they captured with a single tool.

Note that presented product hierarchy can be implemented in standard IP-XACT objects utilizing hierarchical designs without any vendor extensions. For example, we can use *busDefinition* and *abstractionDefinition* for PCIe bus between boards just as we use them for on-chip bus on FPGA. Likewise, we can describe the physical pin tester just as an IP-block in IP-XACT component and we do not need to introduce new XML schemas. However, we do propose certain naming and library guidelines

IV. PRODUCT DATA MANAGEMENT

Figure 7 illustrates the basic capabilities for product data management in IP-XACT: versions, configurations, and views. A new version is needed when object's structure or parameter set is changed whereas configuration stores the assigned values of the various parameters in a design. Views capture the different purposes for the same object. For example, an accelerator may have 3 views: RTL implementation, SW reference implementation, and driver SW. Each is accompanied with different information and files.

A. Versioning and documentation management

IP-XACT components have version identifiers but external version control software, such as SVN or Git, is still mandatory. The idea is to develop IPs under version control and give

978-1-4577-0671-4/11 $26.00 © 2011 IEEE

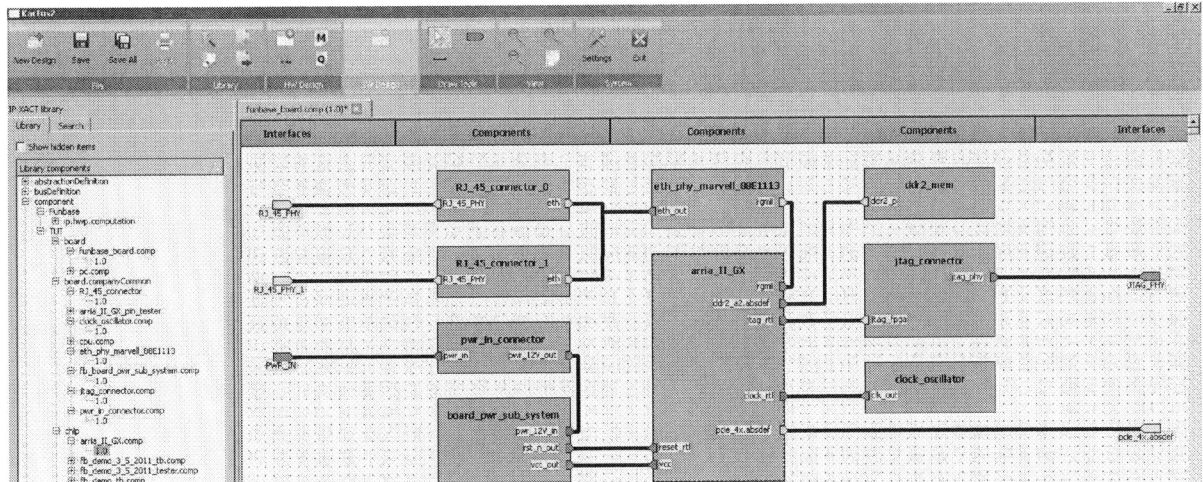

Fig. 5. The structure of Funbase board documented in Kactus.

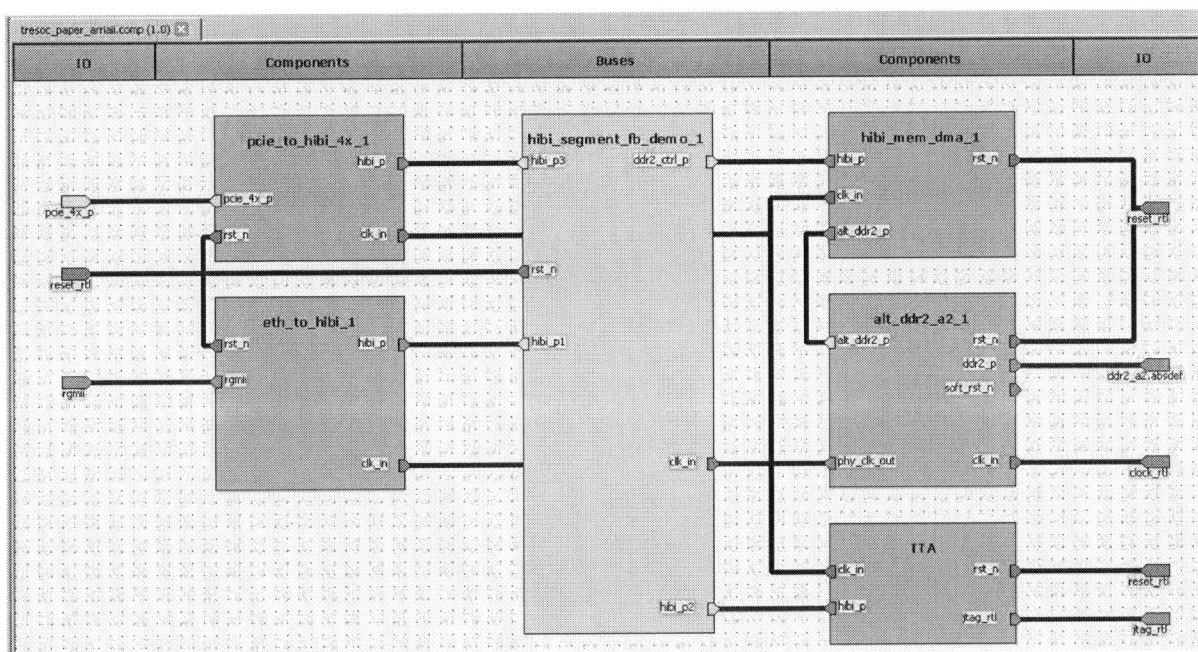

Fig. 6. The SoC architecture.

IP-XACT version label only to frozen components. Version number must always be updated if component is modified. Otherwise, there is no guarantee that other designs using the same component remain valid. Kactus tool helps in this case, because it can list the designs using certain component. The designer must decide case-by-case if she wishes to update them or keep using the original component version.

The documentation will be stored in arbitrary file formats and hence just the files are listed in meta-data, not the actual content. For example, use cases are stored in UML, requirements using TeX, pin maps with synthesis tool's proprietary

format and so on. That way designer can easily locate all the documents related to certain component. Of course, all documents are versioned, too and that is easiest with ASCII-based formats, such as TeX.

B. Use environment

Environments refer to the context in which the product objects are verified and used. It is crucial that the original test environment can be easily set up again in case a problem appears later. Otherwise, the bug fixes are very painful to implement. Similarly, targeted use environment (e.g. home, factory, medical, outdoors) must be well documented. To

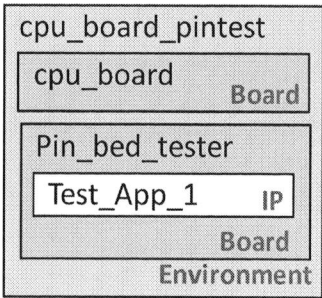

Fig. 8. Processor board is connected to a pin tester.

illustrate the role of environment, an instance of processor board is attached to a pin tester in Figure 8.

The structure of the testbench may differ greatly between levels. For example PCB testing uses some physical signal generators and measurement devices, whereas IP blocks may utilize HDL testbenches and simulation. Moreover, some IPs have only very simple script-based test vectors. However, test-benches that instantiate design-under-verification with specific verification IPs (e.g. protocol checkers, stimulus generation, response checkers) are preferred.

V. RUN-TIME CONFIGURATION CHECK

An important maintenance aspect of PDM is that developers must repeat the error condition in the lab if a bug is discovered in a deployed product. First, they must find out *"what exactly is in there?"*, i.e. determine the versions and parameters of all HW and SW components. In optimal case, the exact configuration of the device is tagged into the versioning system and can be easily restored. In most cases, however, it requires a quite bit of a detective work to find the right files, versions and configurations. Therefore, designer should be able to read out the identification or even the exact configuration from the deployed product.

A. Encoded identification

In simplest case, the identification can be achieved with simple textual label stickers or bar codes on the printed circuit board. PCB can also use a small non-volatile memory to store the identification, using just $64 - 4k$ bits and serial interface [12]. Modern desktop processors offer a special CPUID register, so that software can identify the features that are available, such as the processor family, model, inclusion of streaming or virtual machine extensions [13], [14]. Similar registers can be found in chipset ASICs [16].

We utilize the same idea in an FPGA-based SoC with identifier registers. In addition to debugging, they can be used for more general-purpose bootup routines and to collect dynamic information, such as utilization statistics. During design phase, the static info registers are good for verifying the successful integration of components and network-on-chip instantiation. For example, the first integration test program tries to read all the id registers. That confirms a whole lot of

issues, such as power, reset, clocking, manufacturing, program execution and so on.

However, a single register allows only limited amount information (32b-96b) and needs specialized coding. Processor manuals explain the meaning of codes and there are programs that can read and decipher the CPU IDs into human-readable form, e.g. [15]. This is good approach as long as the stored properties and selected codes remain constant for a long time.

B. SW-based solution

In embedded devices, however, the needed information is very case-dependent and decoding programs become hard to maintain. Therefore, a *human readable system diagnosis* is preferred and it can be implemented as a constant ASCII array in a SW header file. The main CPU just prints it out upon user's request. The used connection is product-specific.

Since the memory space is limited, the designer can use some standard compression to store the top-level VHDL file or master header file in C. In some cases, a sub-set suffices, for example just the generic parameters of VHDL, but that necessitates automation to avoid typing errors and tedious maintenance during development. However, it can be rather easily incorporated into system level design tools, such as Kactus, or extracted from the HDL description with scripts, for example using grep + Python + gzip.

VI. APPLYING DETAILED TAXONOMY FOR REUSABLE COMPONENTS

IP components are traditionally identified by natural-language- like name. VLNV forces some formalism but, unfortunately, there is no common agreement on how to use its fields in practice. On the other hand, disciplined categorization of IPs helps maintenance and finding the appropriate ones. The most obvious criterion is purpose, for example, computation versus storage. Additionally, certain nonfunctional constraints need to be considered when searching for IP.

For example, OpenCores community lists over 500 open source hardware projects and divides them into 15 categories, such as arithmetic, communication controller, and SoC [18]. Furthermore, user can easily filter projects based on their implementation language, license type, and development stage. Design & Reuse portal lists IPs in 3 categories: silicon IP, verification, and software [19]. Due to large number, there are 6-16 sub-categories (e.g. analog, peripheral, processors), several sub-sub-categories (e.g. CPU, co-processor). Some of the sub-sub-categories are duplicates (for example *DSP core* is filed under *Digital Signal Processing* and *Processors &Micro-controllers*) or sound very similar (*CPU* vs. Microprocessor).

There is a delicate trade-off between details and practicality. Too many categories makes the classification exact, but hard to use and leads to very few components in each category. Another issue is strictness: whether a component can belong to many categories at the same time. Finally, the tool should allow easy-to-use navigation, searching, and sorting function-ality. As this is a major research topic itself, we only briefly outline our profile proposal here.

978-1-4577-0671-4/11 $26.00 © 2011 IEEE

TABLE I
EXAMPLES OF RESERVED KEWORDS AT LIBRARY

Level	Category	Function	Configurability
product	hw_platform	computation	fixed_function
board	hw_application	communication	parameterizable
chip	sw_platform	storage	programmable
soc	sw_application	interface	
ip	independent		

A. Keywords

Classification starts by identifying the criteria and allowed values (keywords). An extension to IP-XACT is proposed in [17] to help searching for proper components from libraries. They proposed XML schemas that include additional descriptions for verification, non-functional properties, and behavior (or purpose) of the component. The search finds matches in both the names and semi-formal use cases, and scores similarities to create a rank order. We adopt somewhat similar approach. Some of the reserved keywords are listed in Table I. The three rightmost columns are examples mainly for IP level. We believe that a handful of allowed words is a good compromise between usability and expressiveness.

Since we do not wish modify the XML schema, there are two basic ways to store the keywords: encoded into VLNV and as separate, prenamed parameters of a component.

B. VLNV naming

In the following, each VLNV field is composed of keywords which are separated by dots. Table I shows the reserved words in each subcategory. Note that columns in Table are not dependent on each other. IP-XACT standard for string characters is followed.

The format for the VENDOR is defined as *vendor.[project | organization]*. For example, *tut.funbase* (according to project) or *tut.dcs* (according to department).

Product hierarchy is reflected by LIBRARY that is defined as *level.category.function.configurability*.

NAME and VERSION are freely selectable by the vendor.

C. Separate classification parameters

The drawback of the above method is that VLNV becomes awkwardly long and that component can belong only to single category. Components have model parameters that can be mapped, for example, to the generics in VHDL. One can easily augment all IPs with a set of parameters that will be used just for classification. For example, include parameter "function" and obtain the legal value from Table I.

This seems more natural way in XML-based tools. Kactus provides search and sorting based on these parameters. All tools can easily show them to user even if they cannot provide any additional functionality. Furthemore, designer can put the same IP into two categories in unclear cases.

VII. CONCLUSIONS

IP-XACT is developed for IP and SoC level HW descriptions with plenty of metadata elements for the structural description and associated information like source files, their languages and tools. Unlike typical extension proposals, we have kept the metadata standard untouched and apply product data management to IP-XACT through usage conventions, model views, and parameters. As a result, we can use same XML parsers and other tools for managing HW and SW objects as well as HW/SW mappings at different hierarchical product levels. Hence, the method should be easy to adopt.

Our work-in-progress tool, Kactus2, can be used to both capture embedded product designs and manage configurations. Future work includes algorithm development for change detection and management, and further development of general IP taxonomy for the VLNV library and name fields. Evaluating the applicability and comparison to other tools are also left for future work.

REFERENCES

[1] A. Sangiovanni-Vincentelli. Quo Vadis SLD: Reasoning about Trends and Challenges of System-Level Design. Proceedings of the IEEE, 95(3):467-506, Mar. 2007.

[2] W. Wolf, A.A. Jerraya, G. Martin, Multiprocessor System-on-Chip (MPSoC) Technology, IEEE Trans. Computer-Aided Design of Integrated Circuits and Systems, vol.27, no.10, pp.1701-1713, Oct. 2008

[3] J. Browne, Leveraging Your IP in the New Economy, keynote presentation at IP-SOC, Mar. 2010.

[4] F.R. Wagner et al., Strategies for the integration of hardware and software IP components in embedded systems-on-chip, Integration, the VLSI Journal, Sep. 2004, Vol. 37, Iss. 4, pp. 223-252.

[5] D. Densmore, R. Passerone, A. Sangiovanni-Vincentelli, A Platform-Based Taxonomy for ESL Design, IEEE Design&Test of Computers, Sep/Oct 2006, vol. 23 no. 5, pp. 359-374

[6] IEEE Standard for IP-XACT, Standard Structure for Packaging, Integrating, and Reusing IP within Tools Flows, IEEE Std 1685-2009, pp.C1-360, Feb. 2010.

[7] A. Kamppi et al., Kactus2: Environment for Embedded Product Development using IP-XACT and MCAPI, accepted to Euromicro DSD, Aug-Sep. 2011.

[8] T. Kangas et al., UML-based Multi-Processor SoC Design Framework, Transactions on Embedded Computing Systems, May 2006, Vol.5, Issue 2, pp. 281-320

[9] W. Krujitzer et al., Industrial IP Integration Flows based on IP-XACT Standards, in: DATE, Mar. 2008, pp.32-37.

[10] T. Arpinen et al., Evaluating UML2 Modeling of IP-XACT Objects for Automatic MP-SoC Integration onto FPGA, in: DATE, Apr. 2009, pp. 244-249.

[11] A.P. Dahlqvist et al., Product Data Management and Software Configuration Management - Similarities and Differences, The Association of Swedish Engineering Industries, 2001, 181 pages.

[12] Printed Circuit Board Identification (PCB ID) and Authentication, Maxim Integrated Products, [online], Available: http://www.maxim-ic.com/products/1-wire/pcb_id_authentication.cfm, 2011.

[13] Intel processor identification and the CPUID instruction,Intel Corporation, Application Note 485, Jan. 2011, 124 pages.

[14] CPUID Specification, Advanced Micro Devices, Publication #25481, Sep. 2010, 38 pages.

[15] CPUID X86 technical resources, web page, [Online] Available: http://www.cpuid.com/

[16] AMD RS690 ASIC Family Register Reference Guide, Advanced Micro Devices, Technical Reference Manual, Rev. 3.00o, 2007, 422 pages.

[17] C. Trummer et al., Searching Extended IP-XACT Components for SoC Design Based on Requirements Similarity, IEEE Systems Journal, Vol. 5, No. 1, Mar. 2011, pp. 70-79.

[18] OpenCores community, web page, [Online] http://www.opencores.org

[19] Design & Reuse - Catalyst of collaborative IP Based SoC design, web page, [online] http://www.design-reuse.com

978-1-4577-0671-4/11 $26.00 © 2011 IEEE

Increasing Energy Efficiency of Automotive E/E-Architectures with Intelligent Communication Controllers for FlexRay

Christoph Schmutzler, Abdallah Lakhtel, Martin Simons
Daimler AG
HPC: 050/G007, 71059 Sindelfingen, Germany
Email: {christoph.schmutzler, martin.simons}@daimler.com

Jürgen Becker
Institute for Information Processing Technology (ITIV)
Karlsruhe Institute of Technology (KIT), Germany
Email: becker@kit.edu

Abstract— When a modern vehicle is in use, its interconnected *Electronic Control Units* (ECUs) are effectively always on. By temporarily deactivating those ECUs, whose functions are not needed, we can contribute to a vehicle's energy efficiency and lower its fuel consumption as well as its emissions. FlexRay is an automotive bus system typically used to interconnect high-performance ECUs. FlexRay, however, does not support the deactivation of single nodes. In this paper, we introduce the concept of *Intelligent Communication Controllers*, which allow us to perform a demand dependent deactivation of FlexRay ECUs. We also outline an FPGA based prototypical implementation.

I. INTRODUCTION

The number and feature set of *Electronic Control Units* (ECUs) has been steadily increasing over the past generation of cars. Starting at just 7 ECUs and 100 communication signals in the 1984 Mercedes-Benz E-Class, the current E-Class is equipped with up to 70 ECUs and uses over 6000 signals for communication between ECUs.

From the early beginning, idle ECUs had to meet strict power targets (approx. < 6 mW per ECU) when the vehicle was inactive, i. e., when it was locked and the engine was not running. When the engine and electrical generator were running, however, a car's energy consumption was not considered as important. Hence, for a long time, energy consumption of active ECUs was usually not subject to optimization efforts. Due to the rising number of ECUs, this lead to an ever increasing electric base load of an active car.

Because the electric energy consumption of a car has a direct impact on its fuel consumption, it has become increasingly important in the past years [1], [2]. Facing increasingly challenging emission standards, rising fuel prices, and the emergence of electric powertrains, car manufacturers are now exploring ways to increase the energy efficiency of cars [3].

However, current automotive bus systems only provide very limited support for improving a vehicle's energy efficiency. Therefore, we are working on the concept of *Intelligent Communication Controllers* (ICCs) for FlexRay. ICCs allow us to send temporarily not required ECUs to sleep on an ECU-selective basis. This is currently not possible for FlexRay.

Even when the ECU is "sleeping", i. e., its components are in an energy efficient state, or are not powered at all, an ICC is still able to

- filter bus-traffic for wake-up reasons;
- store important messages that are required after the ECU resumes normal operation;

- and send static cyclic status messages.

To validate and test the concept, we developed a *Field Programmable Gate Array* (FPGA) based ICC that we will present in this paper.

In Section II, we introduce the notion of demand dependent deactivation of ECUs based on the current vehicle state, and explain why current automotive bus systems and other energy saving mechanisms do not provide the required flexibility to use this approach. We then introduce the concept of ICCs for the FlexRay bus in Section III. Our FPGA-based ICC implementation is described in detail in Section IV. In Section V we conclude and outline our further research agenda.

II. ELECTRIC/ELECTRONIC-ARCHITECTURES

Electronic components in modern cars, such as ECUs, and networked sensors and actuators, are interconnected by a complex heterogeneous network. A car's *Electric/Electronic-architecture* (E/E-architecture) is a logical and physical representation of this network. The logical representation describes functional components, and the communication behaviour and dependencies between those components. The physical description maps functional components and their communication onto ECUs and buses. Depending on bandwidth, timing and safety requirements, the automotive bus systems CAN, LIN, MOST and FlexRay are being used to connect ECUs. A good overview of the different bus systems and E/E-architectures can be found in [4], [5]. In this work we focus on FlexRay, because it is becoming increasingly important for bandwidth- and safety-related applications.

An E/E-architecture's energy efficiency can be improved by focussing on local, function-dependent optimizations of individual ECUs, e. g., by optimizing actuators, or by introducing means to completely deactivate ECUs while they are not required. The latter saves the effort of individual local optimizations.

Local optimizations are especially suitable for ECUs that have components that consume a lot of power, such as the electric power steering pump, the compressor of the air conditioning system, or the power distribution of the seat heating ECUs. However, for many ECUs that do not have specific power-intensive components, local optimization can quickly become a time- and cost-consuming task when compared to the achievable energy savings. Here, a demand dependent deactivation of temporarily not required ECUs is a worthwhile approach.

978-1-4577-0671-4/11 $26.00 © 2011 IEEE

A common drawback of today's automotive buses is that they only provide a reliable support for a bus wide sleep: as soon as any node starts to communicate, all sleeping nodes connected to the same bus become active. They also do not provide mechanisms for an ECU-selective activation of sleeping nodes: a node can only request a bus wide wakeup. Of course, this severely restricts the potential for deactivating currently not needed ECUs. [6], [2] further discuss the supported sleep mechanisms for the different bus systems and arising problems with respect to energy efficient design of E/E-architectures.

Although there are several bus independent mechanisms for an ECU-selective sleep, such as *clamps* and dedicated *wake-up lines*, none of them provide the necessary flexibility for a truly demand depended deactivation with a reasonable increase in system complexity and cost [2]. Therefore, we are working on concepts that allow for an ECU-selective sleep with existing bus systems.

Since MOST and LIN are typically used for applications that are better suited for a bus-wide deactivation of ECUs, both are not the primary targets for the concept of ECU-selective deactivation. For CAN, new *Partial Networking*-Transceivers are being developed, that allow for a selective deactivation [6], [7]. For FlexRay, no similar mechanisms are available or planned.

FlexRay uses a shared bus medium, called channel. Nodes communicate by sending *Frames* which contain up to 256 bytes of payload. Frames are identified by a unique Frame-ID. For a detailed description of FlexRay we refer to the FlexRay specifications [8], or dedicated literature [5].

III. Intelligent Communication Controllers for FlexRay

Figure 1a shows the simplified structure of a regular ECU. It consists of a *Microcontroller* (μC), a *FlexRay Communication Controller* (CC) and a FlexRay transceiver. The transceiver converts analog signals (BM, BP) into digital values when receiving, and digital into analog values when sending. The CC encapsulates the FlexRay protocol towards the μC, which controls communication.

The transceiver controls a voltage regulator (omitted from Figure 1a) with its *Inhibit-pin*. The voltage regulator powers the μC and CC. In normal mode, the Inhibit-pin is enabled. A regular ECU disables itself by switching the transceivers state machine into sleep mode. Now the Inhibit-pin is disabled, the voltage regulator is switched off, which in turn disables the rest of the ECU. If the transceiver detects a wakeup pattern, it enables its Inhibit-pin and the ECU is powered again. Because the transceiver often erroneously detects the pattern in regular communication, current ECUs are effectively always on during bus communication.

Because of the described limitations of FlexRay, we are proposing the concept of *Intelligent Communication Controllers* (ICCs), which allow ECUs to go to sleep when there are not required. Based on bus traffic, and, therefore, vehicle state, the ICC detects when the sleeping ECU needs to resume operation. Other nodes do not need to send an explicit request for reactivation.

The ICC can also send predefined messages, e. g., containing status information. An absence of these messages could

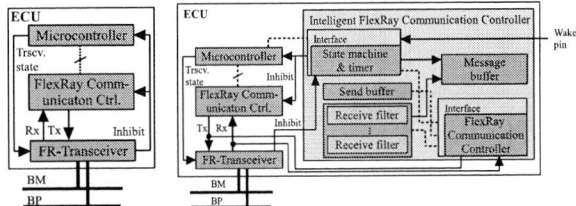

(a) Regular ECU. (b) ECU with ICC. The Inhibit-pin is now controlled by the ICC.

Fig. 1. Simplified structure of an ECU with, and without an ICC.

lead to errors in other nodes. In summary, other nodes can not detect the ICC sleep mode based on just bus traffic.

In an ICC equipped ECU (cf. Figure 1b), the Inhibit-pin is controlled directly by the ICC. The ECU can now go to sleep if it is not required based on the current vehicle state. Prior to going to sleep, the μC configures the receive filters and send buffers of the ICC. It then hands over communication to the ICC and instructs the ICC to disable the Inhibit-pin. It now

1) filters incoming messages for wakeup reasons;
2) performs timeout monitoring for configured frames;
3) stores relevant messages that can be read back directly after wake-up;
4) sends static status messages;
5) and checks for local IO-events (e. g., pressed buttons).

The demand for an ECU can be determined based on the state of the vehicle (e. g., speed or ignition state) and environmental conditions (e. g., temperature, or light conditions).

The parking assist system, for example, is fully active up to a vehicle speed of about 20 km/h. It signals the driver how much space is available in front and back of the car. Above 20 km/h it stops actively indicating the measured distance, but it keeps scanning for parking spots up to a speed of approx. 30 km/h. Even though most internal components can be disabled above that speed, the μC always remains powered because of constant bus communication. Using the ICC, however, the μC could go to sleep. Wake-up conditions are vehicle speed and the ignition state. Because the ICC can continue to send static status frames, other ECUs can not detect the sleeping node.

Other examples that could make use of the ICC are the trailer module, which could also be disabled above a certain speed; or the Nightvision ECU, which could be disabled if the light sensors indicate that nightvision is not required based on outside light conditions.

An even greater amount of ECUs could be disabled during the recharge of the high-voltage battery of hybrid or all-electric vehicles. The recharging process requires cyclic communication between a few ECUs. With current bus systems, all ECUs on the same bus would be woken up. The ICC could be used to disable those ECUs during the charging process.

Safety and reliability concerns are addressed by defining appropriate ICC configurations that contain only safe and valid static data to be sent and by considering all possible wakeup events.

IV. Concept Validation and Experimental Results

To test the ICC concept we developed an FPGA-based ICC implementation that resembles the previously described ICC architecture. We are using an Altera Stratix III (EP3SL150F1152C2) FPGA. However, whereas in Figure 1

978-1-4577-0671-4/11 $26.00 © 2011 IEEE

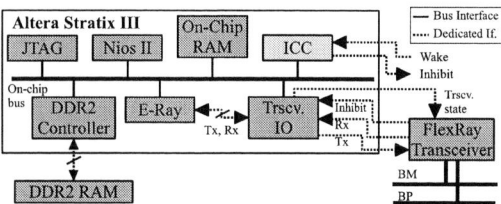

Fig. 2. SoC design including the prototypical ICC implementation.

the FlexRay-CC and μC are depicted separately, and the ICC is using its own CC, many commercially available μCs already include an CC. External FlexRay-CCs are not commonly used. Hence, it is very likely that an ICC would also be integrated on the μC, instead of being made available as a dedicated component. An integrated ICC would reuse the existing CC on the μC. All other components on the μC could then be disabled, e. g., with power or clock gating mechanisms.

Our System-on-Chip (SoC) design was created with Altera's Quartus II design software. We use several standard IP blocks that are provided by Quartus II, such as a JTAG-debug interface, a Nios II Softcore CPU (i. e., a synthesizable CPU using only FPGA resources), a small On-Chip-RAM and an interface to an external DDR2-RAM, as shown in Figure 2. The components are interconnected with Altera's *Avalon Interface* [9].

We are using the *E-Ray* FlexRay Communication Controller from Robert Bosch GmbH [10]. The E-Ray core is used both by the CPU and ICC. The FlexRay transceivers are controlled by a custom interface. In part, it is directly connected to the E-Ray, but also provides access to the transceiver's state machine and Inhibit-pin via the bus interface.

Our ICC implementation is integrated into the system. The resulting SoC design is comparable to the architecture of available μCs that use an integrated FlexRay core: μCs also use a multiplexed On-Chip bus that interconnects the different system components. Furthermore, the Bosch E-Ray IP is used in many μC families from different vendors.

The ICC extensions could be transferred without major changes to existing platforms for upcoming μC architectures. Of course, the required development effort, verification of the new component, and additional chip space are only justifiable if there is a broad demand for the ICC functionalities.

A. Test Setup

The software running on the Nios II mimics the communication behaviour of an existing FlexRay node. Communication behaviour of all other ECUs is simulated with *Vector's CANoe* network analysis, simulation and testing tool [11]. CANoe is running on a regular PC and uses an USB-FlexRay interface that connects to the rest of the system, in this case the ICC-equipped FPGA-based FlexRay node.

By using CANoe to simulate the network, we can test for correct ICC behaviour without having to integrate the ICC node into an existing car. Furthermore, we can control the provided stimuli, whereas creating specific, reproducible bus traffic in a car is often difficult and very time consuming.

B. ICC implementation

Figure 3a shows the internal structure of the ICC. The main components are the transmit buffers, the filtering and timing

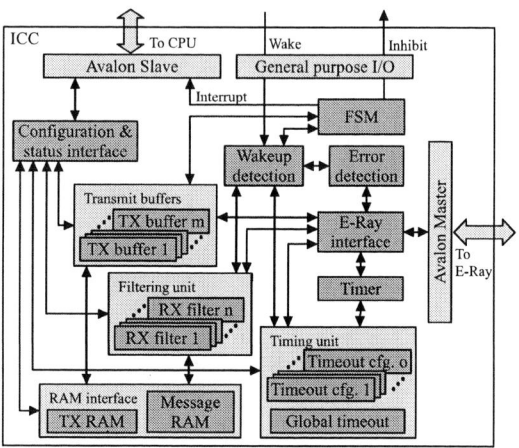

(a) Internal ICC structure.

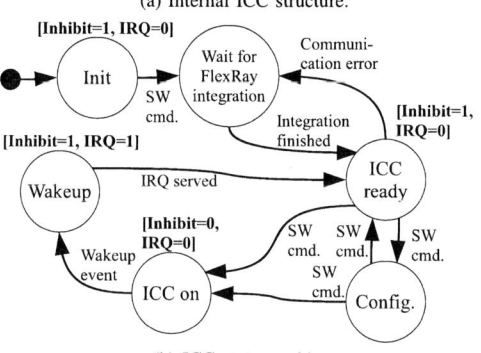

(b) ICC state machine.

Fig. 3. Schematic and state machine of the ICC.

units and the interface to an internal memory. The memory contains the payload and configuration of static messages that are to be sent. It is also used to store messages that have passed the receive filters. The number of transmit buffers, receive and timing filters are configurable.

The E-Ray interface polls the E-Ray CC for new frames and stores received frames in an internal buffer. It then notifies the filtering and timeout unit about the received frame.

For each new frame, the filtering unit loops through all filter configurations. If the received frame matches the configured frame ID, it compares a segment of the payload (up to 32 bit) with the configured compare value and arithmetic compare operation. If the result is positive, the frame is stored in the internal RAM. Depending on the configuration, the filtering unit also signals that a filtering wakeup event occurred.

The timeout unit cyclically loops through all configured timeout configurations and checks if a timeout occurred, i. e., if a frame has not been received for a certain amount of time. If the arrival of a new frame has been signalled, the matching timeout filters are updated accordingly.

The wakeup detection unit raises a wakeup event based on the state of the different filtering units, as well as the error detection block that observes the state of the E-Ray. It also monitors a dedicated *Wake-pin*, that can be used to generate local wakeup events (e. g., press of a button).

The transmit buffers are controlled by a *Finite State Machine* (FSM). Each transmit buffer configuration contains a static FlexRay frame and its cycle time. If the cycle time of a buffer has expired, the FSM request the transmission of the buffer via the E-Ray interface and resets the cycle time. The FSM also controls the Interrupt- and Inhibit-pin based on the current ICC state and wakeup events. Since all components require the internal ICC state, the FSM is connected to all other blocks (omitted from Figure 3a).

The internal memory can be accessed by the ICC status interface to configure the payload of sent frames and to read back stored messages after a wakeup. The configuration interface encapsulates the settings of the different ICC components.

The ICC state machine is shown in Figure 3b. After a reset, the ICC enables its Inhibit-pin and enters the **Init** state. It then pauses until the CPU instructs it to switch to the **Wait for Integration** state. Now the ICC waits until the FlexRay-CC has successfully established communication and then switches to **ICC Ready**. If communication is interrupted, it falls back to **Wait for Integration**. When resuming from a wakeup, a pending interrupt is cleared in **ICC Ready**.

When put into the **Configuration** state, old status information (e. g., stored messages, or the last wakeup reason) is cleared, and the internal RAM is made accessible for configuration. The ICC can then be switched back to the **Ready**, or the **ICC on** state. In **ICC on**, the Inhibit-pin is disabled to power off connected components. The ICC now filters incoming and sends static messages according to its configuration. In case of a wakeup event (e. g., a matching filter or a communication error), it switches to **Wakeup**, enables the Inhibit-pin and raises an interrupt. When the software has served the interrupt, the ICC switches back to **ICC ready**, where stored messages can be read back.

C. ICC resource utilization

For synthesis and fitting of our design we used Altera's Quartus II 10.1. For synthesis, we selected the "Extra" effort level and the optimization technique "Balanced" for speed and area. The relative resource utilization of the ICC components is shown in Figure 4.

Because we are using an encrypted E-Ray core, i. e., we can't access internal information directly, we have to buffer complete frames in the filter unit, and constantly update and store several E-Ray status registers in the E-Ray interface. This requires a large amount of registers (filter unit: 26%, E-Ray interface: 21%). The use of an unencrypted FlexRay CC that is directly accessible by the ICC would significantly reduce resource utilization. Using a fixed amount of filters and only offering advanced filtering options for a limited number of filters would further reduce the 38% of combinational logic currently used by the filter unit.

Memory consumption for filter, timeout and send buffer configuration is already optimized towards the use of dedicated memory instead of logic registers. Timer logic and timer and filter status, however, are still using a lot of dedicated registers.

V. CONCLUSION AND OUTLOOK

In this article, we introduced the concept of ICCs for FlexRay and described its prototypical implementation. ICCs allow temporarily not required ECUs to go to sleep to save power. This is currently not possible for FlexRay ECUs: if any node starts communication, all other nodes on the same bus wake up. This effectively means that all ECUs are always on

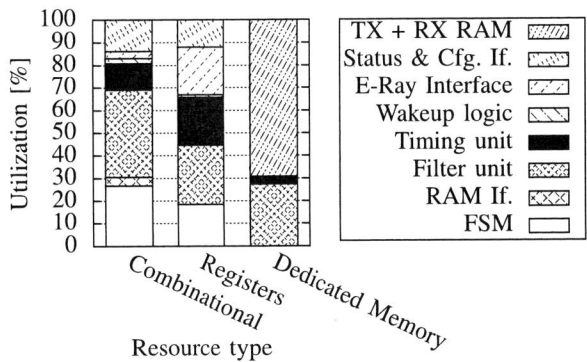

Fig. 4. Relative resource utilization of the ICC components.

when the car is being used, which severely restricts the energy efficiency of current E/E-architectures.

As a next step, we are working on an optimized ICC variant that uses a static number of filters and send buffers. The resulting resource figures can then be used to evaluate the feasibility of implementing an ICC in upcoming μCs.

Based on the FPGA ICC implementation that we presented in this work, we are currently developing a demonstrator that will use an adapted embedded OS that makes uses of the ICC. With the help of the demonstrator, we are planning to measure the reaction time upon wakeup events. We are also planning to extend the development tool chain to include support for ICC configuration.

ACKNOWLEDGEMENT

We thank the Robert Bosch GmbH, in particular Thomas Lindenkreuz and Christian Horst, for providing us with their FlexRay IP implementation for research purposes.

REFERENCES

[1] C. Koffler and K. Rohde-Brandenburger, "On the calculation of fuel savings through lightweight design in automotive life cycle assessments," *The International Journal of Life Cycle Assessment*, vol. 15, pp. 128–135, 2010.

[2] C. Schmutzler, A. Krüger, F. Schuster, and M. Simons, "Energy effcient automotive networks: State of the art and challenges ahead," *International Journal of Communication Networks and Distributed Systems, Special Issue on Optimization Issues in Energy Efficient Distributed Systems*, 2011, to be published.

[3] T. Weber, V. Lauer, D. Mann, and M. Simons, "The Comprehensive Energy Management - From the Conventional Powertrain to the Full e-Drive," in *4. VDI-Tagung: Baden-Baden Spezial 2010 Elektrisches Fahren machbar machen*, 2010.

[4] N. Navet, Y. Song, F. Simonot-Lion, and C. Wilwert, "Trends in automotive communication systems," *Proceedings of the IEEE*, vol. 93, no. 6, pp. 1204–1223, 2005.

[5] D. Paret, *Multiplexed networks for embedded systems : CAN, LIN, Flexray, Safe-by-Wire...* Wiley, 2007.

[6] C. Schmutzler, A. Kruger, F. Schuster, and M. Simons, "Energy efficiency in automotive networks: Assessment and concepts," in *High Performance Computing and Simulation (HPCS), 2010 International Conference on*, July 2010, pp. 232–240.

[7] B. Elend and H. Huber, "Reduktion des CO$_2$ Ausstoßes durch Teilnetzbetrieb in der Fahrzeugvernetzung," in *2. Elektronik automotive congress*, May 2010.

[8] *FlexRay Communication System, Protocol Specification Ver. 2.1 Rev. A*, FlexRay Consortium, Dec. 2005. [Online]. Available: www.flexray.com

[9] *Avalon Interface Specifications*, Altera Corporation, May 2011. [Online]. Available: www.altera.com/literature/manual/mnl_avalon_spec.pdf

[10] *Product Information E-Ray IP Modul*, Robert Bosch GmbH, July 2009. [Online]. Available: www.semiconductors.bosch.de/media/en/pdf/ipmodules_1/flexray/bosch_product_info_eray.pdf

[11] *CANoe*, Vector Informatik GmbH. [Online]. Available: www.vector.com/vi_canoe_en.html

AN AUTOMATIC EXPERIMENTAL SET-UP FOR ROBUSTNESS ANALYSIS OF DESIGNS IMPLEMENTED ON SRAM FPGAS

Uli Kretzschmar, Armando Astarloa, Member, IEEE, *Jesús Lazaro, Jaime Jimeńez, Aitzol Zuloaga*

Department of Electronics and Telecommunications
University of the Basque Country
Bilbao, Spain

ABSTRACT

This paper introduces an experimental test-flow for evaluating the susceptibility of SRAM based FPGA designs to SEU (Single Event Upsets). Using this method it is possible to cover both SEUs and MBU (Multiple Bit Upsets) in the configuration memory of Xilinx FPGAs for applications based on tiny soft microprocessors.

The introduced test-flow imposes a minimal effort to the system developer and achieves a good estimation on the percentage of critical bits in the configuration memory of a design. This flow is executed for a design using multiple tiny soft microprocessors and the reliability values extracted by the test-flow are compared to non-experimental estimation techniques.

1. INTRODUCTION

Nowadays, more and more electronic devices, such as FPGAs, are being used in safety critical systems. When a system failure may result in death or serious injury to people, the system failure rate is of great importance. In this context, functional safety standards, such as IEC 61508 [1], use the λ parameter to quantify the safety integrity level of the design.

These standards are used as guide for designing systems in a way to prevent dangerous failures or to control them when they arise, with the aim of reducing the system failure rate. Therefore a lower system failure rate will lead to a higher safety level. The required safety integrity level for a specific application depends on the frequency of system failures and the severity of their consequences.

The technology of SRAM based FPGA is well known and commercially available since a long time. It offers a short time-to-market, low costs for low-volume productions and the possibility of reconfiguring the device in the field.

But together with these advantages comes a higher susceptibility to SEUs induced by heavy ions, protons and ground level radiation. As all the configuration information of the FPGA, e.g. interconnections and look-up-tables, is stored in SRAM cells, a SEU in these cells can have a significant impact on the functionality of the complete system. Like this a SEU can change the proper connection of a signal or the performed logical operation of a look-up-table, which in the worst case changes the FPGAs behaviour. Furthermore, continuously shrinking process geometries does not only increase the number of configuration bits on a FPGA that could be affected by a SEU but also the probability of a SEU to occur.

When designing safety critical applications an estimation needs to be made to decide whether it is necessary to perform special measures to guarantee the proper operation of the device over its lifetime. The unit metric for λ parameter is FIT (Failures In Time). Information on the λ for Xilinx FPGAs can be found in the Rosetta experiment [2]. This information is given in FIT per MB of memory, which can be configuration memory or Block RAM memory. To get the information on the realistic λ of a FPGA-design not only the size of the memories, but also the number of critical bits are an important measure.

This paper presents a practical approach to obtain realistic λ values for a given design by injecting SEUs into the configuration bitstream. Using this approach, it is possible to use the data from the Rosetta Experiment to estimate the overall FIT rate and MTBF (Mean Time Between Failure) for the actually implemented design. The test-flow was applied on a Virtex-5 FPGA and the resulting expected failure rates are compared between traditional estimation and emulation.

The concept-proof application introduced in this paper is used as an example to show the execution of this practical approach for experimental robustness analysis as well as several non-experimental estimation methods. It implements multiple tiny processors of the type PicoBlaze [3]. The software of the different PicoBlaze processors is stored in the internal dedicated memory of the FPGA and the processors itself are implemented using general purpose FPGA logic. Many of this kind of ultra low area consuming soft processors can be implemented in a FPGA device to obtain multiprocessor systems.

Due to their simplicity and flexibility they are starting to be adopted for industrial applications, like for example the FPX KCPSM module [4], Software Defined Radio processing [5] or motor control cores.

The remainder of this paper is organized into five sections. Section 2 presents an introduction into error estimations based on the number of critical bits of FPGA designs. This papers test-flow is introduced in Section 3 as an alternative more reliable source for the estimation of the number of critical bits of a design. Section 4 introduces the used concept-proof application. The results of both presented error estimation methods, experimental- and the non-experimental, are shown in Section 5 and finally the paper ends with the conclusion for the test-flow and on the gained results.

2. RELATED WORK

In [6] calculations are presented for getting an estimation based λ using the data of the Rosetta experiment.

$$\lambda = CB * RTSER \qquad (1)$$

In equation (1) λ is the application specific failure in time rate, where a λ of one corresponds to one failure in 10^9 hours of system operation. *RTSER* stands for Real Time Soft Error rate, a value that is provided for each FPGA in [7], depending on the used device family. Finally *CB* stands for the number of critical bits. There are several ways of finding an approximation on the number of critical bits and since the *RTSER* is fixed, finding a good estimation on the number of critical bits (*CB*) is the key for a good λ estimation. An alternative measure for the robustness is the MTBF (Mean Time Between Failure), which is indirectly proportional to λ and measured in years.

$$MTBF = \frac{10^9}{\lambda * 365.24 * 24} \qquad (2)$$

2.1. CB estimation based on FPGA properties

Obviously *CB* can be set to the device size which can serve as a worst case estimation.

$$CB = deviceSize \qquad (3a)$$
$$CB = FC * BPF \qquad (3b)$$

The Xilinx configuration memory is organized in frames, where each frame consists of a certain number of frames BPF (bits per frame). The device size is calculated by the product of the BPF and the number of frames on this device (FC).

A less pessimistic estimation on the number of critical bits is given in [8]. There an approximation on the number of configuration bits per logic feature of the FPGA, as shown in Table 1, is used.

Table 1. Configuration bits per device feature (Virtex-5)

Abbr.	Device Feature	Configbits approx.
LS	1 logic slice	1,181
BR36	1 blockRAM (36 Kb)	1,170
BR18	1 blockRAM (18 Kb)	585
IOB	1 I/O block	2,657
DSP	1 DSP48E slice	4,592

$$CB = a * LS + b * BR36 + c * BR18$$
$$+ d * IOB + e * DSP \qquad (4)$$

Using this method the number of critical bits is calculated using equation (4) where a, b, c, d, e stand for the number of used logic features (logic slice, 36k block ram, 18k block ram, I/O block, DSP48E slice) and LS, BR36, BR18, IOB and DSP for the configuration bits per feature.

Another option to estimate the number of critical bits is to use an estimation provided by the Xilinx development tool-suite. The Xilinx bitstream generator *bitgen* [9] provides an option to generate a file (.ebd) containing the information of which bits in the bitstream are critical to the correct function of the FPGA. This file is organized as a mask for the bitstream.

$$CB = EBD_{one} + FC * BPF * 0.03 \qquad (5)$$

The number of ones in the .ebd file EBD_{one} represents the number of critical bits. Attention needs to be paid, because 3 percent of the device size needs to be added to this number, because the .ebd file does not comprise of the complete device size.

2.2. CB estimation by SEU emulation

Another way of getting estimations on the number of critical bits of a design is to inject SEUs on the actual FPGA by changing its configuration bits via one of the programming interfaces. After the SEU is injected it needs to be verified, that the behaviour of the design has not changed.

One way of injecting the SEUs in run-time is shown in [6]. In this method the so called SEU controller in conjunction with the ICAP port of Virtex-5 FPGAs. In this case the ICAP port is used to flip one or more bits of the configuration memory within runtime which is equivalent to a SEU. The benefit of this methodology is, that SEUs occur some time after device startup, just like in SEUs provoked by radiation. Disadvantage of this methodology is, that the SEU controller itself requires resources on the FPGA, which itself are vulnerable to SEUs.

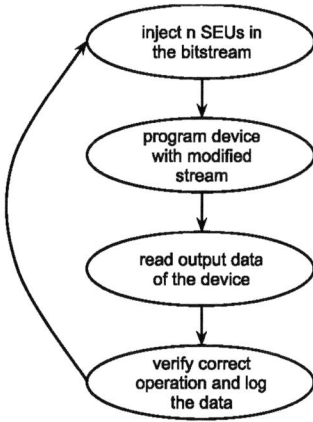

Fig. 1. SEU injection and test-flow

Another way of injecting SEUs into the configuration memory of a FPGA is to manipulate the bitstream prior to the programming and then writing a modified bitstream to the device. This manipulation can be done on a host PC as in [10] or using additional hardware as in [11]. This bitstream manipulation prior to device configuration also allows evaluating the effects of SEUs in the configuration of the device under test without additional resource requirements.

3. SEU INJECTION AND TEST-FLOW

The methodology for SEU injection and test-flow presented in this paper is summarized in Fig. 1. The basic idea of this flow is to program the device with a bitstream, where a SEU or MBU was injected beforehand and to check if the design is still behaving as expected.

In the first step one or more bit positions of the original bitstream are inverted. This faulty bitstream gets programmed to the FPGA holding the design to be investigated in the second step. In the third step the output of the design is read in order to verify the proper operation of the design despite of the injected SEU or MBU in the first step. Finally in the fourth step the read data is verified and the corresponding bit positions of the injected error are marked to be critical or not. The benefit of this four step flow is that all steps can be automated by using scripts running on a host PC controlling the overall flow. This allows a large number of SEUs and MBUs to be tested. Generally it is not feasible to have a test run for each bit position of the bitstream, but it is desirable to have as many test runs as possible, because like this the bit positions covered by the tests becomes more representative for the whole device.

Each individual step of the test-flow is explained in the following in more detail.

3.1. SEU injection and device programming

Since the bitstream generated by the Xilinx *bitgen* generator is in a well defined format, it is easily possible to invert one or more of its bits that are going to be written to the configuration SRAM.

The Xilinx bitstreams are packet oriented. There are two types of packets [7]. Each packet is partitioned into header and payload. In packets of *type 1* the header defines how many words (32bit) are going to be read or written to what register. *Type 2* headers can be used following *type 1* headers and allow specifying a bigger word count, while neither changing the read/write setting nor the destination register. To inject SEUs into the configuration memory only bits have to be changed, that are in the payload of a package, which is written to the FDRI packet register of the Virtex-5 FPGA. It needs to be ensured, that the CRC checking is disabled when generating the bitstream using bitgen.exe.

A custom set of PHP scripts have been developed for the bitstream manipulation. The selection of the bit position for injecting the SEU is done using the Mersenne Twister [12] to avoid issues with the quality of the random numbers for the SEU position selection.

In order to get a completely automatic test environment the programming of the FPGA needs to be automated as well. For Xilinx FPGAs this is achieved by using the command line interface of the *IMPACT* [13] programming tool.

3.2. Reading and verifying the device data

In safety critical applications finding a way to verify the proper device operation is a complex task. All sections of the hardware need to be used and their proper operation needs to be ensured. This requires, that the test running after the programming of the bitstream needs to be as comprehensive as possible in terms of hardware usage. This work uses tiny soft core micro-controllers for testing the hardware. These micro-controllers have the advantage of not having to externally apply test vectors. Instead the software running on these can be used to achieve good test coverage.

The SEU injection test-flow proposed in this work uses a UART channel to receive a selection of registers representing the internal state of the FPGA design on the host PC. The data for a failure free design can be determined by running the FPGA with the desired system without any SEUs in a so-called golden run. After that for each run with an injected SEU the UART data is compared to the data of this golden run. Choosing UART for the communication does not only reduce the extra hardware needed to a very low level (there is only 1 pin used and sending UART requires only a small state machine, a counter and a shift-register) but it also eases the automatic evaluation of the test runs.

If the output is the equal to the golden run, the SEU or MBU positions are marked as non-critical. If the output dif-

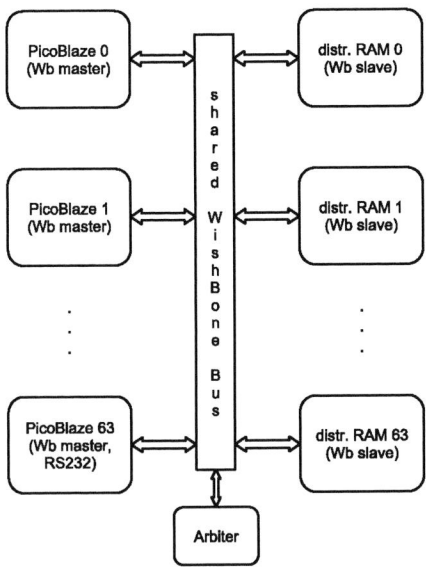

Fig. 2. Concept-Proof Application block diagram

fers, the locations of the injected SEUs are stored for later validation of the critical bit positions.

This fact that micro-controllers are used as test-vector generators is no restriction to the introduced test-flow. If no micro-controllers are available for testing, than for each test-case it would be necessary to use the test PC for providing the test-vectors, which could also be done using UART.

The UART reception in this work was done using a C++ program, the checking, logging and test run management was done using PHP scripts running on the host PC. Like this an automated testing of a high number of SEU or MBU positions was possible without the need of manual interaction.

All bits marked as critical were re-validated in a separate run, which re-tests all the recorded findings to avoid non-critical bits from being marked as critical due to test setup issues.

4. CONCEPT-PROOF APPLICATION

The Concept-Proof Application serves as an example on how to apply the test-flow introduced in the previous chapter as well as on how to calculate non-experimental λ values. A block diagram of this application is shown in Fig. 2.

This application consists of a shared Wishbone reference bus architecture with 64 masters and 64 slaves. The Wishbone slaves are Xilinx distributed RAM cells with a Wishbone wrapper as used in the Wishbone library. The masters are built from 64 PicoBlaze micro-controllers. All 64 of these PicoBlaze micro-controllers have a Wishbone master

interface, whereas the 64th has an additional UART transmit hardware.

In general it is important to achieve a high resource utilization of the FPGA when applying the SEU injection test-flow. Otherwise many injected SEUs will affect configuration bits of unused resources, which will most likely cause no error. By having a homogeneous application like the Concept Proof Application, which is dominated by the PicoBlaze micro-controller it is possible to obtain a realistic approximation of the SEU susceptibility of this very module.

4.1. Initialization and Application Dataflow

This application does not need any special hardware for setting up test-inputs, since the code memories of the different PicoBlazes contain all the information needed. Other applications, which do not have an internal stimulus provided by a tiny soft micro-controller, do have to ensure a proper initialization, that all test runs can be done using the same inputs. This could for example be done by receiving test vectors from the test PC.

After applying a reset the Application will start to run and the PicoBlaze masters 0 to 62 will begin executing their programs and by this writing in an interleaved manner a predefined sequence of values to the different distributed RAMs. Actually all masters will request the bus at the same time, but the implemented round-robin arbiter will guarantee this interleaving. Starting slightly after the other masters, master 63 will start reading the data of all 64 slaves and transmit the read data via the UART interface to the PC.

There is either of two test programs running on all the PicoBlaze masters 0 to 62:

- *TA*: Test program A, which triggers the Wishbone transfers using a minimal set of 38% of the PicoBlazes instructions

- *TB*: Test program B, which triggers the Wishbone transfers using 88% of the PicoBlazes instructions to ensure usage of a high percentage of the PicoBlazes logic.

Both *TA* and *TB* are causing the same activities on the Wishbone bus and by this the overall application is equivalent. The difference between the two test programs is only visible inside each of the masters.

These two versions of the test programs serve as an example to underline, that not only the logic design, but the actual application influences the robustness against SEUs. Test program B fulfils the requirement of testing as much logic as possible, which was shown in Section 3, whereas test program A only fulfils the application needs.

5. TEST RESULTS

The presented test results were achieved using the ML507 Xilinx-5 development board [14], which populates the Virtex-5 XC5VFX70T FPGA device. Programming was done using a Xilinx JTAG programmer.

The software used on the Windows XP based host PC was ISE12.3/ISE13.1, PHP5.3.4 CLI and MinGW 4.4.1.

5.1. Theoretical λ calculation

The term theoretical λ will be used for the evaluations using the *CB* estimation methods presented in Section 2. Corresponding theoretical numbers of critical bits are called CB_{theo1} and CB_{theo2}. For the calculation of the theoretical λ it is not needed to distinguish between the two test programs used, since the concept-proof application shows the same device utilization and the same number of critical bits in the .ebd file.

Using the FPGAs device utilization of the logical features as in equation (4) gives the first theoretical estimation on the number of critical bits CB_{theo1}.

$$
\begin{aligned}
CB_{theo1} =& 5080 * LS + 0 * BR36 + \\
& 64 * BR18 + 4 * IOB + 0 * DSP \quad (6) \\
=& 6.048 Mbit
\end{aligned}
$$

The second theoretical number of critical bits CB_{theo2} is obtained by using the information of the .ebd file, according to equation (5).

$$
\begin{aligned}
CB_{theo2} =& 2552190 + 21080 * 1312 * 0.03 \\
=& 3.382 Mbit
\end{aligned} \quad (7)
$$

5.2. Estimation of λ based on emulation

Using the test-flow described in this paper six test runs have been executed, where a SEU, a two bit MBU or a three bit MBU were injected into the concept-proof application which was running test program A or test program B. Each test run consists of 10000 injected SEUs or MBUs. The measurement results are summarized in Table 2.

Assuming, that the percentage of measured critical bits $perc_{CB}$ equals the percentage on critical bits for the whole device an emulated estimation of the number of critical bits CB_{emu_A} and CB_{emu_B} can be calculated for the respective test program.

$$
CB_{emu_A} = FC * BPF * perc_{CB_A} \quad (8a)
$$
$$
CB_{emu_B} = FC * BPF * perc_{CB_B} \quad (8b)
$$

Table 3 summarizes the estimations on the robustness for the different estimations for $CM_{theo1}, CM_{theo2}, CM_{emu_A}$

Table 2. Results of executed SEU and MBU tests

	SEU		MBU 2bit		MBU 3bit	
	TA	TB	TA	TB	TA	TB
SEUs injected	10000		10000		10000	
critical	171	232	335	458	463	692
$perc_{CB}[\%]$	1.71	2.32	3.35	4.58	4.63	6.92

Table 3. Results for the three robustness analysis methods

	theoretical				emulation	
	theo1		theo2		emu1	emu2
	TA	TB	TA	TB	TA	TB
CB[Mbit]	6.048		3.382		0.473	0.642
λ	792		443		62.0	84.1
MTBF	144		258		1841	1357

and CM_{emu_B} using a RTSER of $131\frac{FIT}{Mb}$ of the Virtex-5 family.

It can be seen, that the emulation of SEUs (CM_{emu_A} and CM_{emu_B}) showed the lowest λ. The estimation based on the .ebd file generated by *bitgen* (CM_{theo2}) is more pessimistic and the highest λ estimations, and by this the most pessimistic estimation, is achieved using only the device utilization for the estimation (CM_{theo1}).

The test-flow presented in this paper is superior the two presented theoretical ways of determining the SEU robustness of a design. This is due to the fact that an actual application does usually not use all the hardware in all actual possible combinations, whereas tools like *bitgen* can not exclude this combinations from the analysis. Test programs A and B demonstrate this issue inherent to theoretical analysis. Both applications have exactly the same device utilization and exactly the same number of critical bits in the .ebd file. This leads to equivalent theoretical numbers of critical bits resulting in the non-experimental estimation. But the experimental λ estimation shows, that test program A shows a lower λ than test program B. This is caused by the fact, that test program A uses less instructions and by this not all the logic of the PicoBlaze micro-controller. But this lower λ is only true if these instructions are never used, not even in rarely executed code.

For each injected SEU there can be two results. Either this SEU produces the verification step to fail or not. According to this the corresponding bit will be marked critical or not by the test-flow. Attention needs to be paid to the bits marked as non-critical, because these belong to either of two categories:

1. The injected SEUs did actually not have an effect on the concept-proof application or

2. The injected SEUs did have an effect the functionality of the concept-proof application, but the test-flows verification did not find the error.

The first category represents the desired result of a robustness analysis, which means that injected SEUs did actually not have any impact on the systems behaviour. One reason for bits having no impact on the functionality can be unused device resources due to the device utilization. Another reason can be only partly used device resources, like e.g. a LUT16, that uses only 3 inputs and outputs. Also there could be complete sections of logic, which are not used at all. The PicoBlaze micro-controller is a good example. This supports its complete instruction set, meaning lots of sub-functions like e.g. subtracting and shifting. If a program does not do shift-operations at all, the whole shift logic will never be used. But not all of this logic can be cut out in the optimization step of the design synthesis, because the synthesis tool can not analyse the codes flow to 100%. This was demonstrated by emulating the concept-proof application with test program A and test program B. Other reasons for SEUs in the bitstream not leading to errors are redundancies and possible other effects of the Xilinx bitstream.

The second category SEUs that did have an effect, but this effect was not detected can be seen as a disturbance to the test-flow. When testing the system it is very hard to achieve input vectors that reach all the logic implemented.

To keep the number of errors belonging to the second category as small as possible, the test program and UART readout must be structured in a way to test as much of the hardware as possible. The concept of using UART to verify the correct operation of the system does not impose a limitation of the testing. Within the design of the test program it needs to be determined how many bytes of internal data need to be output to the host PC to ensure the correct device operation. The assessment of the bytes needed to be output via UART needs to be done in the development process and is highly application dependent.

6. CONCLUSION

A test-flow for determining the number of critical bits of a design is introduced in this paper. Due to the high degree of automation by using scripts for injecting SEUs in the configuration bitstream, programming of this bitstream with subsequent functionality testing via UART it is possible to test a big number of SEUs without the need of manual interaction. The hardware overhead of this flow is at a very low level, so the impacts of SEUs effecting this parts of the FPGA is kept very small.

The testing of the given concept-proof application has no need for external inputs, since the implemented micro-controllers are used for ensuring high test coverages. Using different test programs together with the concept-proof ap-

plication it was shown, that the development of a suitable test program is a key for good robustness estimations.

7. REFERENCES

[1] International Electrotechnical Commission, "Functional Safety of Electrical/Electronic/Programmable Electronic Safety-Related Systems , internacional standar IEC61508," http://www.iec.ch, 2010.

[2] Xilinx Corp., "Device reliability report, fourth quarter 2010," Xilinx Documentation, http://www.xilinx.com, Feb. 2011.

[3] K. Chapman, "PicoBlaze 8-Bit Microcontroller for Virtex-E and Spartan II/IIE Devices," Xilinx Application Notes, http://www.xilinx.com, Feb. 2003.

[4] H. Fu and J. Lockwood, "The FPX KCPSM Module: An Embedded, Reconfigurable Active Processing Module for the Field Programmable Port Extender (FPX)," Technical Report. Applied Research Laboratory Department of Computer Science. Washignton University, Jul. 2001.

[5] N. Palermo, "A 9 MHz Digital SSB Modulator," http://www.microtelecom.it/ssbdex-e.htm, Sep. 2002.

[6] Ken Chapman, "Virtex-5 SEU Critical Bit Information Extending the capability of the Virtex-5 SEU Controller," Xilinx Documentation SEU lounge, http://www.xilinx.com, Feb. 2010.

[7] Xilinx Corp., "Xilinx-5 FPGA Configuration User Guide." Xilinx Documentation, http://www.xilinx.com, Aug. 2010.

[8] Ken Chapman and Les Jones, "SEU Strategies for Virtex-5 Devices," Xilinx Documentation, http://www.xilinx.com, May 2009.

[9] Xilinx Corp., "Command Line Tools User Guide," Xilinx Documentation, http://www.xilinx.com, Sep. 2010.

[10] Zachary K. Baker and Joshua S. Monson, "Fault Injection Into SRAM-Based FPGA For The Analysis Of SEU Effects," in *PROCEEDINGS 2003 IEEE International Conference on Field-Programmable Technology (FPT)*, 2003.

[11] M.Alderighi and F.Casini and S.DÁngelo and M.Mancini and A.Marmo and S.Pastore, "A Tool for Injecting SEU-like Faults into the Configuration Control Mechanism of Xilinx Virtex FPGAs," 2003.

[12] Makoto Matsumoto and Takuji Nishimura, "Mersenne Twister A 623-Dimensionally Equidistributed Uniform Pseudo-Random Number Generator," *ACM Transactions on Modeling and Computer Simulation Special issue on uniform random number generation*, vol. 8, 1998.

[13] Xilinx Corp., "SEU Strategies for Virtex-5 Devices," Xilinx Documentation, http://www.xilinx.com, 2002.

[14] ——, "ML505/ML506/ML507 Evaluation Platform User Guide," Xilinx Documentation, http://www.xilinx.com, Oct. 2009.

A Coarse-Grained Reconfigurable Protocol Processor

Mohammad Badawi and Ahmed Hemani
Royal Institute of Technology
Stockholm, Sweden
{badawi,hemani}@kth.se

Abstract—Trade-off between flexibility and performance became an important factor for characterizing modern protocol processing architectures. While some solutions tend to be more flexible and less computational efficient like GPPs, other solutions like custom ASIC devices provide high computational efficiency while loosing the ability to cope with the diversity of current and evolving protocols. We propose a reconfigurable protocol processor that is flexible and highly adaptable to the needs of the required protocol with the ability to operate individually or as a multi-core integrating processors. We show how a common protocol processing task that consumes one third of RISC CPU time can be performed on our processor at high speed and low energy cost.

I. INTRODUCTION

Protocol processing in modern communication applications found to increasingly demand improved hardware architectures, not only for meeting speed and energy requirements, but also for flexibility to cope with protocols diversity. Since flexibility is a trade-off of performance [1], this trade-off became an important characterizing factor of protocol processing architectures. Targeting protocol processing domain, RISC based architectures including single-core RISC, single-core RISC with accelerators and RISC multi-core have been heavily utilized. Although their simplified instruction set leads to higher flexibility, some protocol tasks may require large sequence of instructions to be performed, which affect performance, energy and instruction store area (as shown in Section II: Related Work).

In Order to achieve high flexibility as well as computation efficiency (computed tasks per unit energy), we propose a coarse-grained reconfigurable and small sized protocol processor. It is designed to be deployed as a pool of resources operating within a multi-application computational fabric called *Dynamically Reconfigurable Resource Array* [2]. It is also designed to be configured for handling the protocol processing task as an individual processor or as a group of integrating processors, as detailed later in this paper. The protocol processor we propose is composed of modules that are configured at run time. Upon configuration, modules and their internal resources and processing kernels can be enabled or disabled for suiting the required protocol task, resulting in high flexibility and adaptability. On the other hand, disabling unnecessary resources saves energy and improves the computation efficiency of the processor.

In this paper, we present our protocol processor describing its architectural properties and reconfigurability. We also demonstrate mapping of protocol tasks which shows the high adaptability of the processor. The protocol processor we propose is the focus of (Section IV). For more clarity and comprehension, we provide a brief illustration of the DRRA fabric in (Section III) highlighting protocol processor scope within the fabric.

For evaluating our proposed protocol processor, we report in (Section V: Evaluation Results), results for area, energy and speed of some of the processor's kernels that are heavily used in protocol processing.

We draw the conclusions of this research paper in (Section VI) and in (Section VII) we state our future work.

II. RELATED WORK

During the last two decades, a number of protocol processor variants brought forth by industry as well as academia [1] [3] [4]. In this section we consider three selected architectures describing their architectural properties and mentioning their difference to our proposed architecture.

The IXP1200 [5] handles protocols at layers 2-4 according to OSI model and it provides the advantage of separating data plane processing from control plane processing. It adopts parallelism at thread level utilizing six RISC based cache-less *Micro-Engines* for data plane processing. Each of these *Micro-Engines* supports four threads sharing a local 2K instruction store. IXP1200 uses different RISC processor with a local cache for handling control plane processing and it provides a set of interfaces to host processor, physical layer and memories. Additionally, the IXP1200 provides dedicated HW units for queuing and hashing while some of its variants and not all provide HW units dedicated for error checking. For improved performance, the IXP2800 [6] was produced as the next generation similarly for serving core, edge and access applications but with sixteen *Micro-Engines*. Each of the *Micro-Engines* is additionally connected to its neighbors by *neighbor-to-neighbor* connections and having an 8K instruction store shared among its threads. In these two processors, instructions and variables can be shared among threads running on the same *Micro-Engine*. This is different from our proposed architecture were each processor is independent of its microcodes and state variables. As a result of sharing in these two processors, the size of the instruction store and number of

978-1-4577-0671-4/11 $26.00 © 2011 IEEE

instructions shared between threads become critical to processor's performance and programmability [7]. Additionally, the RISC nature of the *Micro-Engine* puts a constraint on the minimum size of the required instruction store.

In [8], T. Henriksson presented a shared memory dual protocol processor architecture in which protocol processing is partitioned into *Inter-Packet* and *Intra-Packet* tasks. According to [8], *Inter-Packet Processor* is assumed to be a traditional RISC micro-controller for handling irregular tasks meant for control and management. The *Intra-Packet Processor* is a programmable non-pipelined data-flow processor performing tasks that can be existing in different protocol layers and can be processed independently from packet context information. *Intra-Packet Processor* core does not have general purpose registers and thus, when the core encodes an instruction that involves computation of intermediate results it dispatches this task to an accelerator. There are five loosely-coupled accelerators used for CRC calculation, IP and UDP checksum calculations, packet length counting, and memory management. Other *Intra-Packet* tasks including demultiplexing and address matching are not accelerated and they are performed within the core instruction flow as as set of single-cycle compare and branch instructions. Since the core and accelerator are supposed to work in parallel, parallelism can be exploited at the instruction level. However, utilizing the core instruction flow and accelerators in parallel is not always possible and it has impact on processor's performance. That is, when an instruction is dispatched to an accelerator, other instructions need to fill the core flow until accelerator's result is returned. On the other hand, the flexibility of the architecture is restricted by these custom accelerators. This is different from our proposed architecture were configurable modules are utilized to achieve flexibility.

Based on a patented data-flow architecture, Xelerated [9] provides different processors targeting *Unified Fiber Access*, *Metro*, and *Core* Applications. Xelerated's data-flow architecture is a one clock-cycle per stage pipeline formed of programmable elements. Each of these stages is optimized for one or more packet processing operations and it can be one of three types of elements. The first is a packet processing core with a local instruction store. It is called the *Packet Instruction Set Computer PISC* and up-to 400 of it can be in a pipeline. The second is an *Execution Context* associated with every packet traversing the pipeline and is formed of registers, flags and the first 256 bytes of the corresponding packet. The third is an I/O unit. This data-flow architecture shown to have simple programming and application mapping. However, the local instruction stores and the context that has to traverse the pipeline remain considerable issues and we tend to avoid in our proposed architecture.

III. DRRA

Dynamically Reconfigurable Resource Array (DRRA) is a coarse-grained reconfigurable computational fabric. In contrast to other coarse-grained reconfigurable architectures which targets single application or accelerating hot-spots and inner

loops [10], the DRRA targets hosting multiple complete applications. The DRRA is not only different in hosting compute intensive parallel functionality, but also in hosting control intensive protocol processing functionalities. As illustrated by Figure 1, the DRRA is composed of pools of resources organized in regular tiled fashion, and mainly customized to three layers of functionalities: 1) the reactive application layer, 2) the control and memory intensive protocol processing functionality, and 3) the data-parallel functionality of a typical PHY layer. Figure 1 also shows applications as composed of these three layers and how they are mapped to their respective pools of resources. The application layer is a low-power general purpose processor and our preliminary investigation suggests that being computationally light weight and application layers of multiple applications can be merged into a single processor. This will also simplify the inter-application communication as it is likely to be initiated at application layer. The protocol processing layer consists of a pool of agile reconfigurable protocol processors that can be allocated to handle the task either individually or as an integrating group. The data-parallel PHY layer consists of a pool of morph-able Data Path Units as well as controlling logic and it has been under development in our group and results have been reported in [2]. The multi-bank multi-port memory is a shared resource between the data-parallel PHY layer and the protocol processing layer, it has been under development in our group and results have been reported in [11]. A separate processor is used as the runtime system controller. It utilizes the system Level interconnects and it is responsible for creating and configuring private execution environments customized to the needs of the demanded applications. In other words, when an application is demanded, the runtime management system creates from the pool of resources a private execution environment having its local memory and state variables and customizes it to serve the application. Upon completion, the resources will be disabled and returned to the list of available resources. In this paper, we elaborate the scheme of the pool of protocol processors.

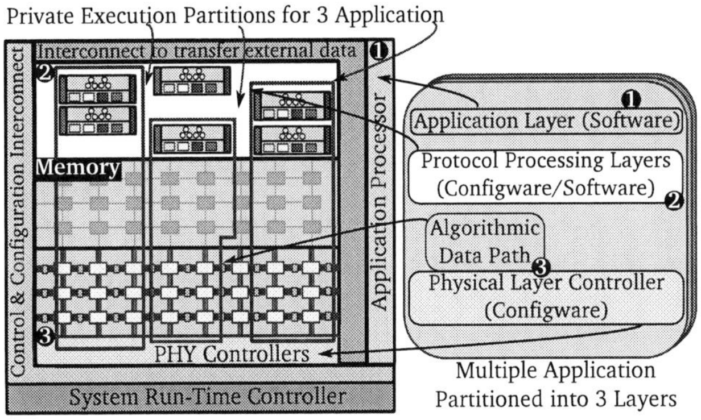

Fig. 1. DRRA Conceptual view

IV. PROPOSED ARCHITECTURE

In this section, we provide a description of our proposed protocol processor that is reconfigurable, small and simple enough to be deployed as a pool of resources within the DRRA fabric, and built to work either individually or in a group integrating to handle large protocol processing task. We also demonstrate mapping of protocol tasks.

A. Architecture

The protocol processor we propose is a coarse-grained reconfigurable architecture performing protocols within layers 2 to 6 according to *Open Systems Interconnection* model and mainly targeting *Access* and *End-User* nodes. It exploits parallelism at packet level and it assumes two ordered levels of configuration. The first level of configuration defines the operation and connectivity of the processor. The second level of configurations is used to finely customize processor's internals to suite required protocols.

Figure 2 shows the organization of the protocol processor, enclosed in the dotted line and composed of five modules: *Interfaces, Data Processing Path, Control Processing Path, Micro-Code Storage*, and *Booting and Organization Unit (BOU)*. We next describe the role and functionality of each of these modules as well as their configuration levels.

1) Interfaces: This module is a set of level-1 configurable interfaces responsible for inter processor communication. The protocol processor utilizes these interfaces to communicate with other DRRA resources, specifically, application layer, physical layer, and the run time controller. The protocol processor also uses these interfaces to communicate with other protocol processors in case where multiple protocol processors are allocated and configured to integrate in handling the protocol task.

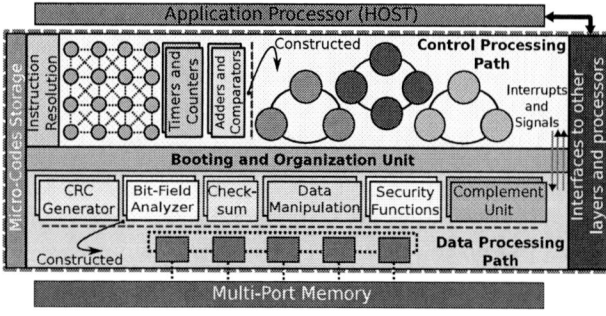

Fig. 2.   Proposed Protocol Processor

2) Data Processing Path: This module is responsible for high-speed packet processing and it consists of a group of low-power configurable packet processing kernels. Each of these kernels is specialized in a particular protocol operation but commonly found in different protocols. Unlike [5], [6] that use identical RISC based processing elements and [8] that uses RISC processor with intra-packet accelerators, we provide kernels for the most common protocol operations and we apply both mentioned configuration levels to them in order to suite the required protocol task. Using level-1 configuration, kernels that are required for the protocol task are determined and enabled and the rest of the kernels remain disabled in order

to save energy. These kernels are used for CRC generation, bit-field analysis, checksum computation, data manipulation, security functions and complementary functions.

The CRC Generator is used in several Link-layer protocols like Ethernet, ATM or in others like Bluetooth but with different generator polynomials and message lengths. The second configuration level configures and customizes the CRC generator to the needed polynomial and length.

Checksum is the one's complement addition needed in different protocols including IP, TCP and UDP. Level-2 configuration is used to select checksum for each of these protocols.

Bit-field analyzer handles pattern matching, flag searching and table look-up. It also has level-2 configuration determining which fields and flags are to be matched and searched.

Data manipulation is configured with level-2 configuration and it is used to edit exact fields within the packet, such as field filling during encapsulation/framing or decrementing Time-To-Live and updating checksums during packet forwarding.

Security functions kernel provides a set of encryption/decryption and hashing functions such as AES, RC4 and md5. These functions can be required for IPsec, WLAN security and other authentication protocols and can be selected using level-2 configuration accordingly.

The Complementary Unit is more general kernel performs a number of shift, compare and other logical operations and it can be configured with level-2 and used to provide support for any other kernels if the case needed.

Each of these kernels is connected to memory and to both neighbors via configurable connections of a bit-width varying between 1 and 4 bytes. Level-2 configuration is used to enable/disable any of these connections and to set its bit-width. Connections to memory are bidirectional while the ones between kernels are unidirectional and they can be used to forward a calculated result and/or to by-pass a data loaded from memory or received for previous kernel to next kernel in order to reduce total memory accesses. When the required Kernels are enabled and configured, they are constructed as a data-flow fashioned pipeline. The data-flow based architecture is adopted by [8] and [9] and it is the architecture whose resources are synchronized by the availability of the data instead of constant time slots. In our case we adopt the data-flow since processing time on each kernel can be different depending on the packets themselves and also, the memory response time might be variable. It is worth mentioning the number of kernels from each type is selected by the end-user engineer at implementation time giving a flexibility on deciding protocol processor's limits.

3) Control Processing Path: This module is responsible for control processing such as sequencing, tables management, and handling interrupts resulting from cases like erroneous packets and packet fragments. This path consists of the instruction resolution, timers/counters, adders/comparators, and a set of level-1 configurable finite state machines. Depending on the protocol task to be performed, one or more state machines can be configured for suiting the task or handling possible

978-1-4577-0671-4/11 $26.00 © 2011 IEEE   104

interrupts. Some protocol tasks shown to have more interaction with the application layer which might require the use of off-loading instructions. In this case, the instruction resolution is used for mapping the coarse-grained off-loading instructions to their corresponding micro-codes. Similarly, the maximum size of the state machines and the number of timers/counters and adders/comparators blocks are selected by the end-user engineer at implementation time since they participate in deciding protocol processor's limits.

4) Micro-Code Storage: This module holds all information sets needed during processing. These information sets include the micro-codes of the off-loading instructions, the micro-codes of the interrupt handlers, tables, and constants such as HW and IP addresses.

5) Booting and Organization Unit: This module is the first module to be enabled when the protocol processor is requested and it is responsible for enabling and configuring other protocol processor's segments for both configuration levels. Through its interface, it receives the configuration macros, the micro-codes of off-loading instructions and the interrupt handler micro-codes from the run-time controller of the DRRA fabric. Then it construct the data and the control processing paths, it configures needed interfaces, and it fills the micro-codes storage with the micro-codes, tables and constants.

B. Mapping Demonstration

As we mentioned before, the protocol processor is configured and customized for suiting the inquired task. When the run-time detects an application with protocol processing tasks, it requests a protocol processor and it sends configuration data. *BOU* within the protocol processor receives the request and constructs the needed resources within the processor and it fills the micro-code storage. Depending on the inquired protocol task, interrupts and off-loading instructions can be different, which shows the reason and the advantage for adopting micro-coded instructions instead of hardwired instructions. To demonstrate how protocol parts are mapped to processor's resources, we consider the case of TCP client connected to an Ethernet network. This client task can be mapped to a single protocol processor but for showing more complete image, we will map the TCP/IP functions to one processor and the data-link layer functionality for Ethernet to another protocol processor. Since number of kernels is decided by the end-user engineer at implementation time, we assume that the protocol processor is implemented with two Checksum kernels and one kernel form each other type.

A typical TCP client C program implemented with Berkeley Sockets Application Programming Interface [12] can be represented with the pseudo-code given in Listing 1. At compilation, with the aid of compilation directives the compiler can distinguish which parts of the task are mapped to which processor. Consequently the compiler generates configuration macros for both processors as well as any involved off-loading instruction as shown in Listing 2. The pseudo-code in Listing 1 is then translated into Listing 2 as follows: The first and second

lines appear in Listing 1 which are used to create a socket for TCP protocol and to check creation success are translated to the configuration macros appearing at lines 1-5 in Listing 2. Line 3 in Listing 1 which is used to associate the socket with local values is translated to configurations macros at lines 6 and 7 in Listing 2. Lines 4-7 in Listing 1 which are used to establish the connection and to communicate with the server are translated into the off-loading instructions at lines 8-10 appearing at Listing 2. Finally, line 8 in Listing 1 which is used to terminate the socket is translated into configuration macro at line 11 in Listing 2.

Listing 1 TCP Client using Berkeley Socket API

#Directive 0: Ethernet ($HW\ Address$) $\rightarrow Processor\ i$
#Directive 1: TCP/IP $\rightarrow Processor\ j$

1: Create Socket ($protocol\ type = TCP$)
2: Check Socket Creation Status
3: Set $port\ number$ and $IP\ address$
4: Connect Socket ($Port,\ IP,\ via\ Ehernet\ Interface$)
5: Check Socket Connection Status
6: Write ($Socket,\ Packet = Request,\ Packet\ Length$)
7: Read ($Socket,\ Packet = Responce,\ Packet\ Length$)
8: Close Socket

Listing 2 TCP Client using Protocol Processor

1: Construct ($PID = i,\ map = Ethernet,\ Mem\ Space\ i$)
2: Construct ($PID = j,\ map = TCP/IP,\ Mem\ Space\ j$)
3: Link ($[i,\ PHY],\ [i,\ j], [j,\ Application\ Processor]$)
4: Configure Kernels ($PID = i,\ Connection\ settings$)
5: Configure Kernels ($PID = j,\ Connection\ settings$)
6: Initialize ($i,\ HW\ Address,\ ARP\ Table$)
7: Initialize ($j,\ Port\ number,\ IP\ address$)
8: Connect ($j,\ Port,\ IP$)
9: Send ($j,\ Data,\ Data\ Length$)
10: Receive ($j,\ Data,\ Data\ Length$)
11: Destruct ($i,\ j$)

While the configuration macros and off-loading instructions are extracted from the code at compilation time, resolving the actual processor's identification number and its memory space are dependent on the availability of resources and they are done at run-time by the run-time controller which initiates the configuration. The run-time controller starts the configuration by sending configuration macros of Listing 2 through *Configuration Interconnect* to the *BOU* of both protocol processors, resulting in the structure shown in Figure 3.

The first *Construct* macro is sent to the *BOU* of processor *i* which a) enables the CRC Generator, the Bit-Field Analyzer, and the Data Manipulation b) build the state machine for controlling send and receive to/from PHY with CSMA/CD protocol c) sets the processor's memory boundaries as stated by the macro. The second *Construct* macro is sent to the *BOU* of processor *j* which a) enables both available Checksum, the Bit-Field Analyzer, and the Data Manipulation b) build

---

978-1-4577-0671-4/11 $26.00 © 2011 IEEE

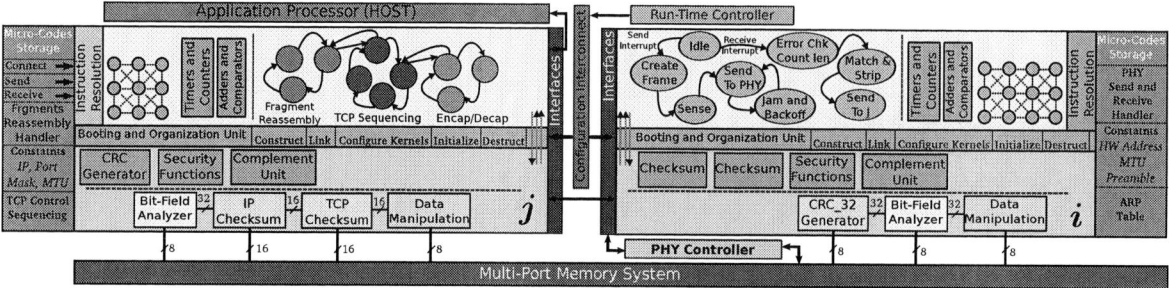

Fig. 3. Configuration of TCP/IP Over Ethernet

the state machines for controlling encapsulation/decapsulation, fragments reassembly, and TCP sequencing c) enable the Instruction Resolution d) sets the processor's memory boundaries as stated by the macro. Each processor has its own memory space but they can access the memory space of each other since they belong to the same private execution environment.

The *Link* macro is sent to the *BOU* of both processors for configuring their interfaces and setup the connections. The macro results in connecting processor j to the application processor, processor j to processor i, and processor i to the PHY controller.

While *Construct* and *Link* macros are level-1 configurations, the macro *Configure Kernels* is level-2 configuration, it has to be subsequent and it is used for setting the internals and connections of enabled Kernels. The first *Configure Kernels* macro is sent to the *BOU* of processor i for a) setting the CRC generator polynomial to 32, b) setting the Bit-field analyzer to count length and to check *Type* field, c) setting Data manipulation for Ethernet framing, and d) configure kernels connection as in Figure 3. The second *Configure Kernels* macro is sent to the *BOU* of processor j for a) setting one of the checksum kernels to IP checksum and the other to TCP checksum, b) setting Bit-field analyzer for matching flags patterns and fields of IP Packets as well as TCP segments, c) setting the Data manipulation for TCP/IP encapsulation, and d) configure kernels connection as in Figure 3.

Once both processors are configured, they send success signals to the run-time controller which then sends *Initialize* macros. The first and second *Initialize* macros are sent to *BOU* of processor i and of *BOU* processor j respectively, filling their *Micro-Code Storage*. In this demonstration, the *Micro-Code Storage* of i will contain a static ARP table, a set of constants including the HW (MAC) address and a value of maximum transmission unit (MTU), and the interrupt handler code of "receive from j and send to PHY" and "receive from PHY and send to j" interrupts. On the other hand, the *Micro-Code Storage* of processor j will contain the a micro-code driving the state machine for send/receive with TCP sequencing, a set of constants including IP, port and MUT, the interrupt handler code of fragments reassembler, and lastly, the micro-codes of the off-loading instruction involved in this task.

Once both processors are completely configured and ini-

tialized, application processor starts off-loading the task to processor j as coarse-grained off-loading instructions. When these instruction reach processor j, the *Instruction Resolution* maps them to the corresponding pre-loaded micro-codes. The micro-code sequence of the *Connect* instruction results in building a request packet to establish a connection with the server, sending its memory address to processor i and wait. Processor i receives the address as a "receive from j and send to PHY" interrupt. It builds an Ethernet frame, send it to PHY Controller and wait. When the response comes from server, the PHY Controller send the frame memory address to the i. Processor i receives the address as a "receive from PHY and send to j" interrupt. It checks the frame for errors, strip the packet from the frame and it sends the packet's memory address to j. Processor j checks the packet, initialize the sequencing, sends acknowledgment packet to the sever as done before and return success signal to application processor. Application processor then sends the off-loading instruction *Send* that carry the memory location and length of data to be sent to the server. Processor j encapsulates it in one or more packets, depending on its size and the MTU, and it sends to server passing through i. When the application processor wants to read the data from the server, it sends *Receive* instruction to processor j. Processor j in its turn reassembles packet fragments from the received packets, re-request any erroneous packets, sort the packets and strip the data from these packets. By the end of reception, processor j returns to the application processor the memory address and length of the received data.

When the task is completed, the run-time controller sends *Destruct* configuration macro to the *BOU* of both protocol processors in order to disable them. Meaning that, both processors as well as their memory and connection space will be return to the list of available resources.

V. EVALUATION RESULTS

In this section, we select two of the protocol processor kernels as a case for performance evaluation. We show area, speed, power, and energy of the preliminarily implementation of the *CRC Generator* and the *Checksum* which found to occupy around one third (32%) of RISC CPU time during TCP/IP operation according to [13]. We also show required configuration time which holds for all other processor's kernels.

978-1-4577-0671-4/11 $26.00 © 2011 IEEE

The CRC kernel supports CRC-8, CRC-16-CCITT, CRC-32-IEEE 802.3, and CRC-40-GSM. It has a configuration register that is used for selecting the desired polynomial as well as the width of the output port. It consumes 8-bit input from memory each cycle and it outputs the check polynomial by the end of the input packet.

The Checksum kernel can be configured to perform IP or TCP/UDP checksum. It has a configuration register that is used for selecting desired checksum as well as the width of the output port. For IP Checksum configuration, it calculates the Checksum for IP header as well as the Pseudo header that is forwarded for TCP/UDP Checksum. For TCP/UDP Checksum configuration, it calculates the Checksum over the segment and it includes the Pseudo header coming from IP Checksum.

We synthesized both kernels using Synopsys Design Compiler with TSMC 90 nm technology and we show synthesis results in Table I. We performed gate level simulations with 500MHz frequency using ModelSim and we back-annotated the synthesized net-lists to estimate power consumption. Table II shows the throughput, estimated power, energy per word and the configuration used during the simulation.

TABLE I
SYNTHESIS RESULTS

	Max. Frequency GH_z	Cell Area μm^2
CRC Generator	1.05	3068
Checksum	1.04	2898

TABLE II
POWER ESTIMATION

	Throughput $Gbps$	Power mW	Energy pJ	I/O width
CRC-32 Generator	4	1.2447	2.489*	8/32
IP Checksum	8	0.6874	1.374**	16/16
TCP Checksum	8	0.6813	1.362**	16/16

* Per 32-bit word   ** Per 16-bit word

Required configuration time is similar for all kernels. It consists of three cycles as illustrated by the timing diagram in Figure 4. In *cycle 0* the *Data Processing Path* configuration register is written to determine the kernels to be enabled. In *cycle 1* the kernels are enabled and in *cycle 2* their internal configuration register is written.

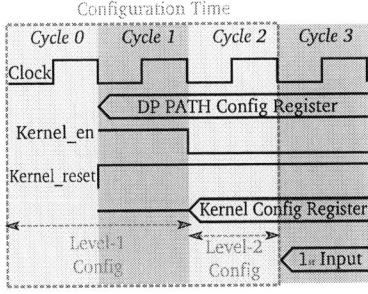

Fig. 4. Required Configuration Time

VI. CONCLUSION

In this Paper, we have presented our coarse-grained protocol processor illustrating configuration, mapping and adaptability. While application layer is kept on a light-weight GPP and the physical layer is kept on dedicated controller, the protocol task within the application is mapped to the protocol processors. Starting from standard C application, we were able to extract the protocol processor task and to generate configuration macros and off-loading instructions for both protocol processors integrating in handling the task. By adopting microprogrammed codes, we were able to avoid the extra area required for RISC instruction store and to avoid performance degradation due to shared instructions and variables. By exploiting specialized kernels, we were able to perform common protocol functionality at high speed and low energy cost, where such tasks consume one third of the RISC CPU time.

VII. FUTURE WORK

The memory system is not described in this paper and we plan to consider it separately describing how processor-memory interaction is done and showing its performance. We are also working on more optimized version of the processors kernels. We next plan to perform a complete benchmarking of the processor using specialized benchmarks where we provide the comparisons against other systems including RISC based and custom ASIC.

REFERENCES

[1] N. Shah, "Understanding network processors," Master's thesis, Electrical Engineering and Computer Sciences Department, University of California, Berkeley, 2001.
[2] M. Shami and A. Hemani, "Control scheme for a cgra," in *Computer Architecture and High Performance Computing (SBAC-PAD)*, Oct. 2010, pp. 17–24.
[3] S. A. Virtanen, "On communications protocols and their characteristics relevant to designing protocol processing hardware," 1999.
[4] M. Ahmadi and S. Wong, "Network processors: Challenges and trends," in *The 17th Annual Workshop on Circuits, Systems and Signal Processing, ProRisc*, 2006, pp. 223–232.
[5] Intel Corporation, *Intel IXP1200 Network Processor Family, Hardware Reference Manual*, 2001.
[6] ——, *Intel IXP2400/IXP2800 Network Processor Family, Hardware Reference Manual*, 2003.
[7] N. Shah, W. Plishker, K. Ravindran, M. Gries, S. Weber, A. Mihal, C. Kulkarni, M. Moskewicz, C. Sauer, and K. Keutzer, "Successfully deploying the asip," in *Building ASIPS: The Mescal Methodology*. Springer US, 2005, pp. 179–225.
[8] T. Henriksson, "Intra-packet data-flow protocol processor," Ph.D. dissertation, Department of Electrical Engineering, Linköpings Universitet, SE-581 83 Linköping, Sweden, 2003.
[9] [Online]. Available: http://www.xelerated.com
[10] R. Hartenstein, "A decade of reconfigurable computing: a visionary retrospective," in *Design, Automation and Test in Europe*, 2001, pp. 642–649.
[11] M. A. Tajammul, M. A. Shami, A. Hemani, and S. Moorthi, "Noc based distributed partitionable memory system for a coarse grain reconfigurable architecture," *VLSI Design International Conference*, pp. 232–237, 2011.
[12] W. R. Stevens, B. Fenner, and A. M. Rudoff, *Unix Network Programming, Volume 1: The Sockets Networking API*, 3rd ed. Addison-Wesley, 2004.
[13] N. Maruyama, T. Ishihara, and H. Yasuura, "An rtos in hardware for energy efficient software-based tcp/ip processing," in *Application Specific Processors (SASP)*, June 2010, pp. 58–63.

moviTest: A Test Environment Dedicated to Multi-Core Embedded Architectures

Teodor Tite*, Adelina Vig*, Nicolae Olteanu**, Cristian Cuna**

* "Politehnica" University of Timisoara, Department of Computer and Software Engineering, Timisoara, Romania
** Movidius SRL, Department of Development Tools, Timisoara, Romania
E-mail: Teodor.Tite@dsplabs.cs.upt.ro, Vig_Adelina@yahoo.com,
Nicolae.Olteanu@movidius.com, Cristian.Cuna@movidius.com

Abstract—The major shift towards multi-core hardware brought additional complexity to the software development process and made software testing even more important. Movidius Test Environment (moviTest) is a test environment which addresses these challenges, a novelty, especially in the embedded field. It can be used for generating/running automatic tests, which target heterogeneous multi-core architecture ASICs or simulators. The tests may be intended for both hardware and software validation. The environment offers key features for writing test cases, aimed to validate heterogenous parallelized software. The proprietary test script language allows – by usage of specific directives – the inclusion of target-specific assembly language and/or C/C++ language code and the loading/running of the resulting binaries on the desired heterogeneous targets. To support the different build processes, of each language and associated target, the user may specify multiple tool-chains which will be used by the environment. If the tests do not use all core targets, and they do not share the same resources, of the targeted embedded system, the scripting language provides the possibility to overlap multiple tests to make use of all targets simultaneously, thus considerably decreasing the testing time. Other important features like multi-core debug information collection are also present in moviTest. Experimental results show the usefulness of the previously enumerated features.

I. INTRODUCTION

One of the most important subjects that have often shaped the orientation of software development has been the evolution of the underlying hardware. Major breakthroughs in hardware technology had a large impact on software. Even the first major software crisis has been related to the transition to integrated circuits, which boosted tremendously hardware performance [10]. During the time, hardware development has been focused on faster instruction execution and higher clock frequencies and thus, no significant change in the intuitive sequential-structured software development was needed to gain full advantage of hardware reserves [11].

Special software tools, designed for parallelized software development, are at an early stage and generally immature, but are a major research direction towards increasing development productivity of parallelized software. Tools and environments for testing multi-core hardware and software represent even more a rarity and there is much room for research. For embedded systems, multi-core hardware has just entered the market and therefore, development tools for parallelized software are even more an exception. A very interesting example is

pTest [1], a testing tool, specially dedicated for validating parallelized software, on embedded multi-core hardware. It is mainly focused on detecting parallel computing-specific errors, like synchronization errors between threads, and does not cover all the levels needed by software development.

II. RELATED WORK

One example of the most powerful test environments, which are widely used in industry, is IBM's Rational Test RealTime [7]. This environment provides a variety of testing features like runtime analysis, component testing and system testing for languages, used in embedded systems, like C, C++ and Ada. The environment possesses a powerful test script API, enabling stub responses to a variety of input and detailed data definitions for data intensive tests like those found in image processing. With the help of the test script API, so-called – in Test RealTime documentation – virtual tester behaviour may be obtained. Virtual tester scripts define the input data to be gathered by the tested system and, also, define the expected output data which will be used for comparison with the output obtained by the system under test.

Another example of testing tool is Time Partition Testing (TPT), developed by PikeTec GmbH. This tool is largely used by car manufacturers because of some key features, which will be presented in next paragraphs. TPT supports the following – testing process typical – steps: test modelling, test execution, test assessment and test documentation [6]. An important feature of this testing tool is the support for graphical modelling of test cases. This enables very intuitive, productive test case creation. Another advantage of the graphical approach is reduced redundancy. In scripting language only-based test cases, many similarities may arise, a problem which is solved by TPT graphical approach, using joint structures. The main component of Time Partition Testing is the TPT Virtual machine, which may be integrated into many test environments. This component provides a major advantage for testing embedded systems: portability. The Virtual machine may operate on different platforms, keeping all differences isolated into a platform adapter [5].

III. AN AUTOMATED TEST ENVIRONMENT

A. Test environment caracteristics

Movidius Test Environment (moviTest) is an environment which can be used for generating/running automatic tests on Movidius specific targets: simulators and boards containing ASIC or FPGA. Movidius-designed targets are heterogeneous multi-core VLIW SoC

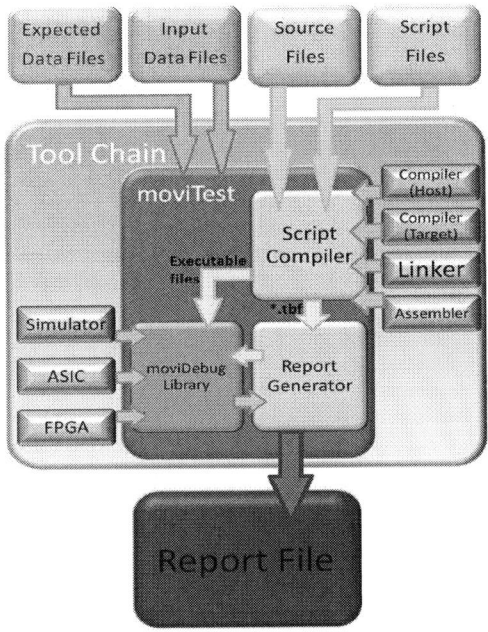

Figure 1. moviTest structure and testing process.

communicate with target platforms: simulators, ASIC or FPGA. This file contains the main actions for running simulation, taken from the test script file like the ones presented in the example below. After that, the test batch file and the other temporary data files are processed by the *test report generator*, resulting in the final .html format test report file.

Automatic testing is performed using test scripts, which are sequences of commands describing the test actions (initializations of registers, memory locations, system registers, testing different parameters in the running system, etc.). The processing of the test scripts is performed in several steps (as shown also in Fig. 2):

1. The test scripts are compiled using the internal *test script compiler*, which – during the *source generation part* – takes the script file, along with additional source files as input and creates the final source files needed for the test script.

2. The second step is accomplished by the *binary generation (compile, assembly and link-edit) part* of the *test script compiler* in which the test environment invokes the tool-chain for each of the cores involved in the test in order to build the binaries (detailed in Fig. 3), which will be used in the following steps.

3. The third step is the generation of a set of debugger pseudo-commands – a *test batch file* (**.tbf*) – with the purpose of conducting the entire test process and associated operations.

4. Finally the **.tbf* file is then interpreted by the *test report generator* which uses the *moviDebug library* (multi-core debugger) to load, save, set and get values (registers or memory locations) which will serve as terms in data comparison operations, performed in order to generate the final test report as a browse-able .html file after running the steps described at 3.

The environment collects debug information during compile and execution time, enabling support for usual debugging features like breakpoints, hovering, jumping over code sections and other.

architectures, which gives 20 GFLOPs of processing power, with low power consumption profile.

Testing is available at all software levels, from assembly language code to C/C++ language code, for all the cores inside the ASIC. As heterogeneous architectures are supported, different tool chains may be used for each type of the cores. Thus the binaries are built, deployed and executed to one of the execution platforms which can be one of: the SoC simulator, the ASIC or an FPGA containing just parts of the SoC (used in early hardware development), depending on the user's needs.

The general structure of the test environment is presented in Fig. 1. The input files are processed by the *test script compiler* component, and thus, generating temporary files needed for the *test report generator* component. As stated in the above paragraph, the *test script compiler* may use different tool-chains, for generating executables for the host computer and for each cores of the heterogeneous target platform. One of the temporary files is the *test batch file* (**.tbf*) which is written in an intermediate script language that contains so-called pseudo-commands for the *moviDebug library* component (the multi-core debugger library), which is used to

Although almost entirely written in C/C++ language, moviTest has been developed portably, to be run on both Windows (built with Visual C++), and Linux (built with GCC) environments. Its source code comprises over 20000 lines of C and C++ language code along with about 200 windows batch file lines (shell script for Linux build) used for enabling flexible configuration of the test environment.

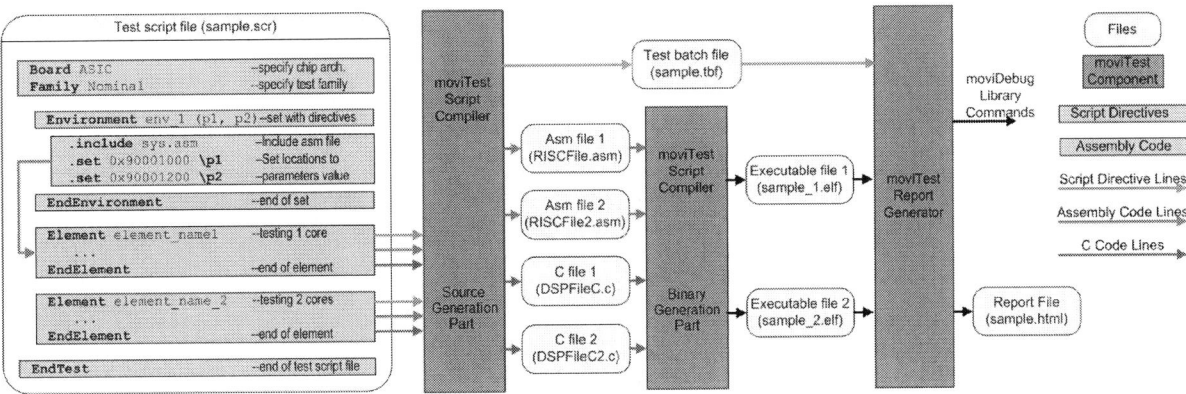

Figure 2. Test script compilation and test report generation processes.

978-1-4577-0671-4/11 $26.00 © 2011 IEEE

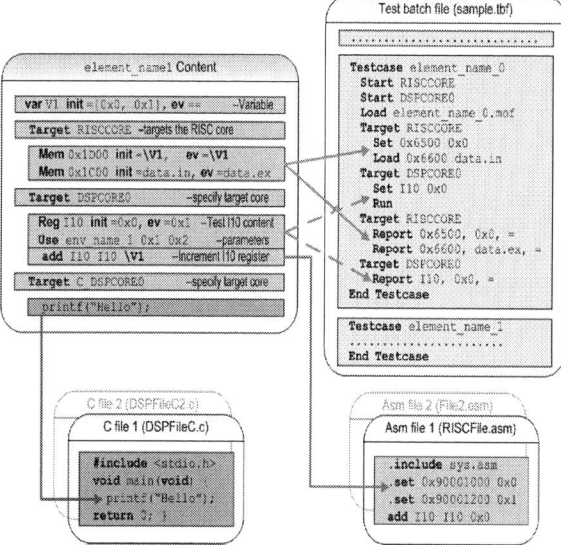

Figure 3. Expanding the test script into resulting files.

B. Test script language caracteristics

The main block of a test script file represents a *test element*. As shown in Fig. 3, a *test element* contains one or more *test cases* which are sequences of data set/test, source code and environmental usage. A test element may contain source code, test environment variables declarations and usage, debugger commands, build instructions, etc. Each test element may expand into multiple test cases. The number of test cases generated for a test element is determined by the number of values which a test script variable can have, and for each of these values a separate test case (with all variables replaced by the current index of value entry) is processed in a sequential manner, from the first to the last.

On the other hand, in contrast to classical test environments, the number of test sections is determined by the number of used targets (cores). For each of those targets, a different section of source code and directives (those after each "Target" directive) is processed in a concurrent way, therefore executing in a multi-core manner. As explained earlier, different tool-chains are called for each type of target (core), in order to obtain the associated binaries from each section of source code.

By usage of specific directives, it is possible to include target-specific assembly language and/or C/C++ language code and the loading/running of the resulting binaries on the desired heterogeneous targets.

C. Comparison to related test environments

Although it lacks graphical modelling [5], moviTest uses instead 3 levels of scripting language, to avoid test script redundancy. The first step of the 3-level scripting language, used for the input *.scr* file, enables the writing of almost redundancy-free scripts. This is achieved though high level directives, e.g. *Environment*, which groups script directives, used for multiple test elements with the same environment settings, or the support for intervals, which makes possible to use the same script directives with different data, according to the intervals. In contrast to the previously presented environments, expected data may be acquired from specific portions of data files and

the result of comparison may be obtained using an expected tolerance which is very useful when comparing floating point operations, or when a slight deviation from the expected data is acceptable for some test cases. Despite it lacks virtual tester feature [6], C/C++ helper functions may be included in the test script for generating input data, or expected data, which will be then used for comparison with the obtained test data. These data-generating C functions are compiled with a user specified compiler – Visual C++ and GCC are supported by the environment – and executed on the host machine (PC) on which the test environment is run.

The second scripting language step is reached when the directives of the scripting language are expanded, by an internal component of the test environment, into intermediate low-level scripting directives, saved into the *.tbf* file (Test Batch File). This process can be observed in Fig. 2 and Fig. 3. The *.tbf* file contains a set of commands, targetting mainly the heterogeneous multi-core debugger, with various purposes like: setting the environment, initialization of the system, loading the test and the additional input files, running the test and waiting until the test is complete, retrieving the output data by saving different memory areas or register content, compare the obtained results with the expected ones etc. The user may manipulate the script language also at this level (although more redundancy at this stage) and run the *.tbf* file as input, together with the previous generated temporary files; this is useful in some cases as it provides more granularity of the test process and it can also serve for some debugging purposes of the test process flow.

Finally at the 3rd step, the content of the *.tbf* file is translated into debugger batch directives to communicate with the *moviDebug library* component (Fig. 2).

moviTest provides automatic results analysis, report documentation generation, similar to the above presented environments, but also includes useful features like automatic summary mail delivering, automatic report files upload on a server.

Although not as powerful as the TPT Virtual machine, portability issues are managed using a separate layer, between the hardware part and the testing part (Fig. 1). The environment may integrate alternative versions of this layer, to support other architecture types, or simply extend the present version.

IV. EXPERIMENTAL RESULTS

An example is shown in Fig. 4, which presents more clearly the multi-core-oriented advantages of the environment's test script API. The test script file written for this example, contains 2 *test elements*, organized in an overlapped test approach, with the intent to reduce the total time of testing. After compiling the test script file, the test batch file, contains the final 2 *test cases*. The first *test case* presents how the environment makes possible overlapping of independent tests. The 1st process (in Fig. 4 shown in dark green), which contains 4 threads, runs concurrently with 4 other single threaded processes, which are completely independent executed on the remaining 4 DSP cores. As told in the previous paragraphs, the 4-threaded test project is managed by dedicated C language code, executed on the RISC core. The second *test case* contains only single threaded projects, which were only organized together, to reduce testing time.

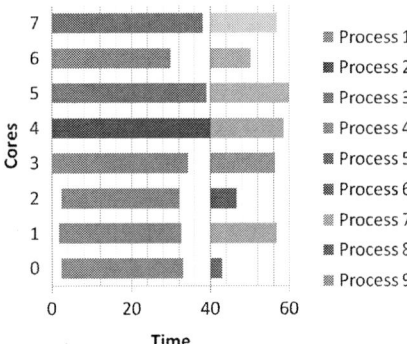

Figure 4.   Multiple tests execution example

This experiment shows the capability of the test environment to run multi-threaded tests along with other independent single-threaded tests, in such a manner that all available DSP cores are exploited. In contrast, the sequential approach, the total testing time is the sum of all overlapped threads' execution times, far higher than those presented in Fig. 4.

The evolution of the assembler, compiler, linker, debugger and simulator, which are part of the Movidius tool-chain, is currently validated using regression tests. With each new added feature, one or more corresponding test cases are added to the regression test collection, most of them employing all stages of the tools chain.

The older, sequential test approach, using Perl scripts has been replaced with the new approach using moviTest test environment. Initially the test cases where run sequentially also with moviTest to avoid conflicts between test cases. The test cases have been, since then, rewritten and organized into fewer test elements, each comprising 8 test cases, which will be run concurrently on the 8 DSP cores, for the purpose of reducing total testing time. This new regression resembles the last presented approach from Fig. 4.

Time measurements have been performed to survey the performance gains between the former and the latter approach, which are presented in Fig. 5. The first two test setups are differentiated by the hardware target to be tested and the third equals the second one, but with all simulation optimizations turned on. As shown in the figure, the total testing time has been reduced for all test setups, by a quantum of about 80%, or about 5 times faster than previous. These performance gains are achieved through parallel execution on the 8 DSP cores simultaneously and through reduced TCP/IP traffic (i.e.

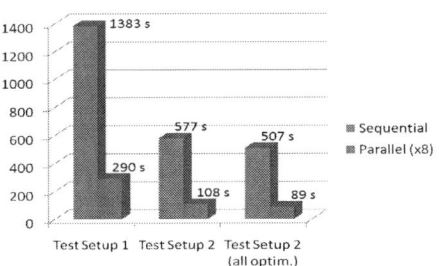

Figure 5.   Performance gains through reorganizing test elements to run in bunches of 8 parallel processes (in seconds)

the connection between the debugger and either the simulator, or the debug server to which the execution platform is connected).

V.   CONCLUSION AND FUTURE WORK

The presented test environment addresses the heterogeneous multi-core domain of research, supporting key features like for enabling efficient testing of heterogeneous multi-core/parallelized systems and provides a solution for the problems of multi-core systems we detected, as for example, the feature of supporting multiple and heterogeneous hardware targets and the associated tool-chains, or the feature of full traceability and determinism during the test life cycle; also the preservation of the intermediate files produced after the compilation of the test script provides easy reproduction and debugging of some eventually failing test cases, decreasing the fixing time. There is plenty of room for improvement also in the moviTest environment, but it is a step toward multi-core automated testing.

ACKNOWLEDGMENT

This work has been funded by the EU Structural Funds Falx Daciae - SUIM 499/11844, POS CCE O2.1.1 research program.

REFERENCES

[1] Shou-Wei Chang, Kun-Yuan Hsieh, Jenq Kuen Lee, *"pTest: an adaptive testing tool for concurrent software on embedded multicore processors"* , Proceedings of the Conference on Design, Automation and Test in Europe (DATE'09), European Design and Automation Association 3001 Leuven, Belgium, Belgium ©2009

[2] N.S. Eickelmann, D.J. Richardson, *"An evaluation of software test environment architectures"*, 18th International Conference on Software Engineering (ICSE'96), Berlin, Germany, 25-29 March

[3] Leitner, A., Ciupa, I., Oriol, M., Meyer, B., Fiva, A., *"Contract Driven Development = Test Driven Development - Writing Test Cases"*, Proceedings of ESEC/FSE'07: European Software Engineering Conference and the ACM SIGSOFT Symposium on the Foundations of Software Engineering 2007, (Dubrovnik, Croatia), September 2007

[4] E. Larsson, Z. Peng, *"An integrated system-on-chip test framework"*, Proceedings of the conference on Design, automation and test in Europe (DATE'01), IEEE Press Piscataway, NJ, USA ©2001

[5] Eckard Lehmann, *"Time Partition Testing: A Method for Testing Dynamic Functional Behavior"*, DaimlerChrysler AG, Research and Technology

[6] Time Partition Testing, Systematic automated testing of embedded systems, PikeTec GmbH, http://www.piketec.com

[7] IBM Rational Test Real Time ftp://ftp.software.ibm.com/software/rational/docs/v2003/ test_realtime/rtrt-tut.pdf

[8] Larry Cantwell, *"Architecting An Automated Test Environment"*, OnPath Technologies A TesLA Whitepaper

[9] Mojtaba Mehrara, Thomas Jablin, Dan Upton, David August, Kim Hazelwood, and Scott Mahlke, *"Multicore Compilation Strategies and Challenges - An overview of parallelism and compiler technology"*, IEEE SIGNAL PROCESSING MAGAZINE, NOVEMBER 2009

[10] Edsger W. Dijkstra, *"The Humble Programmer"* (EWD340), Communications of the ACM.

[11] Bryan Schauer, *"Multicore Processors – A Necessity"*, ProQuest Discovery Guides, September 2008

[12] Andrew Richards, *"Multicore Tools Issues"*, Codeplay Software Ltd., November 2009

Mismatch Characterization of High-Speed NoC Links using Asynchronous Sub-sampling

Sebastian Höppner, Dennis Walter, Georg Ellguth, René Schüffny

Faculty of Electrical Engineering and Information Technology

Technische Universität Dresden

Dresden, Germany

Emails: {sebastian.hoeppner, dennis.walter, georg.ellguth, rene.schueffny}@tu-dresden.de

Abstract—**This paper presents asynchronous sub-sampling techniques to measure delay mismatch of clock and data lanes in high-speed serial network-on-chip (NoC) links. The techniques allow the use of low quality sampling clocks to reduce test hardware overhead for integration into complex MPSoCs with multiple NoC links. It enables compensation of delay variations to realize high-speed NoC links with sufficient yield. The proposed techniques are demonstrated at NoC links as part of an MPSoC in 65nm CMOS technology, where the calibration leads to significant reduction of bit-error-rates of a 72 GBit/s (8 GBit/s per lane) link over 4mm on-chip interconnect.**

Index Terms—asynchronous subsampling

I. INTRODUCTION

High-speed serial point to point connections are proven to be energy and area efficient solutions for on-chip data transmissions over long distances in the range of some mm [1] in networks-on-chip (NoCs). As link data rates increase, the influence of process variations gets more severe especially in sub-100nm CMOS technologies. To achieve high data rates and maximize yield calibration techniques have to be employed to compensate static mismatch variations. In [2] a calibration strategy with optimal sizing of compensation delay elements has been proposed. However, measurement access to on-chip signal characteristics is required.

Asynchronous sub-sampling where a high-speed signal is periodically sampled by a low-speed clock is widely used for on-chip measurement purposes. In [3] on-chip oscilloscopes using asynchronous sampling clocks are presented which allow measurement of high-speed signals but require low jitter sampling clocks. This is a mayor drawback for complex MPSoCs where measurement signals have to be distributed over longer distances with negligible circuit and area overhead. In [4] measurement of static skews of periodic signals is proposed using statistical averaging. This approach eases integration due to relaxed requirements for sample clock quality. Using this method periodic on-chip signals whose period is in the range of sample clock jitter can be characterized.

This work presents an asynchronous sub-sampling technique to provide measurement access to delay characteristics of multiple high-speed NoC links in a 65nm MPSoC. Therefore a low frequency asynchronous sampling clock with relaxed jitter requirements can be used which simplifies integration and scalability.

II. NoC LINK ARCHITECTURE

We consider MPSoC network-on-chip point-to-point connections over large distances (e.g. some mm) as shown in Fig. 1. Data is transmitted differentially with low voltage swing to achieve high energy efficiency. The link architecture shown in Fig. 2 uses high speed serialization and deserialization (SERDES) where each NoC link contains a single high-speed clock lane and several data lanes with a serialization factor of m. Each lane consists of two differential lines. The clock is shifted by $90°$ at the transmitter to center its sampling edge to the data eye for double-data-rate (DDR) signaling. Both clock and data lanes contain delay cells for selective delay of rising and falling edges with 8 steps to allow compensation for delay mismatch as presented in [2]. At the receiver side the clock is amplified by a time-continuous amplifier whereas the data levels are detected by clocked sense amplifiers for high energy efficiency [5], [6].

For the proposed measurement technique, the low frequency sub-sample clock is applied to each NoC link receiver. A multiplexer selects the sampling clock for the data lanes, which is the received clock signal in functional operation and the sub-sample clock in measurement mode. During operation and measurement the data lanes are sampled by the closely located RX sense amplifiers with a clock signal from the *same* physical net for low relative skew. In functional mode the sampled data is fed to the deserializers (DES) and for measurement the sampled data is routed directly to an output pin of NoC link. A typical MPSoC contains multiple NoC links which are usually distributed over the whole chip area. For low effort and test overhead the sampling clock is provided over a standard digital input pad and routed conventionally over the chip. This induces significant amount of jitter. The sample data outputs are multiplexed to standard digital chip output pads. Since there is only one global sample clock, and the sample data of several NoC links is multiplexed, this technique is well suited for MPSoCs with multiple NoC links and limited routing and I/O resources. The delay configuration data is stored locally in the NoC transmitter and can be set via JTAG.

III. MEASUREMENT TECHNIQUE

A. Sub-sampling Algorithm

For delay measurements periodic signals are sent on the links, which are realized by repeating identical data patterns.

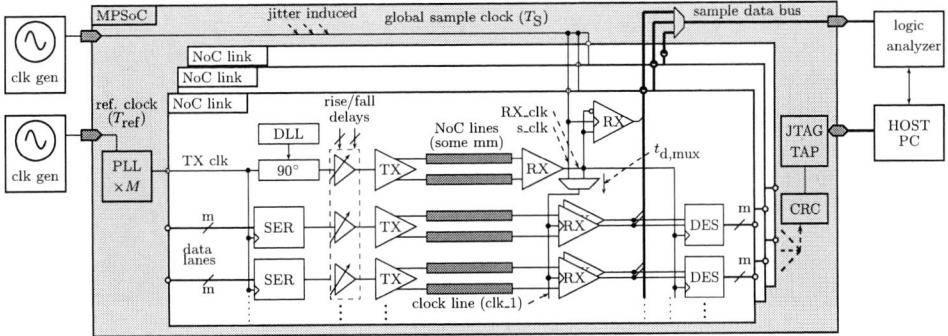

Fig. 2. NoC Link system and measurement setup for an MPSoC with multiple NoC Links

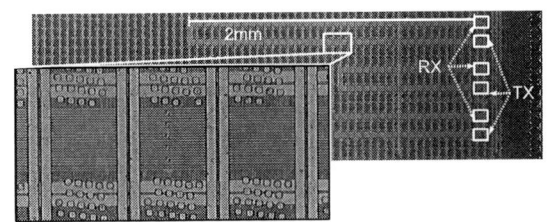

Fig. 1. Chip photo of 65nm prototype MPSoC with high-speed NoC links routed in the upper metal layers between power mesh

The sub-sample data stream is captured for all data and clock lanes of one NoC link.

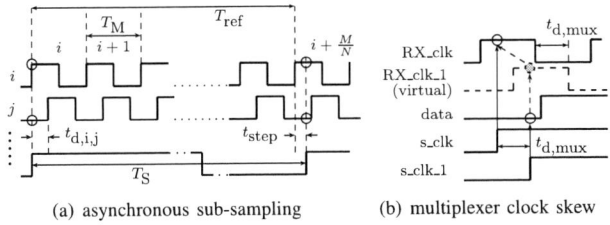

(a) asynchronous sub-sampling   (b) multiplexer clock skew

Fig. 3. Measurement signal waveforms

As illustrated in Fig. 3(a), T_S is the period of the sample clock. The measured signal period is $T_M = T_{ref} \cdot \frac{N}{M}$, where T_{ref} is the global reference clock period, M the PLL frequency multiplication factor and N the data pattern multiplier (e.g. $N = 8$ for 0xFF00 pattern). Asynchronous sub-sampling is achieved by detuning T_{ref} vs. T_S. The effective sub-sampling timebase reads

$$t_{step} = |T_S - T_{ref}|. \tag{1}$$

The occurrence of samples within a specific time t_{bin} is summarized in a histogram of n_{bin} bins with $t_{bin} = \frac{T_M}{n_{bin}}$. To avoid bins with no data points falling in the condition $t_{step} < t_{bin}$ should be fulfilled. A cumulative density function histogram (CDF) $h(t)$ is constructed based on the method proposed in [3]. Using the algorithm in Fig. 4, the effective time of each sample within the histogram t_{eff} is calculated and it is assigned to the corresponding bin.

When starting the capturing process the relative phase of the sample clock and the measured signal is unknown, i.e. the location of the rising and falling data edges within the effective time range $[0; t_{eff}]$ randomly varies within several measurements. Thus an initial histogram $h'(t_{eff})$ is constructed and $h(t_{eff})$ is estimated by centering $h'(t_{eff})$ within $[0; t_{eff}]$. This allows separation of rising and falling signal edges. Fig. 5 shows a measured example histogram.

```
1  for all lanes do
2  │    h(1 : n_bins) = 0
3  │    for i = 1 to n_samples do
4  │    │    t_eff = (i · t_step) mod T_M
5  │    │    bin = floor(t_eff/t_bin) + 1
6  │    │    h(bin) = h(bin) + sample(i)
7  │    end
8  end
```

Fig. 4. Sub-sampling histogram algorithm

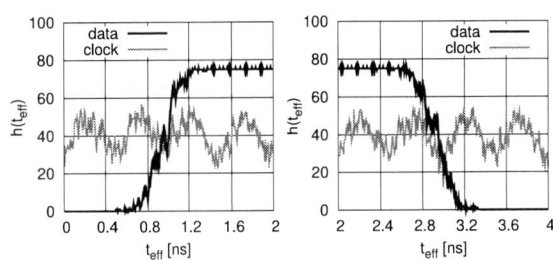

Fig. 5. Example $h(t_{eff})$, $M = 40$, $N = 8$, $T_{ref} \approx 20.000$ns, $T_S \approx 20.005$ns, data pattern 0xFF00

As shown in Fig. 2, there is a multiplexer with delay $t_{d,mux}$ between the amplified functional clock (RX_clk) and the one used to trigger the clocked data amplifiers (clk_1). Due to the sub-sample clock (s_clk) taking the same physical path between sampling RX_clk and the data lanes, RX_clk is sampled as if it were virtually delayed by $t_{d,mux}$ (Fig. 3(b)). Hence, this delay can be ignored and introduces no additional skew between clock and data histograms.

B. Sample clock frequency error and jitter

As shown in Sec. III-A, the exact difference t_{step} of the measured signal and the sample clock period have to be known to construct $h(t_{\text{eff}})$ accurately. Due to clock generator uncertainty or temperature drift, a step size error Δt_{step} between the expected and the real step size exists. This error causes subsequent periodic samples to be assigned to the wrong bin if the condition $n_{\text{samples}} > t_{\text{bin}}/\Delta t_{\text{step}}$ is fulfilled. This is especially the case for large n_{sample} as desired for good accuracy of delay measurement, as shown in Sec. III-C, and for high time step errors Δt_{step}. As illustrated in Fig. 6(a), the measured signal slope in $h(t_{\text{eff}})$ decreases with increasing n_{samples} and Δt_{step}. For estimation of Δt_{step} we propose to consider the standard deviation $\sigma(h(t'_{\text{eff}})$ as optimization criterion, which is maximized if the $t_{\text{step}} + \Delta t_{\text{step}}$ estimation fits the real value of $|T_S - T_M|$. Fig. 7 shows a measured $\sigma(h(t'_{\text{eff}})$ characteristic.

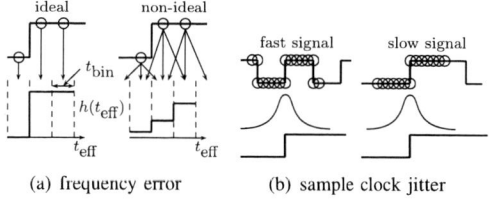

(a) frequency error   (b) sample clock jitter

Fig. 6.   Non-ideal effects in estimation of $h(t_{\text{eff}})$

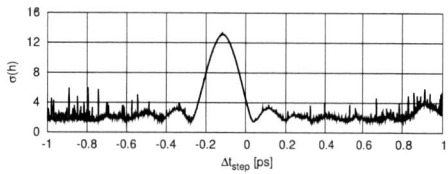

Fig. 7.   Frequency estimation criterion versus reference period offset

In contrast to [3] this work considers clocks with significant jitter due to clock generation (e.g. PLL jitter) and distribution on-chip (e.g. supply noise, crosstalk) as well as long-term accumulated jitter in the signal to be measured. As shown in Fig. 6(b), the separation of rising and falling edge samples is not possible if the sample jitter is larger than the period of the measured signal. Therefore, the effective data period is enlarged by a periodic data pattern of length N, whereas the clock edges can not be separated properly as shown in Fig. 5.

C. Delay difference of data channels

The link architecture shown in Fig. 2 relies on a relative delay matching between the channels. After separating the $h(t_{\text{eff}})$ histograms for rising and falling edges the arithmetic difference technique presented in [4] is employed to determine the delay difference (skew) between data lanes i and j.

$$t_{\text{d},i,j} = e \cdot \frac{T_M}{n_{\text{samples}}} \cdot \sum_{k=1}^{n_{\text{bins}}} \left(h_i(t_{\text{eff},k}) - h_j(t_{\text{eff},k})\right) \quad (2)$$

where e denotes the edge type (rise: $e = 1$; fall: $e = -1$).

D. Clock duty cycle

Due to sample clock jitter, the rising and falling edges of the high speed clock signal cannot be separated within $h(t_{\text{eff}})$.

However, duty cycle d can be estimated by averaging $h(t_{\text{eff}})$ and is related to the rising and falling edge delays by

$$d = \overline{h(t_{\text{eff}})} = \frac{1}{2} + \frac{t_{\text{d,f}} - t_{\text{d,r}}}{T_{\text{clk}}} \quad (3)$$

E. Clock to data skew

For estimation of clock to data delay ideal clock $(h_{\text{clk,est}}(t_{\text{eff}}))$ and data $(h_{\text{data,est}}(t_{\text{eff}}))$ signals are constructed within the t_{eff} timebase with arbitrary delay. Their delay difference with respect to the measured functions $h_{\text{data}}(t_{\text{eff}})$ and $h_{\text{clk}}(t_{\text{eff}})$ are estimated using the maximum of their cross correlations [7] $CCF(h_{\text{clk}}, h_{\text{clk,est}})$ and $CCF(h_{\text{data}}, h_{\text{data,est}})$. Fig. 8 shows an example measurement result with aligned clock and data estimation signals. The relative clock to data delay, represented by the skew of $h_{\text{clk,est,aligned}}$ and $h_{\text{data,est,aligned}}$, is the sum of the 90° delay shift $(T_{\text{clk}}/4)$ for DDR signaling and the clock to data delay mismatch $t_{\text{d,c2d}}$.

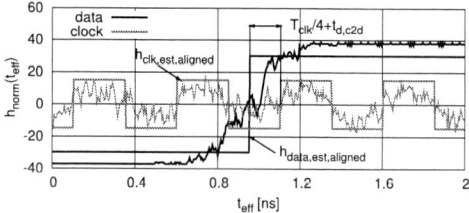

Fig. 8.   Estimation of clock skew by cross correlation with ideal waveforms

IV. EXPERIMENTAL RESULTS

An MPSoC prototype chip has been manufactured in 65nm CMOS technology (Fig. 1). It contains several high-speed serial NoC links, which include the proposed measurement circuitry, with data rates up to 72 GBit/s (8 GBit/s per lane).

Fig. 9(a) shows the measured jitter histogram of the effective sample edge. It is estimated by binning $h(t_{\text{eff}})$ in larger bins and calculating its first derivative with respect to t_{eff}. The high effective sample edge jitter of $\sigma(t_{\text{S,rise}}) \approx 155\text{ps}$ is caused by accumulated long-term jitter of the NoC signals and sample clock jitter. The latter is caused by input pad noise and crosstalk due to large distance routing on the MPSoC die.

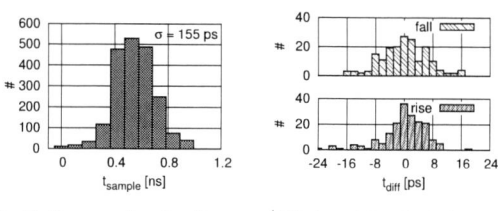

(a) effective sample edge jitter   (b) data delay deviation

Fig. 9.   Measured timing variations, 4mm link, $V_{\text{DD}} = 1.1\text{V}$

A. Delay Mismatch Compensation

Fig. 9(b) shows the histogram of the measured delay differences of the data lanes to mean average data delay for a 4mm NoC-link (9 data lanes) across 20 dies. The standard deviation is $\sigma(t_{\mathrm{D}}) \approx 6.4$ps at $V_{\mathrm{DD}} = 1.1$V. A 2 mm link with identical TX/RX circuitry shows $\sigma(t_{\mathrm{D}}) \approx 2.9$ps at $V_{\mathrm{DD}} = 1.1$V. The delay mismatch within the NoC link is compensated by rise/fall delay elements, as shown in Fig. 2, which have 8 delay settings with ≈ 6ps step size at $V_{\mathrm{DD}} = 1.1$V. Data-to-data delay mismatch is measured as shown in Sec. III-C and the compensation values are determined by an algorithm similar to the one presented in [2]. Fig. 10 shows the measured data delay difference before and after compensation for one particular die. The remaining difference is below 4ps. Clock duty cycle adjustment is performed by the methods presented in Sec. III-D leading to a difference value of rising and falling clock delay compensation. By this the delay compensation parameter space has been reduced from 20 dimensions to 1. The remaining parameter is clock to data skew, which can be estimated by the method shown in Sec. III-E. However, this estimation is limited to an accuracy of t_{step} and is further degraded by clock jitter. For fine skew compensation, bit-error-rate (BER) measurements are performed with a few clock skew compensation values in the estimated region to find the optimum value. Tab. I summarizes the compensation results.

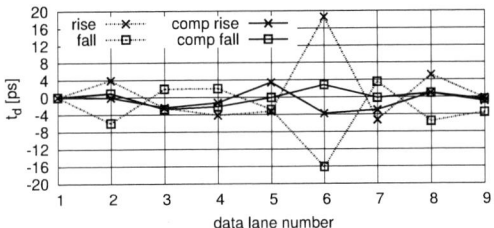

Fig. 10.   Data delay difference to lane 1 (4mm link, $V_{\mathrm{DD}} = 1.1$V)

To illustrate the benefits from delay calibration, BER measurements of the same NoC-link as in Fig. 10 have been performed before and after calibration. Fig. 11 shows the results for two different link speeds for different supply voltage levels V_{DD} and DC signal swing V_{swing}. Due to the capacitive line drivers, AC signal swing scales with V_{DD} and constant V_{swing} has to be matched accordingly to prevent inter-symbol-interference (ISI). At low-speed (unit-interval UI = 250ps) the adjustment of delay variations in the ps range does not lead to significant improvements, whereas at high link speeds (UI = 125ps) the proposed compensation technique enables low BERs over a wide range of V_{DD} and V_{swing}.

TABLE I

NoC LINK DELAY COMPENSATION RESULTS AND DELAY SETTINGS

	uncompensated	compensated	delay setting
data rise	see Fig. 10		[4 3 4 5 5 0 5 3 4]
data fall	see Fig. 10		[1 3 0 0 2 6 0 3 2]
clock duty	0.42	0.52	fall-rise=3
clock skew	from Sec. III-E and BER meas.		rise=1 (fall=4)

V. CONCLUSION

Measurement techniques based on asynchronous sub-sampling have been presented which allow mismatch characterization and compensation of high-speed serial NoC links.

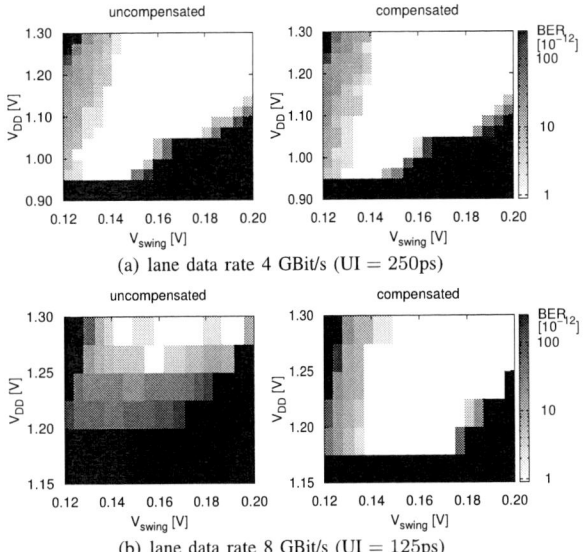

(a) lane data rate 4 GBit/s (UI = 250ps)

(b) lane data rate 8 GBit/s (UI = 125ps)

Fig. 11.   Measured BER results of 4mm NoC-link at 36 GBit/s and 72 GBit/s

This enables high bandwidth operation and maximizes yield. These methods require only minimum test hardware effort and have relaxed requirements for frequency accuracy and jitter of sampling clocks. The techniques have been demonstrated successfully by measurement and calibration of high-speed NoC links up to 8 GBit/s per channel over 4mm on-chip lines in 65nm CMOS technology. The proposed techniques lead to significantly reduced bit-error-rates at high data rates.

ACKNOWLEDGMENT

This work is supported by the German Ministry of Education and Research BMBF under grant number 13N10788 (CoolBaseStations). The authors are responsible for the content of this publication.

REFERENCES

[1] J. Park, J. Kang, S. Park, and M. Flynn, "A 9-Gbit/s serial transceiver for on-chip global signaling over lossy transmission lines," *Circuits and Systems I: Regular Papers, IEEE Transactions on*, vol. 56, no. 8, pp. 1807 –1817, aug. 2009.

[2] S. Höppner, D. Walter, H. Eisenreich, and R. Schüffny, "Efficient compensation of delay variations in high-speed network-on-chip data links," in *System on Chip (SoC), 2010 International Symposium on*, 2010, pp. 55 –58.

[3] J. Schaub, F. Gebara, T. Nguyen, I. Vo, J. Pena, and D. Acharyya, "On-chip jitter and oscilloscope circuits using an asynchronous sample clock," in *Solid-State Circuits Conference, 2008. ESSCIRC 2008. 34th European*, 2008, pp. 126 –129.

[4] B. Amrutur, P. K. Das, and R. Vasudevamurthy, "0.84 ps resolution clock skew measurement via subsampling," *Very Large Scale Integration (VLSI) Systems, IEEE Transactions on*, 2010.

[5] R. Ho, T. Ono, R. Hopkins, A. Chow, J. Schauer, F. Liu, and R. Drost, "High speed and low energy capacitively driven on-chip wires," *Solid-State Circuits, IEEE Journal of*, vol. 43, no. 1, pp. 52 –60, jan. 2008.

[6] E. Mensink, D. Schinkel, E. Klumperink, E. van Tuijl, and B. Nauta, "Power efficient gigabit communication over capacitively driven rc-limited on-chip interconnects," *Solid-State Circuits, IEEE Journal of*, vol. 45, no. 2, pp. 447 –457, feb. 2010.

[7] C. Knapp and G. Carter, "The generalized correlation method for estimation of time delay," *Acoustics, Speech and Signal Processing, IEEE Transactions on*, vol. 24, no. 4, pp. 320 – 327, aug 1976.

978-1-4577-0671-4/11 $26.00 © 2011 IEEE

Impact of Proactive Temperature Management on Performance of Networks-on-Chip

Tim Wegner, Martin Gag, Dirk Timmermann

Institute of Applied Microelectronics and Computer Engineering, University of Rostock,
Richard-Wagner-Str. 31, 18119 Rostock-Warnemuende, Germany
tim.wegner@uni-rostock.de

Abstract— With the progress of deep submicron technology power consumption and temperature related issues have become dominant factors for chip design. Therefore, very large-scale integrated systems like Systems-on-Chip (SoCs) are exposed to an increasing thermal stress. On the one hand, this necessitates effective mechanisms for thermal management. On the other hand, appliance of thermal management is accompanied by disturbance of system integrity and degradation of system performance. In this paper we propose to precompute and proactively manage on-chip temperature of systems based on Networks-on-Chip (NoCs). Thereby, traditional reactive approaches, utilizing the NoC infrastructure to perform thermal management, can be replaced. This results not only in shorter response times for appliance of management measures and therefore in a reduction of temperature and thermal imbalances, but also in less impairment of system integrity and performance. Simulations show that proactive management achieves improvements of nearly 150 % regarding reduction of average temperature inside a 3×3 NoC compared to identical reactive approaches, while mitigating additional delay for packet transmission by more than 50 %.

I. INTRODUCTION

The emergence of nanotechnology is accompanied by cumulative power densities and switching activities per unit area. Therefore, increasingly complex and highly integrated systems like SoCs have to contend with well-known challenges. Amongst others, this concerns heat dissipation, leading to high circuit temperatures and possibly strongly unbalanced on-chip temperature distributions. In the light of a growing number of transistors per chip, which are increasingly susceptible to environmental influences and deterioration, this issue is topical more than ever. As a consequence, thermal stress and physical effects exponentially depending on temperature [1] threaten the integrity of Integrated Circuits (ICs) and have major influence on operability, lifetime and performance. The relationship between temperature and deterioration is illustrated by the Arrhenius model [2], describing the influence of temperature on the velocity of chemical reactions. For this reason, monitoring and control of on-chip temperature distribution are important tasks to secure system functionality and ensure high performance.

Typically, monitoring of on-chip temperature is performed by collecting temperature-related data (e.g. by using integrated diodes). In order to react to undesirable temperatures this data has to be transfered to a component responsible for data evaluation and determination of appropriate reactions (i.e. thermal management). Then instructions are sent to the concerned components. For NoC-based systems, commonly the NoC infrastructure is used for this communication. Despite the importance of thermal management reactive approaches impairs system performance, since the utilization of the NoC presents an intrusion into the system and the induced traffic curtails the availability of the NoC for regular communication. Another drawback is the comparatively long response time of thermal management caused by transmission delay, when using the NoC. Since two transmissions (i.e. reporting temperature and sending instructions) are necessary, an already highly congested NoC additionally exacerbates thermal management. Hence, we propose to predict the on-chip temperature profile based on a model that is realized as part of a Thermal Management Unit (TMU), instead of reverting to physical sensors. By means of the made predictions, the TMU is able to immediately initiate execution of instructions for thermal management. Such a TMU can be implemented in software running on a core of the SoC or it is an inherent part of a core implemented in hardware. Thereby, response time for thermal management is shortened by avoiding transmission of temperature-related data to the TMU and the traffic load of the NoC is reduced freeing up communication capacities for regular data traffic. Prerequisites are that predictions can be accomplished rather fast without inducing unreasonable calculation effort generating additional heat. To ascertain to which extent proactive thermal management influences system performance and on-chip temperature distribution, this approach is compared to a setup reverting to reactive management and to a setup without any thermal management.

The remainder of this paper is organized as follows. In section II an overview over existing work regarding modeling of on-chip temperature and approaches for reactive and proactive management strategies is given. In section III the environment for the simulation of proactive and reactive thermal management of NoC-based systems is introduced. In section IV experiments focusing on the impact of proactive and reactive management on system performance and temperature are conducted. Finally, in section V conclusions are drawn.

II. RELATED WORK

Numerous investigations have already been conducted in the field of modeling thermal behavior [3]–[6] of ICs by exploiting the equivalence of electrical and thermal energy flows [7], since this approach implicates some worthwhile consequences

978-1-4577-0671-4/11 $26.00 © 2011 IEEE

regarding effort for thermal management. In [3], electrical RC-circuits are used to model the thermal behavior of an entire chip. Variability of modeling granularity allows for a trade-off between modeling accuracy and speed. Temperature of the functional blocks is computed by using values for average power dissipation. In [4] this approach is tailored to the simulation of the thermal behavior of on-chip networks. For this purpose, the model of equivalent RC-circuits is extended by the integration of heat spreading angles. Temperature estimation is performed by capturing the network traffic, using these statistics for estimation of power consumption and computing the temperature profile. The creation of SPICE netlists consisting of RC-circuits in order to model on-chip thermal properties is proposed in [5], [6].

Research that can be related to reactive management strategies for on-chip networks is available abundantly. A general concept of an event-based runtime monitoring service for NoC components using hardware probes is proposed in [8]. In [9] this concept is examined with focus on the integration into an existing NoC and the arising implications. In [10] Guang et al. propose a hierarchical agent framework to realize monitoring services on parallel SoC systems in order to provide for reconfigurability and fault tolerance. An approach specified to reactive monitoring and control of temperature in NoC-based systems is provided by [11], where sensors monitor the temperature of the system components and use the NoC infrastructure to report temperature to a central TMU.

Proactive thermal management can be defined as predicting temperature at runtime and taking appropriate actions instead of monitoring temperature and reacting to changes. Assuming this, investigations in this field are available more sparsely. In [12] autoregressive moving average (ARMA) modeling is used to predict temperature of SoCs by regressing previous measurements from thermal sensors. Predictions are employed for thread allocation in order to balance temperature distribution. An approach using a thermal model based on RC-circuits in order to apply reactive and proactive measures for thermal management is introduced in [4].

Our work is motivated by three issues. Firstly, thermal models eventually depend on offline profiling for the extraction of values for power consumption. This makes these models more suitable for tasks like thermal-aware placement and mapping than for the dynamic modeling of thermal properties of ICs. Secondly, due to their nature reactive strategies for thermal management suffer from long response times because the sending of instructions requires availability of monitoring data. The transmission of this data furthermore impairs system performance due to increased traffic. Thirdly, in many cases proactive approaches are not suited for management of NoC routers and links (e.g. they deploy measures like software-based thread allocation) or they partially still rely on physical sensors [12] and external tools for profiling [4]. For these reasons the main contributions of this work are as follows. We extend a NoC simulation environment by a thermal model, which is based on equivalent RC-circuits and therefore does not rely on thermal sensors, which does not depend on any

external tools and which allows for simultaneous system simulation and thermal modeling. This model is used in conjunction with Dynamic Frequency Scaling (DFS) and task relocation in order to allow for simulation of proactive thermal management. To determine the impact on system performance and on-chip temperature distribution this setup is compared to an analog reactive implementation and a reference system without thermal management.

III. SIMULATION ENVIRONMENT

The simulation environment, developed for evaluation of reactive and proactive thermal management, allows for functional simulation of NoCs based on a 2D mesh topology, wormhole packet switching and XY routing. Amongst others, parameters like NoC size, link width and simulation duration can be specified. The system components, which are connected by the NoC, are represented by Intellectual Property Cores (IPCs). The IPCs are individually configurable concerning generation frequency, length and destination address of packets. The sample period T_S, determining the rate of capturing statistics, can be set, too. To preserve consistency the thermal model was developed by using the SystemC Analog Mixed Signal (AMS) library [20], since the simulation environment itself is based on the SystemC [18] and SystemC Transaction-Level Modeling (TLM) [19] libraries. By deploying the AMS library the dualism of electrical and thermal energy flows can be exploited for modeling, because models of all necessary electrical components are included. This allows for simultaneous system simulation and thermal modeling, while preserving system integrity (i.e. independence from external tools for power tracing). For modeling, the NoC infrastructure is mapped on a regular grid of RC-tiles [6]. The general flow of parallel functional and thermal simulation is depicted in Fig. 1. First the NoC topology is set up ① by analyzing simulation parameters and the deployed strategy for thermal management and configuring the IPCs (represented as sending and receiving components). Subsequently, the equivalent RC-network is established ② according to the specified geometry and modeling parameters. Then, the simulation of the NoC, its thermal behavior and the employed strategy for thermal management is executed ③. Every time the specified sample period T_S expires, the simulation is stalled, NoC component activity statistics are passed to the thermal model for current calculation and temperature output of the thermal model is delivered to the thermal management system. After this, the simulation is continued. During simulation the output of the thermal model is updated every clock cycle. The electrical current I, corresponding to heat flow, which is fed into the RC-network, is calculated by (1).

$$I \cong \left(\sum Trans_{0 \to 1} * E_{Trans} \right) / \Delta T + P_{Static} \qquad (1)$$

$Trans_{0 \to 1}$ is the number of bit transitions from 0 to 1 captured for a particular NoC component, E_{Trans} is the energy a single transition consumes, T_S is the sample period and P_{Static} is the value for static power consumption only relevant for active components. For routers E_{Trans} is set to 1,5 pJ due to energy

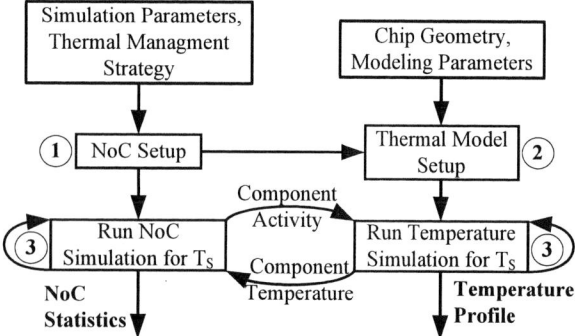

Fig. 1. Flow of simultaneous functional simulation and thermal modeling

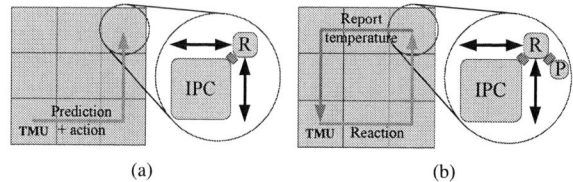

Fig. 2. Management flow and exemplary communication paths for thermal management in a 3×3 NoC: (a) proactive approach, (b) reactive approach (magnified area: components of a NoC-tile a probe is responsible for)

consumption of 0,096 nJ, caused by a 64 bit wide flit crossing a router [15]. Since routers are active elements, P_{Static} for input and output modules as well as FIFO buffers has to be considered [16]. The value of 20 pJ for E_{Trans} of an IPC is unreferenced and only serves to reflect the proportion of IPC to router accounting for the variability of heat generation depending on IPC activity. P_{Static} for an IPC is estimated to be about 100 mW based upon power dissipation of an IBM PowerPC 405 [17] being suitable for integration into a NoC. For NoC links E_{Trans} is set to 11,62 fJ, assuming a wire length of at most 200 μm, random traffic patterns and a transition rate of 50 % [14].

A. Reactive Thermal Management

Reactive thermal management is represented by an implementation of a monitoring and control system, which reverts to event-based monitoring for NoCs [8]. The general flow of reactive thermal management is depicted in Fig. 2 (b). A probe P is attached to every NoC-tile. This probe constantly monitors temperature of all components of its associated tile. This includes the IPC, the router and the 2 links from north to south as well as from east to west (see magnified area in Fig. 2 (b)). In case a temperature change exceeding a threshold T_{Thresh} is detected for one or more components, the probe generates a packet containing the current temperature of the involved components. This packet is sent to the TMU via the NoC infrastructure where it is analyzed and, if necessary, appropriate instructions are sent back to the affected components. The overall scheme of the reactive TMU is illustrated in Fig. 3 (b). All arriving probe packets are stored in the input FIFO and are then processed in the sequence of their arrival. As long as the FIFO is empty the TMU stays in idle mode. In case a packet is available, it is removed from the FIFO and analyzed regarding type (i.e. link, router, IPC), position and temperature value of the involved components. Thereupon, the TMU's internal thermal profile of the NoC is updated and an appropriate reaction (i.e. DFS or task relocation) is determined. Then a packet containing instructions for thermal management is generated and buffered in the output FIFO in order to be transmitted to the concerned NoC components. Reaction policies follow specified values for

step size of DFS as well as maximum and minimum frequency boundaries. Furthermore, an upper temperature limit T_{Bound} and a lower limit for temperature variation ΔT_{Max} between IPCs are defined both triggering IPC task relocation. T_{Bound} serves to reduce hot spots by relocating a task to the IPC with the lowest temperature. ΔT_{Max} is used to balance the thermal profile by relocating a task to the IPC with the biggest temperature variation compared to the affected IPC. The TMU itself is not excluded from this process. Thus, every IPC is a potential TMU, since replacing a whole IPC by a TMU would induce unacceptable overhead. Hence, the TMU can be regarded as being implemented in software. While the TMU is in idle mode or it is currently not located in a particular IPC, this IPC switches to normal operation mode and regularly sends and receives data packets.

B. Proactive Thermal Management

Proactive thermal management does not require probes for temperature monitoring. This accelerates thermal management by reducing response times, excludes packet transmission from probes to the TMU as an error source (e.g. packet loss or data corruption) and redundantizes a set of momentous design decisions (e.g. number and placement of probes). The flow of proactive thermal management is illustrated in Fig. 2 (a). The stage of data transmission from a probe to the TMU is omitted, since the thermal profile of the NoC is directly modeled by the TMU. This avoids additional delay, that would be induced by transmission of monitoring data via the NoC. The detailed scheme of the proactive TMU is shown in Fig. 3 (a). Generally, the TMU periodically updates its internal thermal model of the NoC by analyzing NoC activity statistics, computing the temperature of all components and checking for temperature violations. This corresponds to event-based temperature monitoring executed by the probes of the reactive approach using an identical threshold T_{Thresh}. In case violations are detected, according measures are determined. The measures follow the same policies described for reactive management, again including the TMU itself for possible relocation and therefore turning every IPC into a potential TMU. Once measures are scheduled, instruction packets are generated and buffered in the output FIFO until they are transmitted. Of course this process contributes to heat generation due to additional core activity. As long as no instruction packets are generated, the IPC, in which the TMU is currently located, performs normal operation (i.e. sending and receiving of data

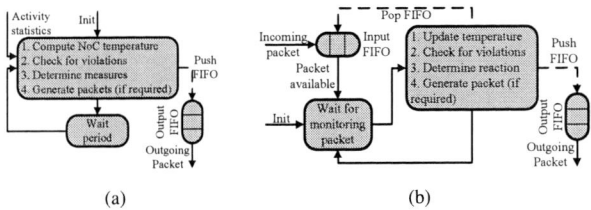

Fig. 3. Thermal Management Unit (TMU): (a) proactive approach, (b) reactive approach

packets). The main challenge to enable a TMU to model the thermal profile of a practical NoC, is to provide the TMU with activity statistics of all network components. For this work it is assumed that this task can be accomplished without occupying the NoC by reverting to system software running on the IPCs, since thermal management might be realized in software, too. Another conceivable approach could be to exploit structures and mechanisms integrated for the purpose of testability (i.e. Built-In-Self-Test, Design for Test) [13]. Admittedly, in case activity statistics have to be transported to the TMU using the NoC, the performance advance of proactive management clearly diminishes. Besides reduced response times and lower traffic load, application of proactive thermal management additionally implies two possible advantages compared to reactive approaches, provided that temperature can be influenced positively. Either, thermal stress and peak temperatures are reduced, when applying identical adjustment measures, leading to increased reliability and lifetime. Or, to achieve identical results, for the proactive approach less effort regarding adjustment measures has to be put in, resulting in lower detraction of overall system performance.

IV. EXPERIMENTS AND RESULTS

In this section the impact of proactive thermal management on temperature and system performance of NoC-based systems is investigated with reference to reactive management. Investigations focus on average temperature, uniformity of temperature distribution, net data throughput $Data_{Net}$, router delay D_R (i.e. time a flit needs to cross a router), delay of packet delivery D_P and the number of delivered packets P_{Trans}. All parameters related to performance refer to values for user data (i.e. traffic for thermal management is excluded). For this purpose, simulations for NoC sizes of 2×2, 3×3 and 4×4 are executed using configuration C1 (see Table I). Moreover, different configurations varying T_{Thresh}, T_{Bound} and ΔT_{Max} are applied to a 3×3 NoC. Since currently simulation of practical periods of time (e.g. a couple of minutes) turns out to be very time consuming, a single run is restricted to 50 ms. In return, to allow for illuminative analysis, all currents injected into the equivalent RC-network are amplified in order to accelerate the occurrence of noteworthy temperature variations. The initial chip temperature and the ambient temperature are set to 60 °C [4] and 45 °C [3]. For the purpose of comparison, results for the reference system without thermal management are depicted in Table II.

TABLE I

CONFIGURATIONS FOR SIMULATION VARYING THRESHOLDS FOR DETECTION OF TEMPERATURE CHANGES T_{Thresh}, TEMPERATURE LIMIT OF IPCS T_{Bound} AND TEMPERATURE VARIATION BETWEEN IPCS ΔT_{Max}

Config	T_{Thresh} [°C]	T_{Bound} [°C]	ΔT_{Max} [°C]
C1	0,2	60,5	0,2
C2	0,2	66,0	0,2
C3	0,2	66,0	2,0
C4	2,0	66,0	2,0

TABLE II

ROUTER DELAY D_R, DELAY OF PACKET DELIVERY D_P, NET DATA THROUGHPUT $Data_{Net}$, THE NUMBER OF DELIVERED PACKETS P_{Trans}, AVERAGE TEMPERATURE T_{Avg} AND TEMPERATURE DIFFERENCE ΔT FOR THE REFERENCE SYSTEM (CONFIGURATION: C1)

NoC size	D_R [cycles]	D_P [cycles]	$Data_{Net}$ [bits/cycle]	P_{Trans}	T_{Avg} [°C]	ΔT [°C]
2×2	5,6	45,9	17,9	≈1,64m	62,7	4,9
3×3	5,7	50,6	38,7	≈3,58m	64,4	13,1
4×4	6,8	60,1	70,8	≈6,55m	66	18

Table III shows the penalties for the performance parameters caused by reactive and proactive management for NoC sizes of 2×2 up to 4×4 using configuration C1. Additionally, improvements induced by the proactive approach are shown. Generally, both reactive and proactive management decrease overall performance for all NoC sizes with degradation growing with larger NoC sizes. For D_R reactive and proactive management cause impairments of at least 223 % (2×2 NoC) exceeding triplication of D_R compared to the reference system. Maximum improvement induced by proactive management is achieved for a 3×3 NoC (29 %). Basically, this also applies to D_P, since larger NoCs are entailed with increased traffic for management due to a higher number of components requiring management. The disproportional degradations for larger NoCs indicate that the NoC precociously gets congested due to additional management traffic. As it can be seen, proactive management, in contrast to reactive management, starts to cause heavy congestion only for larger NoCs, since no additional packets for monitoring data are generated. The observed advances clarify that proactive management is able to considerably relieve the NoC infrastructure. Regarding $Data_{Net}$ and P_{Trans} the proactive approach continuously performs worse than reactive management, although the NoC should have more capacities for regular traffic. This phenomenon can be explained by referring to the design of the TMU. As stated before, the TMU behaves like a regular IPC (i.e. receiving and sending data packets), as long as no management has to be performed. Apparently, this case occurs more often for reactive than for proactive management because in a congested NoC monitoring packets arrive comparatively infrequent. This leads to the reactive TMU being in normal operation mode for most of the time abetting $Data_{Net}$ and P_{Trans} but in turn contributing to congestion, while the proactive TMU primarily stays in management mode.

The above findings lead to the conclusion that for a more

TABLE III

PENALTIES FOR AVG VALUES OF ROUTER DELAY D_R, DELAY OF PACKET DELIVERY D_P, NET DATA THROUGHPUT $Data_{Net}$ AND THE NUMBER OF DELIVERED PACKETS P_{Trans} FOR DIFFERENT NoC SIZES COMPARED TO THE REFERENCE SYSTEM AND IMPROVEMENTS OF PROACTIVE MANAGEMENT COMPARED TO ITS REACTIVE COUNTERPART (CONFIGURATION: C1; MANAGEMENT TRAFFIC EXCLUDED; ABS VALUE / %)

	D_R[cycles]			D_P[cycles]			$Data_{Net}$[bits/cycle]			P_{Trans}		
	2×2	3×3	4×4	2×2	3×3	4×4	2×2	3×3	4×4	2×2	3×3	4×4
Reactive	12,6/ 224	19,2/ 335	27,2/ 398	77,3/ 168	1626/ 3212	≈95k/ ≈159k	1,8/10	2/5	9/13	≈140k/ 8,6	164k/ 4,6	≈803k/ 12
Proactive	12,5/ 223	13,7/ 238	24,2/ 353	69,8/ 152	101/ 200	≈15k/ ≈26k	2,4/ 13,3	5,9/15	12/17	≈205k/ 12,5	≈513k/ 14	≈1,1m/ 17
Improvement	<0,1/ <1	5,5/ 29	3/11	7,5/ 10	1525/ 94	≈80k/ 84	−0,6/ −32	−3,9/ −194	−3/ −30	≈−65k/ −46	≈−349k/ −212	≈−320k/ −40

TABLE IV

PENALTIES FOR AVG VALUES OF ROUTER DELAY D_R, DELAY OF PACKET DELIVERY D_P, NET DATA THROUGHPUT $Data_{Net}$ AND THE NUMBER OF DELIVERED PACKETS P_{Trans} IN A 3×3 NoC FOR DIFFERENT CONFIGURATIONS COMPARED TO THE REFERENCE SYSTEM AND IMPROVEMENTS OF PROACTIVE MANAGEMENT COMPARED TO REACTIVE MANAGEMENT (MANAGEMENT TRAFFIC EXCLUDED; ABS VALUE / %)

	D_R[cycles]				D_P[cycles]				$Data_{Net}$[bits/cycle]				P_{Trans}			
	C1	C2	C3	C4	C1	C2	C3	C4	C1	C2	C3	C4	C1	C2	C3	C4
Reactive	19,2/ 335	12,8/ 224	5,6/ 98	13,7/ 239	1626/ 3212	112/ 221	47/ 92	133/ 262	2/5	7,3/ 19	17,1/ 44	0,9/2	≈164k/ 4,6	≈643k/ 18	≈1,5m/ 42,3	≈62k/ 2
Proactive	13,7/ 238	6,6/ 115	6,1/ 107	10,8/ 188	101/ 200	52/ 103	50/ 99	77/ 152	5,9/ 15	14,9/ 38	16,7/ 43	3/8	≈513k/ 14	≈1,35m/ 38	≈1,5m/ 42,3	≈260k/ 7
Improvement	5,5/ 29	6,2/ 49	−0,5/ −10	−2,9/ −21	1525/ 94	60/ 53	−3/ −7	56/ 42	−3,9/ −194	−7,6/ −104	0,4/ 2	−2,1/ −239	≈−349k/ −212	≈−707k/ −110	−7/ <0,1	≈−198k/ −321

effective management the parameters need to be modified in order to reduce impairment of system performance, while still sustaining positive impact on temperature distribution. For this purpose, different configurations varying T_{Thresh}, T_{Bound} and ΔT_{Max} (see Table I) are applied for proactive and reactive management of a 3×3 NoC. The impact of the different configurations on average temperature and peak temperature difference is illustrated in Fig. 4, while the influence on D_R, D_P, $Data_{Net}$ and P_{Trans} is depicted in Table IV. For configuration C2 T_{Bound} is raised to 66 °C reducing the number of instruction packets for relocating a task from an IPC violating this boundary to the IPC with the lowest temperature. This measure considerably decreases the negative impact on D_R and D_P for both reactive and proactive management, indicating a relaxation of traffic load, with proactive management still outperforming the reactive approach. However, $Data_{Net}$ and P_{Trans} are additionally impaired, since due to relaxed traffic conditions the remaining monitoring and instruction packets reach their destinations much faster. This means that response times are shortened and adjustment measures can be applied more promptly leading to the observed effect and noticeably reduced temperatures as it is depicted in Fig. 4. In detail, proactive management achieves improvements of 150 % and 134 % for average temperature and peak difference compared to reactive management, corresponding to absolute values of nearly 1 °C and 3 °C. For configuration C3 ΔT_{Max} is raised to 2,0 °C. Again, this reduces D_R and D_P for both approaches. The moderate results for proactive management denote that a saturation seems to be reached, because due to the absence of monitoring packets reduced traffic load only affects response times. In contrast, reactive management benefits much more from this measure, since the reduction of traffic load not only influences response times but also transmission times of monitoring packets. Thereby, the time the TMU is in normal operation mode is shortened more drastically for reactive management than for the proactive approach. As it can be seen in Table IV, this leads to a distinct impairment of $Data_{Net}$ and P_{Trans} for reactive management, while proactive management exhibits only slight degradation. These circumstances are also reflected in the results for average temperature and peak temperature differences of the NoC. While reactive management achieves noticeable reductions (see Fig. 4), improvements for proactive management are only marginal. For configuration C4 T_{Thresh} is raised to 2,0 °C. This results in fewer monitoring packets for reactive management and therefore also in fewer instruction packets congesting the NoC, while for proactive management naturally only the latter are reduced. On the one hand, relaxed traffic conditions lead to increased values for $Data_{Net}$ and P_{Trans} partly almost achieving the level of the reference design. This can be attributed to the fact that both TMUs are in normal operation more frequently, improving data throughput and overall number of regular data packets. On the other hand, the increased amount of regular data crossing the NoC leads to exacerbation of D_R and D_P, especially as the remaining management instructions (e.g. DFS) still decrease performance and the routers' operating frequency has a major influence on the delay. Due to a lower sensitivity to temperature changes both approaches achieve the worst results for average temperature and peak difference using configuration C4, partly even acting counterproductive (i.e. temperature rise compared to reference). For both approaches this can be attributed to reactions to temperature changes,

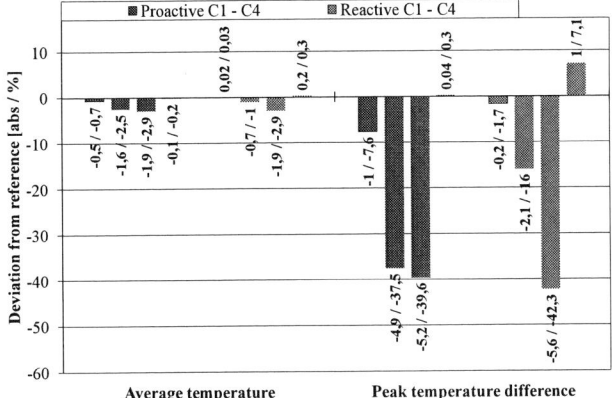

Fig. 4. Reduction of average temperature and peak temperature difference for different configurations normalized to the reference (abs values / %)

which apparently are scheduled too late and therefore apply instructions that are not suitable for the current situation.

The simulations conducted in this section show that parameters for thermal management, system performance, temperature distribution and NoC size are strongly correlated. Results clarify that management parameters have to be individually adapted to different NoC sizes in order to guarantee a certain level of system performance. This applies to both reactive and proactive management. Furthermore, it turns out that reactive management is applicable to large NoCs only to a limited extent, since the management traffic heavily congests the NoC and leads to unacceptable delays. Modification of management parameters facilitates adaptation of thermal management to the performance needs of the underlying system. Generally, results show that for both reactive and proactive management positive effects on temperature distribution can be traded off against performance. In case temperatures and a impairment of delay (D_R, D_P) are required to be as low as possible, parameters for task relocation need to be relaxed. In contrast, if high data throughput ($Data_{Net}$, P_{Trans}) is preferred, thresholds for the detection of temperature violations have to be increased. However, since proactive management exhibits shorter response times and dispenses with monitoring data provided by thermal sensors or probes, the impact of modifications can be predicted more accurately. Therefore, it stands to reason that proactive management reduces the number and the impact of side effects and interdependencies (e.g. negative of effect of raised T_{Thresh} on D_R and D_P), requires less effort for management (i.e. no monitoring data) and needs less fine adjustment of parameters.

V. CONCLUSIONS

In this paper a proactive approach for thermal management of NoCs is proposed. For this purpose, the NoC infrastructure is mapped on a network of RC-tiles. Thereby, the dualism of electrical and thermal energy flows can be exploited in order to model the thermal behavior of a NoC. The RC-model is used to simulate the proactive thermal manage-

ment of NoC-based systems executed by a central Thermal Management Unit (TMU). The TMU uses the temperature model to predict the temperature distribution and triggers appropriate measures, instead of relying on temperature values, which are transmitted by thermal sensors using the NoC. This contributes to the reduction of response times for thermal management and to the decrease of additional traffic congesting the NoC. Comparisons between proactive thermal management and an equivalent reactive implementation show improvements of nearly 134 % and 150 % regarding reduction of temperature imbalances and average temperature inside a 3×3 NoC, while lowering additional routing latency as well as packet transmission delay by more than 48 % and 53 % at the same time. Nevertheless, results for different configurations of management parameters show that in order to achieve practical advancements for on-chip temperature distribution, performance decreases have to be accepted. Furthermore, these parameters have to be individually adapted to particular NoC sizes to sustain sufficient system performance while applying thermal management.

REFERENCES

[1] Failure Mechanisms and Models for Semiconductor Devices, JEDEC publication JEP122F, March 2009
[2] Srinivasan, J., et al.: "RAMP: A Model for Reliability Aware Microprocessor Design", IBM Research Report, RC23048, 2003
[3] Skadron, K., et al.: "Temperature-Aware Microarchitecture, in Proc. of ISCA 2003
[4] Shang, L., et al.: "Thermal Modeling, Characterization and Management of On-Chip Networks", in Proc. of MICRO 2004
[5] Liu, W., et al.: "On-Chip Thermal Modeling Based on SPICE Simulation", in Proc. of PATMOS 2009
[6] Tockhorn, A., et al.: "Modeling Temperature Distribution in Networks-on-Chip using RC-Circuits", in Proc. of DDECS 2010
[7] Krum, A.,: "Thermal Management", in Keith, F., editor, The CRC handbook of thermal engineering, CRC Press, Boca Raton, FL, 2000
[8] Ciordas, C., et al.: "An Event-based Monitoring Service for Networks on Chip", in ACM TOADES 2005, vol. 10, no. 4, pp. 702-723
[9] Ciordas, C., et al.: "NoC Monitoring: Impact on the Design Flow", in Proc. of IEEE ISCAS 2006
[10] Guang, L., et al.: "Hierarchical Agent Monitoring Design Approach towards Self-Aware Parallel Systems-on-Chip", in ACM TECS 2010, vol. 9, no. 3
[11] Atienza, D., Martinez, E.: "Inducing Thermal-Awareness in Multicore Systems Using Networks-on-Chip", in Proc. of IEEE Computer Society Annual Symposium on VLSI 2009
[12] Coskun, A. K., et al.: "Proactive Temperature Balancing for Low Cost Thermal Management in MPSoCs", in Proc. of ICCAD 2008, pp. 250-257
[13] Tran, Xuan-Tu, et al.: "A DFT Architecture for Asynchronous Networks-on-Chip", in Proc. of ETS 2006
[14] Gag, M., et al.: "System Level Power Estimation of System-on-Chip Interconnects in Consideration of Transition Activity and Crosstalk", in Proc. Of PATMOS 2010, pp. 21-30
[15] Ye, T.T., et al.: "Packetization and routing analysis of on-chip multiprocessor networks", Journal of Systems Architecture, vol. 50, issue 2-3, February 2004, pp. 81-104
[16] Cornelius, C.: "Design of complex integrated systems based on networks-on-chip - Trading off performance, power and reliability", Dissertation, 2011
[17] IBM PowerPC 405 CPU Core Product Overview, https://www-01.ibm.com/chips/tech\-lib/tech\-lib.nsf/products/Power\-PC$_$405$_$Embedded$_$Cores
[18] OSCI Core SystemC Language 2.2, http://www.systemc.org/
[19] OSCI SystemC Transaction-Level Modeling Library 2.0.1
[20] OSCI SystemC Analog Mixed Signal Extensions 1.0

978-1-4577-0671-4/11 $26.00 © 2011 IEEE

Bringing Network-on-Chip Links to 45nm

Marco Ferraresi, Giuseppina Gobbo, Daniele Ludovici, Davide Bertozzi
ENDIF, University of Ferrara, 44100 Ferrara, Italy.

Abstract

The literature lacks of a comprehensive overview of achievable NoC link performance when key parameters are swept in the link microarchitecture and in the NoC floorplan. This paper bridges this basic gap while at the same time capturing how link performance is affected by the migration from a 65nm to a 45nm technology node. Finally, it identifies the requirements on EDA tools to keep up with the technology scaling.

I. Introduction

Networks-on-chip (NoCs) build up the communication backbone of virtually all large-scale system-on-chip designs planned for 45nm and below. By removing the dependence of such systems on global on-chip interconnects through path segmentation and retiming, NoCs have relieved the concerns associated with the inverse scaling of interconnect delay in nanoscale technologies.

However, physical synthesis runs qualifying NoC implementation properties on more advanced technology nodes are bringing the interconnect delay concern again to the forefront. In fact, the critical path of the network tends to shift from network internal logic to switch-to-switch links [19], [15]. The higher impact of this trend is expected on embedded computing platforms, where very simple switch architectures are generally designed to meet tight resource budgets [8], [20]. As an effect, performance of an entire network ends up depending on the delay of switch-to-switch links in most cases. Since for the embedded computing domain a full-custom design of such links is often unaffordable, it is very important to understand how effectively automated design tools can handle the synthesis of those links across a wide range of microarchitectural (e.g., repeater insertion, link pipelining) and layout (e.g., link length, non-routable obstructions) parameters as well as synthesis constraints (e.g., target speed). Even more important is to capture how and to which extent quality metrics of NoC links are impacted when moving from current technology generations to newer ones featuring a more aggressive scaling of feature sizes. In fact, while gate delay

certainly benefits from such scaling process, the reverse scaling of interconnect delay paints an uncertain picture for wire-dominated designs like on-chip interconnection networks. Minimizing the implications of this effect on the performance of the network as a whole depends not only on the physical properties of the interconnect, but also on the interaction between them, the optimization strategies of the place&route tool and the mix of cells in the technology library (essential for repeater insertion and link pipelining).

This paper aims at an extensive parametric exploration of physical synthesis properties of NoC links, both for a given technology node (in this case, in 65nm) and across technology nodes (from 65 to 45nm). It is of special interest to this paper to assess whether challenging layout and synthesis constraints as well as scaled technology nodes degrade NoC performance irreparably and/or whether this latter can be restored at a prohibitive cost in terms of resources and/or enabling special features in the EDA toolflow. The more specific questions this paper helps to answer are the following:

- What is the maximum performance that can be achieved by NoC links?
- To what extent and how can link performance be preserved when scaling from 65nm to 45nm technology?
- How predictable is link design (i.e., how closely does post-layout delay track the target delay constraint of the physical synthesis) across technology nodes?
- How sensitive is link performance to the size of the routing channel?
- How many pipeline stages are needed to materialize a given delay on a link of given length?
- How does the need for pipeline stages grow on long links and across technology nodes?

To gain realistic characterization data and scalability insights, this work makes use of a commercial synthesis toolflow and of standard cell libraries from the same technology provider.

II. Related Work

For a 65nm technology, [16] points out the significant gap between post-synthesis and post-place&route performance reports affecting NoC modules when logic synthesis and placement are carried out as two clearly separated stages. Inaccurate wire load models are at the root of this gap, therefore placement-aware logic synthesis tools are advocated. Moreover, although global wires are intrinsically segmented, the maximum interswitch link length still plays a key role in topology design [18] and [19].

In [2] the link design has been studied with a particular emphasis on the bandwidth requirements. Authors in [3] show that the maximal data rate of a NoC link is achieved near the boundary between RC and RLC model validity domains. [4] proposes a timing error tolerant communication system to aggressively design NoC operating speed.

Authors in [5] propose a synthesis approach that utilizes the floorplan knowledge of the NoC to detect timing violations on the NoC links early in the design cycle.

In [6] authors describe several architectural level techniques for link area reduction and a low-voltage link implementation for reducing power consumption. Authors in [13] propose a variable frequency link for power-aware NoC tuned to link utilization.

Authors in [7] propose an equalization technique along with special spacing rules for improving the communication speed and reliability of NoC links.

Other studies in literature address process variation and variability problems of the NoC link. In this direction, the work in [9] proposes a self calibrating link. Another effort in the same field is represented by the work in [11] where authors propose a methodology for the characterization of process variations. Same authors in [12] propose also a new mechanism that deals with process variations in NoC links.

In the 3D stacking domain, an increasing interest arose around fault tolerant link design, like in [10]. The same authors in [14] propose a new physical routing approach for aggregating NoC wires in order to achieve a more controllable and predictable switch–to–switch channel.

As witnessed by the previous work reported in this section, there is an increasing interest in designing and accurately characterizing NoC links. The main reason is that such links, although featuring very simple point-to-point structure, are emerging as key contributors to overall network speed and power and important elements for network reliability. Therefore, it is desirable not to have link quality metrics assessed for specific case studies but explored in a more comprehensive and systematic way (which is the contribution of this paper) and by means of industry-standard design toolflows and libraries.

III. Experimental setup

In order to explore link properties (such as performance and area) as a function of physical synthesis parameters and constraints, we reverted to a flexible experimental setup consisting of a 2-switch test architecture (see Figure 1) with tunable spacing. Due to the modular structure of NoCs, this sub-system enables to capture both link-

level properties and to project implications on the network topology as a whole.

We also include the possibility to instantiate an arbitrary number of pipeline stages along the link itself. Of note, following the trend in [21], [17], a pipeline stage in a NoC environment is not only a simple retiming stage such as a register but also an actual flow control stage that regulates lossless data progress between any two consecutive stages.

Therefore, it is of utmost importance to consider the area implications of such insertion in the link area computation, in addition to the latency implications. For this reason, we developed an analysis framework for the automatic extraction of link area information (both simple buffers/inverters and pipeline stages). Such framework leverages a sign-off tool (i.e., Synopsys PrimeTime) and an in-house made parser.

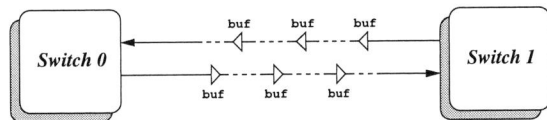

Fig. 1. Simple and flexible switch–to–switch test system to analyze NoC link properties.

In this work we utilized two different standard cell libraries provided by the same industrial technology provider:

- low-power low-Vth 65nm
- low-power std-Vth 45nm

Both libraries share the same nominal voltage, the same process corner (nominal) and the same target (a low power process). They only differ in the feature sizes and in the threshold voltage, but are homogeneous with respect to all other parameters (including number of cells, mix of cell types and set of available driving strengths per cell). The reason behind this choice was that early logic synthesis runs of NoC components (switches and switch sub-blocks) with the two libraries above provided approximately the same critical path delay. More precisely, the std-Vth 45nm library proved only up to 5% faster than a low-Vth 65nm (in practice, feature size and threshold voltage effects offset each other), while it was almost 35% faster than the corresponding std-Vth 65nm. Therefore, *by selecting the above libraries, buffers and inverters for use in NoC links feature approximately the same delay, thus pointing out the neat effect of on-chip interconnects and of their parasitics in 45nm over link quality metrics.* Experimental results on scalability analysis should be read with this choice in mind: results for the 45nm node are pessimistic (in relative terms), and the key take-away we will be aiming at is to understand whether link performance is irreparably degraded moving to 45nm, especially under tight synthesis and layout constraints.

Once the simple system of Figure 1 has been instantiated, placement-aware logic synthesis is performed through

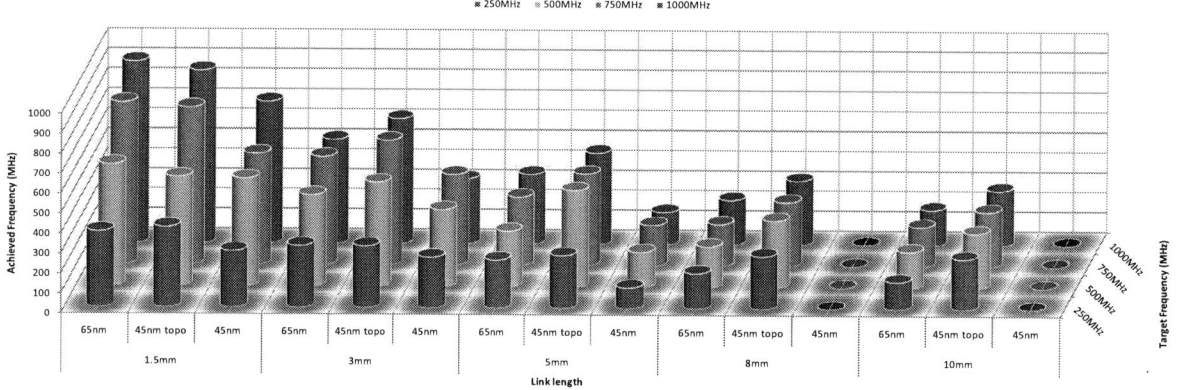

Fig. 2. Switch–to–switch link performance characterization (post-layout results)

Synopsys Physical Compiler. After a very quick initial logic synthesis based on wireload models, the tool internally attempts a coarse placement of the current netlist. Next, it iteratively optimizes the netlist and the placement, based on the actual wire loads implied by the current candidate placement. The outcome is a placed netlist that is optimized also accounting for wire delays, thus mitigating the risk of large deviations between post-synthesis and post- place-and-route performance (design predictability). The final place-and-route step is performed with Cadence SoC Encounter.

Fences are defined to limit the area where the cells of each switch can be placed. Subsequently, the tool automatically places cells without trespassing the fences. At the same time, fence positioning (combined with place&route re-iteration) enables to span different inter-switch spacing scenarios.

The next step is clock tree synthesis. The two switches of the on-chip network under analysis are considered as a single clock domain with a unique clock tree. At this point, the power supply nets are added. We choose the power grid scheme, which distributes power nets from the topmost metal layer of the chip and which minimizes IR drops. After the power nets have been routed, the tool begins to route the logic wires. After an initial mapping, search and repair loops are executed to fix any violations. As a final step, post-routing optimizations are performed, including crosstalk and antenna effect minimization. Finally, a sign-off procedure is run by Synopsys PrimeTime to accurately validate the timing properties of the design.

As an alternative to the aforementioned synthesis flow, in some cases (illustrated later on) we experimented a new approach available in some commercial tools named topographical synthesis. By leveraging such method, it is possible to perform an initial coarse place&route and to generate a DEF file with the exact final cells position. Next, such file containing physical information is reimported in the topographical synthesis tool that is now capable of fully exploiting the exact final position of the logic cells and of sizing their driving strength accordingly.

IV. Inter-Switch Link Delay with Repeater Insertion

The first scenario we analyzed consists of link inference with repeater insertion. By means of many physical synthesis iterations, for a given inter-switch spacing we varied the target frequency (i.e., the speed constraint the tool attempts to meet) and measured the actual achieved speed and cell area on the post-layout netlist with annotated parasitics. The same set of experiments is performed for both the 65nm and the 45nm libraries, and results are reported in Figure 2,

Clearly, the overall performance of the link degrades (by looking the graph from left to right) by incrementing the inter-switch spacing. By looking at the *65nm* row only, it should be observed that even a loose target of 250 MHz cannot be achieved for 8mm links. At 500 MHz, 3mm is the maximum link length while 1GHz is hardly affordable at 1.5mm. These results stress the need for careful engineering of NoC links for high performance applications and point out the infeasibility of many design points (i.e., target speed higher than achieved speed).

By comparing the *65nm* and *45nm* row in each cluster, it is evident that the synthesis tool is not able to materialize the performance achieved with the 65nm library, even for the shorter links. To investigate the root cause for this, we grouped in Figure 3 the logic cells used for a 5mm link inference (synthesized at the target frequency of 1GHz) in each technology based on their driving strength. Cells with the same driving strengths are available in both libraries, but the place&route in 45nm has required a much higher number of those cells, clearly highlighting the effort of the place&route tool in meeting the desired link

978-1-4577-0671-4/11 $26.00 © 2011 IEEE

performance. As a result, physical properties of on-chip interconnects in 45nm are responsible for the observed performance degradation. Clearly, this routing challenge cannot be addressed only by inserting more buffering elements as an afterthought, but the logic synthesis tool should be aware of it from the ground up.

For this reason, we experimented the *topographical synthesis* approach on the 45nm library. It enables the tool to consider the future cell position when synthesizing a logic function (a DEF file generated with a floorplanner must be imported first). The possibility to have floorplanning insights in advance is key to minimize the critical path delay of the link.

By comparing in each "link distance" cluster in Figure 2 the *45nm topo* and *45nm* rows, it is clear that a neat performance speedup is achievable by utilizing a topographical synthesis approach. Performance now tracks results of the 65nm library, and even outperforms them for long links and aggressive target speeds, pointing out that awareness of back-end information for an optimized logic synthesis should be considered an option in 65nm and a feature that cannot be renounced in 45nm.

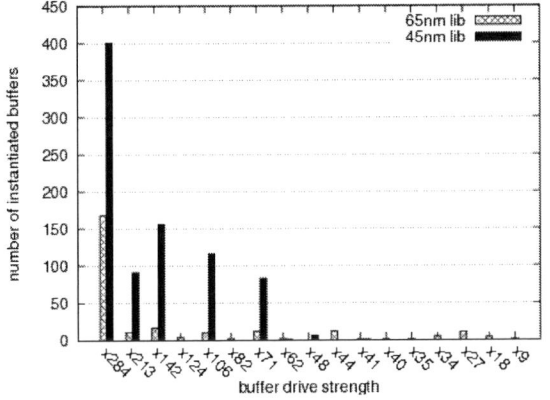

Fig. 3. 5mm Link buffer distribution.

A. Switch–to–switch link obstruction

In order to model a more realistic scenario, the switch–to–switch system of the previous experiments has been studied with a further constraint: the definition of non-routable obstructions. Such obstructions are placed above and below the routing channel and make it progressively narrower. This experiment captures the intricacy of routing many wires on a narrow channel and of positioning repeaters at an ideal spacing in the same conditions. The indications of this experiment can help the designer to avoid expensive oversizing or reserving insufficient channels. As demonstrated in the case study in [15], in this latter case repeaters might be inserted while not materializing any link speed-up. Hence, design curves correlating such speed-up to the routing channel width are essential.

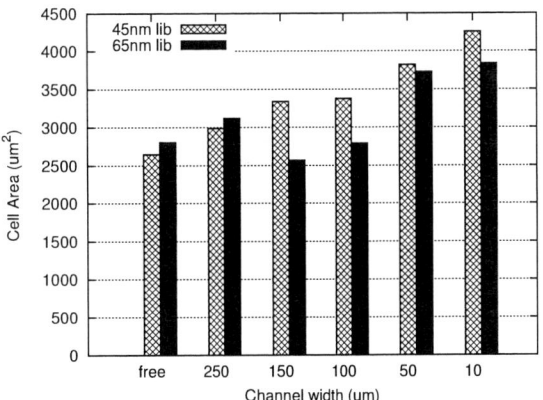

Fig. 4. Link area as a function of channel width at a 3mm switch–to–switch spacing.

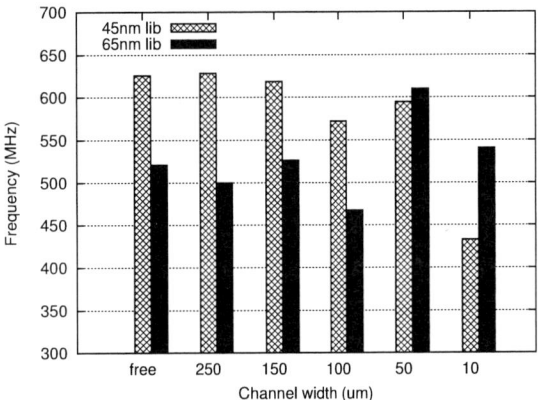

Fig. 5. Link speed as a function of channel width at a 3mm switch–to–switch spacing.

When looking at the 65nm link area results in Fig.4 as a function of the routing channel width, it can be observed that beginning from 150um down area requirements clearly grow. The place&route tool has less margins to perform routing and buffer placement and more/larger buffers are instantiated in an attempt to meet the timing constraints. Nonetheless, optimizations performed by the tool prove not capable of converging to a stable trend of the speed (see Fig.5), which oscillates around 500 MHz. Above 150um, with theoretically more spacing to route and to instantiate buffers at intended places, the tool suffers from the larger design space which challenges the heuristic algorithm it implements. This is proved by the relatively stable speed achieved for those channel widths.

When it comes to the 45nm library (topographical synthesis), the trend turns out to be more regular: cell area linearly increases with channel reduction and operating speed decreases beginning from a 150um channel. This is due to the lower area footprint of 45nm cells, especially

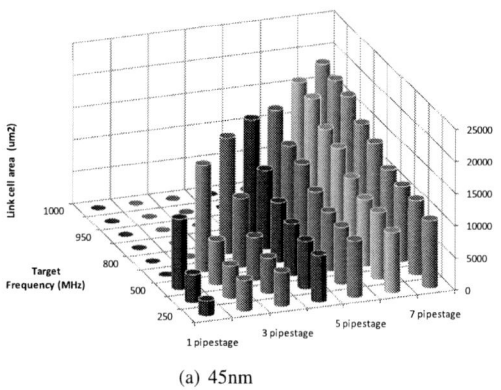

(a) 45nm

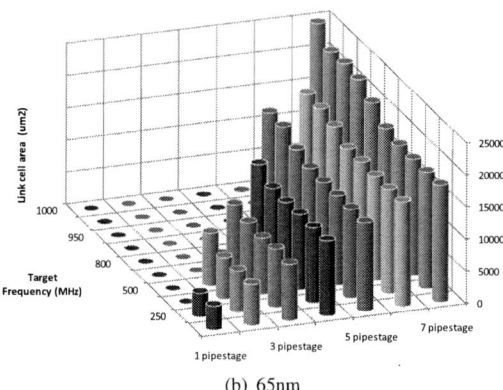

(b) 65nm

Fig. 6. Total link area with pipeline stage insertion at 8 mm.

when compared to the not scaled size of the routing channel: this plays in favor of tool optimizations.

In any case, the above plots point out the significant area penalty of incorrectly sizing the routing channel (too narrow but also too large) or of having it overly constrained by other floorplan blocks. In 65nm technology, an overhead of almost 1000um2 may be incurred, which in a 4x4 mesh topology globally amounts to the area of an extra switch. Another interesting conclusion is that switch height is not necessarily the best sizing option for the routing channel[1].

B. Link pipelining

In this section we study the impact on the performance and the link buffer area of inserting pipeline stages in the simple switch–to–switch system analyzed throughout this paper. First of all, we determine the required number of pipeline stages to break the switch–to–switch critical path, by performing a series of incremental place&route steps. We stop iterating when a conventional (high) target speed of 1 GHz is achieved on the link. The topographical synthesis approach was used for 45nm. Pipeline stages are inserted manually so to break the link into segments of equal length. If we let the tool handle this, all pipeline stages are inserted in the middle of the link, thus vanishing their benefits.

Library	1.5mm	3mm	8mm	10mm
45nm	1	2	7	9
65nm	1	2	7	8

TABLE I. Required number of pipeline stages at various link lengths.

Table I reports the number of pipeline stages needed to sustain the target speed in both 65nm and 45nm.

[1]For the sake of comparison, switch height suggested by Synopsys Physical Compiler for floorplan definition resulted approximately 250um in 65nm technology and 180um in 45nm

Interestingly, the trend is the same for short-to-medium links, thus proving the effectiveness of the topographical synthesis approach in preserving link speed at the scaled technology node. Only for a 10mm link an additional pipeline stage is required in 45nm with respect to 65nm (9 vs 8), however these are highly unlikely wire lengths in commonly used NoC topologies.

We then performed an experiment to assess whether link pipelining can relieve repeater insertion in the intermediate segments to such an extent that meeting a target speed with more pipeline stages on the link results into area savings.

We experimentally assessed that in the 65nm technology this is not the case: the area overhead for a pipeline stage is such that buffer/inverter area of repeaters is marginal. This can be observed from Fig.6(b), which is referred to a link length of 8mm. In this figure, for each number of pipeline stages on the link, we compute the maximum performance, and from there we iterate place&route decrementing the target speed of the whole design. This way, repeaters are taken away from link segments and driving strengths are relieved, till the area saturates to its lower bound. In 65nm, a given target speed should be achieved with the minimum number of pipeline stages.

However, when we moved to 45nm, we collected the results in Fig.6(a). In contrast to the 65nm technology, the design points close to timing violations are inferred with large use of repeaters, confirming the results in section IV and denoting tougher convergence of the tool. Therefore, by inserting one pipeline stage more and demanding the same target speed (e.g., 500 MHz with 1 and 2 pipeline stages), this latter can be materialized at a lower area cost. As a consequence, the designer can choose between two design points spanning the performance-area trade-off.

V. Conclusions

Little work has been devoted so far to understanding the technical issues of implementing a NoC in a 45nm environment. This paper was aimed at shedding light on the challenges and benefits of NoC-based interconnect design at this technology node, especially when compared with the previous node at 65nm. We came to the following conclusions:

- Scaling deeper into the nanoscale regime urges the migration to more refined logic synthesis tools, able to generate results that more closely correlate to physical implementation and therefore reduce iterations between synthesis and physical tools. While these tools (or tool features) are an option in 65nm, their use becomes mandatory in 45nm.

- By reverting to topographical synthesis, link performance in 45nm can be preserved, if we offset logic gate speed-up with standard- vs low-Vth transistors. In other words, NoC link delay will not certainly scale as an effect of the reverse scaling of interconnect delay, but at least it will be kept under control and the same performance achievable in 65nm will be affordable also in 45nm. Without topographical synthesis, this result is not within reach of the 45nm technology.

- As the routing channel between neighboring switches becomes narrower because of non-routable obstructions in the floorplan, the place&route tool is forced to insert more buffers to deal with the more constrained design. This process results into larger cell area in the link and into lower operating speed, since ideal spacing of repeaters is not feasible any more. Incorrect sizing of the channel may lead to a relevant waste of buffering area which may even not materialize the expected multiples in link performance.

- With topographical synthesis, the number of pipeline stages needed to achieve a given speed at a specific link length is the same for 65nm and 45nm. Only for extreme design points (10mm links) there is a marginal difference. Also, since design points close to timing violations are inferred with an extensive use of buffering, the insertion of a pipeline stage can significantly relieve the use of repeater insertion, thus coming up with a more area saving solution (although more costly in terms of latency).

Overall, while NoC link performance can be preserved in 45nm with suitable upgrade of the physical synthesis toolflow, it does not certainly scale like logic gate delay, therefore NoC links remain potential bottlenecks for the performance of the NoC as a whole.

Acknowledgements

This work has been partially supported by the NaNoC European Project (FP7-ICT-248972), by the Hipeac Network of Excellence (Interconnect Cluster).

References

[1] Circuits Multi-Projects, Multi-Project Circuits; http://cmp.imag.fr

[2] Junbok You ; Gebhardt, D. ; Stevens, K.S. ; "Bandwidth optimization in asynchronous NoCs by customizing link wire length", Proc. of ICCD, pp.455. 2010.

[3] Barger, A. ; Goren, D. ; Kolodny, A. ; "Simple Design Criterion for Maximizing Data Rate in NoC Links", Proc. of IEEE workshop on Signal propagation on Interconnects, pp.149,2006.

[4] Tamhankar, R.R. ; Murali, S. ; De Micheli, G. ; "Performance driven reliable link design for networks on chips", Proc. of DAC, pp.749, 2005.

[5] Murali, S. ; Atienza, D. ; Meloni, P. ; Carta, S. ; Benini, L. ; De Micheli, G. ; Raffo, L. ; "Synthesis of Predictable Networks-on-Chip-Based Interconnect Architectures for Chip Multiprocessors", IEEE Transactions on VLSI, Vol. 15:8, pp.869, 2007.

[6] Donghyun Kim ; Kwanho Kim ; Seung-Jin Lee ; Hoi-Jim Yoo "Solutions for Real Chip Implementation Issues of NoC and Their Application to Memory-Centric NoC", Proc. of NOCS, pp.30, 2007.

[7] Lei Li ; Jianhao Hu ; Chun He ; Wanting Zhou ; "Joint equalization technique and special spacing rules for link design", Proc of ICCCAS, pp.1042, 2009.

[8] S.Stergiou et al., "Xpipes Lite: a Synthesis Oriented Design Library for Networks on Chips", Proc. of DAC, pp.559-564, 2005.

[9] Medardoni, S.; Lajolo, M; Bertozzi, D.; "Variation Tolerant NoC design by means of self-calibrating links", Proc of DATE, pp.1402, 2008.

[10] Loi, I. Angiolini, F. Fujita, S. Mitra, S. Benini, L.; "Characterization and Implementation of Fault-Tolerant Vertical Links for 3-D Networks-on-Chip", IEEE Transaction on Computer-Aided Design of Integrated Circuits and Systems, Vol 30:1, pp. 124, 2011.

[11] Hernandez, C.; Silla, F.; Duato, J.; "A methodology for the characterization of process variationin NoC links", Proc. of DATE 2010.

[12] Hernandez, C.; Silla, F.; Santonja, V.; Duato, J.; "A new mechanism to deal with process variability in NoC links", Proc. of IPDPS 2009.

[13] Lee, S. E.; Bagherzadeh, N.; "A variable frequency link for a power-aware network-on-chip (NoC)", Published in the VLSI journal, Vol. 42:4, 2009.

[14] Kakoee, M. R.; Loi, I.; Benini, L.; "A new physical routing approach for robust bundled signaling on NoC links", Proc. of GLSVLSI, 2010.

[15] D. Ludovici, G. N. Gaydadjiev, D. Bertozzi, L. Benini, "Capturing topology-level implications of link synthesis techniques for nanoscale networks-on-chip", Proc of GLSVLSI, pp.125–128, ACM, 2009.

[16] A. Pullini, F. Angiolini, S. Murali, D. Atienza, G. De Micheli, L. Benini, "Bringing NoCs to 65 nm". IEEE Micro Special Issue on Interconnects for Multi-Core Chips, 27(5):75–78, 2007.

[17] A. Pullini, F. Angiolini, D. Bertozzi, L. Benini, "Fault Tolerance Overhead in Network-on-Chip Flow Control Schemes", Proceedings of SBCCI, pp. 224-229, 2005.

[18] F. Gilabert, S. Medardoni, D. Bertozzi, L. Benini, M. E. Gómez, P. López, J. Duato; "Exploring High-Dimensional Topologies for NoC Design Through an Integrated Analysis and Synthesis Framework", Int. Network-on-Chip Symp., pp.107-116, 2008.

[19] D. Ludovici, F. Gilabert, S. Medardoni, C. Gómez, M. E. Gómez, P. López, D. Bertozzi, G. N. Gaydadjiev; "Assessing Fat-Tree topologies for Regular Network-on-Chip Design under Nanoscale Technology Constraints", Proc. of DATE 2009.

[20] M. N. Horak, S. M. Nowick, M. Carlberg, U. Vishkin, " A Low-Overhead Asynchronous Interconnection Network for GALS Chip Multiprocessors", Proc. of NOCS, 2010.

[21] G. Michelogiannakis, J. Balfour, W. J. Dally "Elastic-Buffer Flow Control for On-Chip Networks", Proc. of HPCA-15, 2009.

978-1-4577-0671-4/11 $26.00 © 2011 IEEE

Synchronizing Distributed State Machines In A Coarse Grain Reconfigurable Architecture

Omer Malik
School of ICT
Royal Institute of Technology, KTH
Stockholm, Sweden
Email: omerm@kth.se

Ahmed Hemani
School of ICT
Royal Institute of Technology, KTH
Stockholm, Sweden
Email: hemani@kth.se

Abstract—This work presents methodology for synchronizing distributed FSMs (Finite State Machines) which are generated while implementing different algorithms on a coarse grain reconfigurable architecture. These FSMs interact with each other while executing algorithms and they are dependent upon each other; thus they need to be synchronized with each other for performing correct execution. The algorithms presented in this paper makes appropriate use of different strategies available for synchronizing these FSMs. The tool hides all sorts of low level details from the Programmer. It lets the designer focus on the details of algorithm (at higher level of abstraction) and cycle by cycle timings are resolved automatically.

Index Terms—System Level Synthesis; Distributed FSMs;

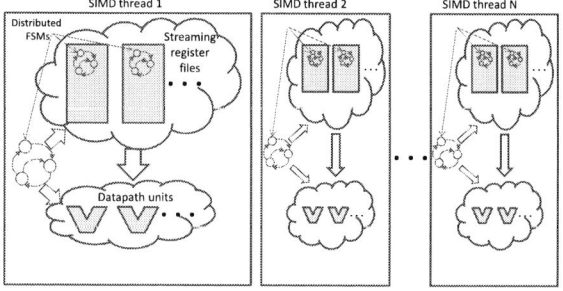

Figure 1. Multiple SIMD threads with distributed FSMs

I. INTRODUCTION

DRRA(Dynamically Reconfigurable Resource Array) is a CGRA (Coarse Grain Reconfigurable Resource Array) organized as regular fabric of morphable datapath units, streaming register files, reconfigurable circuit switch elements and sequencers that enables implementation of local and distributed programmable FSMs that can be chained and/or organized in a hierarchy.

DRRA targets hosting multiple complete DSP sub-systems like modems and codecs. The DRRA fabric can be partitioned dynamically to host specific DSP sub-systems and each such partitions can be sub-partitioned for individual DSP algorithms in the DSP sub-system.

DRRA in general adopts a distributed control model when implementing algorithms. Distributed control logic requires synchronization amongst the individual controllers. DSP algorithms in DRRA are implemented in varying degrees of parallelism in terms of multiple SIMD threads and each SIMD thread in turn has a sequencer, a set of morphable Data Path Units and a set of streaming register files as shown in Figure 1. The streaming register files have programmable FSMs for streaming data with programmable delays and an algorithm can be viewed as consisting of several co-operative FSMs. Depending upon the algorithm they might be interacting with each other or working in parallel. This paper deals with the problem of synchronization amongst distributed controllers when implementing individual DSP algorithms.

Consider the scenario presented in Figure 2, where F represents the set of interacting FSMs for computing an application. The execution of these FSMs are dependent upon each other

and are denoted by set of dependencies D. These dependencies are resolved by some methodology and the output is generated in the form of synchronized interacting set of FSMs F'.

VESYLA is a HLS (High Level Synthesis) like tool used to implement parametric models of commonly used DSP functions [1]. The parameters not only specify the dimensions but also specify the degree of parallelism. Unlike traditional HLS tools, VESYLA generates a distributed control logic and as a results requires synchronization amongst the distributed FSMs. This paper deals with the synchronization problem in VESYLA HLS flow. VESYLA hides all sorts of low level details from the Programmer. It lets the designer focus on the details of algorithm (at higher level of abstraction) and cycle by cycle timings are resolved automatically.

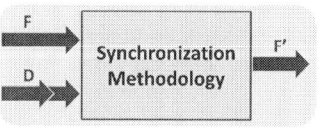

Figure 2. Set of Interacting FSMs with Dependencies

In Section II we review the related work, Section III briefly presents the DRRA fabric, Section IV describes VESYLA, Section V presents the core contribution of this work (Synchronizing distributed FSMs using VESYLA), Section VI presents the experimental results and lastly Section VII presents the conclusions and future work.

978-1-4577-0671-4/11 $26.00 © 2011 IEEE

II. RELATED WORK

In this section we will review some industry standard HLS tools and few schemes for compiling applications on CGRAs and FPGAs.

[2] presents a HLS like technique to map algorithms on a CGRA. Loop un-rolling techniques are applied on the given C code to extract parallelism and then HLS is used to perform scheduling and binding.

[3] is mapping applications on a generic RTL processor architecture. The methodology is based on interactive design environment which enables the designer to gain full control over the design process. The designer can change the synthesis decisions about scheduling, allocation and binding by using a GUI (Graphical User Interface) at any stage.

[4] is HLS tool that takes C code as input and extract parallelism after generating the DFG (Data Flow Graph). Designer can provide/choose constraints for allocation, binding and scheduling through a GUI and the tool generates the target RTL which can also be synthesized on industry standard FPGAs.

In Catapult synthesizes, algorithms are described in ANSI C++ [5]. It allows designers to manipulate various constraints for design space exploration. Designer can view the impact of these constraints and can modify them accordingly. These constraints are applied at different steps of synthesis procedure for loop unrolling, loop parallelization, resource allocation, scheduling etc. Tool generates targeted RTL after considering optimal solution.

Cynthesizer generates RTL from an untimed SystemC code [6]. Designer provides constraints for loop unrolling, clock speed etc. and also specify the target technology file for FPGA implementation.

[7] exploits loop level parallelization for mapping applications on ADRES architecture. Loop level parallelization is achieved by using Modulo scheduling algorithm to achieve optimal performance.

[8] is based on GCC compiler framework and maps C code on the reconfigurable architecture. The goal is to identify and map the most common and regular instructions onto a RFU (Reconfigurable Functional Unit).

[9] transforms application code in to a DFG (Data Flow Graph) and coarse grained operations in the form of hyperops are mapped to the CGRA to achieve code acceleration.

A low level programming language model is proposed in [10] for mapping applications on their respective CGRAs. The disadvantage is that programmer has to micro manage all the details.

Our design flow is also similar to traditional HLS tools but most HLS tools assume a single controller in form of a single FSM, while our tool synthesizes multiple FSMs implemented in the micro-coded sequencer and these multiple FSMs interact with each other at various stages of program execution.

Designer explicitly specify spatial and temporal pragmas to identify the micro architecture implementation on DRRA using VESYLA [1]. By using this technique, the search

space for the tool is reduced as compared to other HLS techniques, which may fail to efficiently make use of the available resources in CGRAs (because they have to cope with large search space).

III. DRRA ARCHITECTURE

In DRRA all resources are organized as a regular, seamlessly connected fabric. DRRA PHY resources, shown in Figure 3 are:

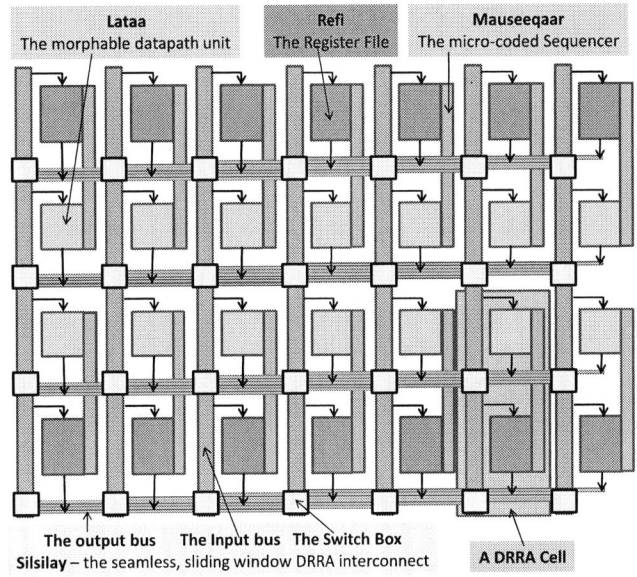

Figure 3. The DRRA PHY layer fabric fragment

a) mDPUs (morphable Data Path Units)[11] are native 16 bit integer units with four 16b inputs and two 16b output. mDPU consist of an arithmetic and logical section. Arithmetic part provides a) MAC, with internal and external accumulation, b) butterfly operations c) add, subtract trees - 4/2 input adders/subtractors, sum-of-difference etc. Logical part consists of logical instructions and single bit shift operations. mDPU can do saturation, truncation/rounding, overflow, underflow check.

b) RFILE - the DRRA Register File [12] is 64 word 16 bit register file with dual read/write ports. RFILE has a AGU (Address Generation Unit) with vectorized, circular buffer and bit reverse addressing. Each of these modes can be executed once or in an endless loop. AGU control these modes by using (initial, middle and repetition) delays. Initial delays are used to delay the start of address generation. Middle delays add a delay between successive addresses and Repetition delays repeat an instruction after some known time intervals. These delays are used for synchronization among DSP functions, I/O and DSP rate change functions. These delays can by dynamically computed at runtime and create elastic streams. Our methodology to synchronize the distributed FSMs is based on successfully computing and allocating the values of these delays to AGU.

c) SEQUENCER is a Micro-coded hierarchical sequencing machine which controls all the resources in a DRRA cell. It sends instructions to the AGUs of the RFILEs, selects mDPU modes, and configure the INTERCONNECTs for proper operations. SEQUENCER also output signals to other SEQUENCERS to/for control communication.

d) INTERCONNECT, the seamless sliding window circuit switched interconnect fabric connects the fabric of mDPUs, RFILEs and SEQUENCERs organized in 2 rows as shown in Figure 3. A circuit switch interconnect, at the intersection of horizontal and vertical buses, controlled by SEQUENCER, selects the programmed outputs from the horizontal buses and loads them onto the vertical buses that feeds the inputs of mDPUs and RFILEs [13]. Each column can reach 3 columns to the right and to the left.

IV. VESYLA

VESYLA is HLS like tool for mapping algorithms on DRRA. VESYLA takes input in form of a C programs and it generates configware that constitutes the FSMs to implement the control flow in the algorithm and the streaming register files to transact data at right time with right periodicity.

A. Overview of DRRA Hierarchical Compiler

We propose developing a library of commonly used signal processing functions using VESYLA. The elements of these libraries are the DRRA implementations of various functions, where each implementation have different architectural style, variation in degree of parallelism and different average energy, latency and area costs. DRRA hierarchical compiler uses a simulink model of an application and it identifies coarse-grained functions with in that application. The DRRA compiler resolves the functions in simulink models with the functions available in DRRA library and if there is a match it records the implementation space - the sweep of architectural styles and degree of parallelism. Once the DRRA compiler knows the possible function implementations, it has to map the application optimally to best meet the global energy and performance constraints (by choosing the optimal function implementation) [14].

B. Overview of VESYLA

Figure 4 shows the VESYLA framework. We will briefly describe the complete framework, but the focus of this paper is the methodology for synchronizing FSMs (box shown in bold dotted lines).

Algorithmic developer is responsible for defining allocation and binding pragmas[15] [1] in the C code. These pragmas sweep the architectural space in terms of degree of parallelism [15]. A unscheduled CDFG is generated after number and types of resources are identified. The CDFG is transformed into a scheduled CDFG during the scheduling phase (Scheduling algorithm is not the focus of this paper and is not discussed here). Data and Resource dependencies are also resolved during this phase. Finally, it synchronizes the FSMs for implementing the control flow. After resolving all issues

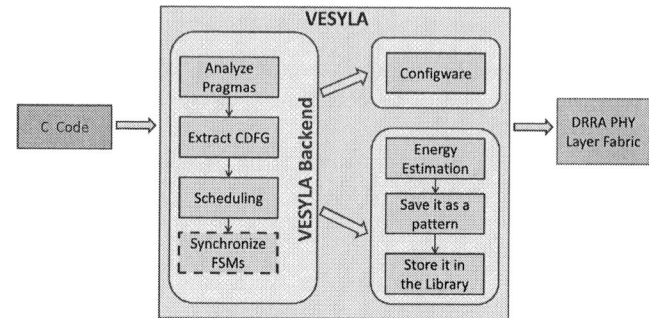

Figure 4.   Overview of VESYLA framework

related to binding, allocation, scheduling and synchronization, it generates configware for programming DRRA. This configware constitutes of distributed control that sequences the design and controls functional, storage units in datapath.

VESYLA serves dual purpose. Apart from generating configware for DRRA, it can build the libraries (discussed in section IV.A) and can also predict the average energy required for implementing any particular style. In other words user can write an algorithm in C language and VESYLA will generate configware for DRRA and will also inform the user about the average energy consumed by that particular implementation [1].

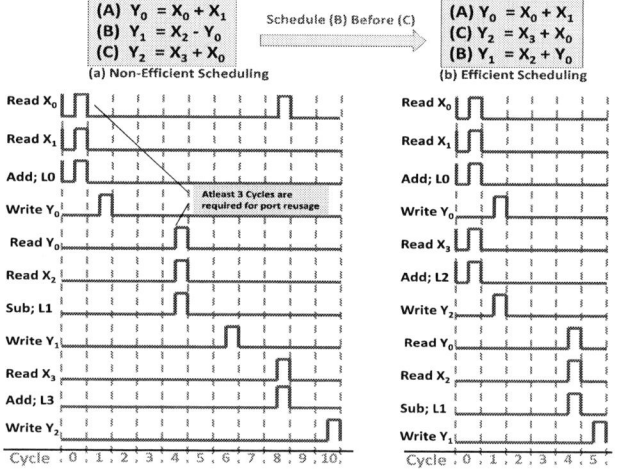

Figure 5.   Scheduling Example

One should consider scheduling and synchronization as two different steps. Consider the algorithm presented in Figure 5(a), which executes three statements (A), (B) & (C). In terms of DRRA component, variables and operators are mapped on to a RFILE and mDPU respectively. Every operator / variable is assigned a unique resource except X_0 and Y_0, which are using the same ports of same RFILE. This condition implies that Read X_0 and Read Y_0 cannot be executed in parallel. This condition leads to a non-efficient scheduling algorithm if the

978-1-4577-0671-4/11 $26.00 © 2011 IEEE   130

statements are executed in the order as shown in Figure 5(a). This non-efficient scheduling scheme can be transformed into efficient scheduling scheme by executing (C) before (B). In this case (A) & (C) can work in parallel and (B) is scheduled later.

Synchronization is always required even if the schedule is non-efficient. Different operations and data streams needs to be synchronized with each other for correct execution of algorithm. For example the read operations X_0 & X_1 must be executed together. Similarly Read Y_0 must wait for statement (A) & (C) to be completed in order to start its execution.

C. From C code to Configware

Consider the C code presented in Figure 6. Comments section in statement 5 - 11 are used as a guideline for allocating and binding the resources. _rfile pragmas are representing the RFILEs and _mdpu pragmas are representing the mDPUs. The generics(row, col, N, M, col_range) make the topological aspect of the implementation relative and the implementation instance can be moved around in the DRRA fabric by changing the values of the generics. M decides degree of parallelism and also specifies the number of SIMD threads.

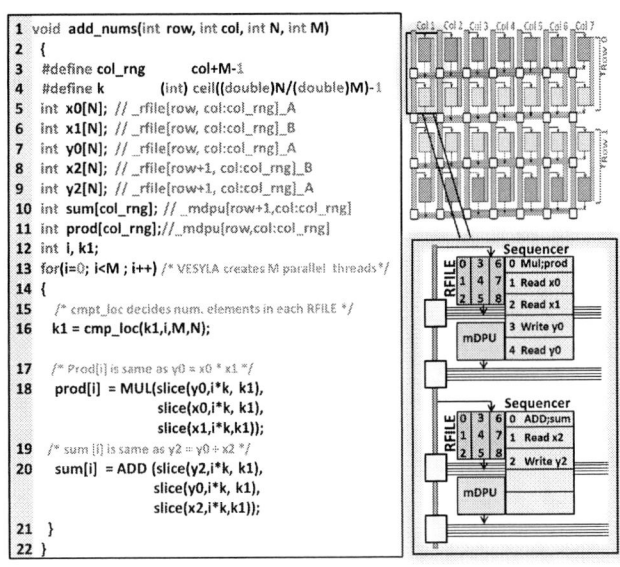

Figure 6. C Code implementation in terms of DRRA components

Figure 6 shows particular implementation instance in terms of DRRA components. VESYLA extracts the values of generics from the C code and decides the exact number of resources required. Let us assume that generic values for row, col, N, M are 0, 0, 18 and 2 respectively. After deciding the number of resources VESYLA binds variables in C code to the resources specified by the pragmas. For example in statement 5, array x0 is mapped on the RFILEs pointed by the row (which is 0) & col_range indices (col_range is calculated by using M number of SIMD threads, so number of columns used here are 2 and these two columns are 0 & 1). As there are two

SIMD threads, so x0 is distributed over two RFILEs and each RFILE is operating on vector of length 9 (M/N = 18/2), where each vector corresponds to addresses 0 - 8 as shown in Figure 6. Lastly VESYLA schedules the statements and generate configware for programming the SEQUENCERs.

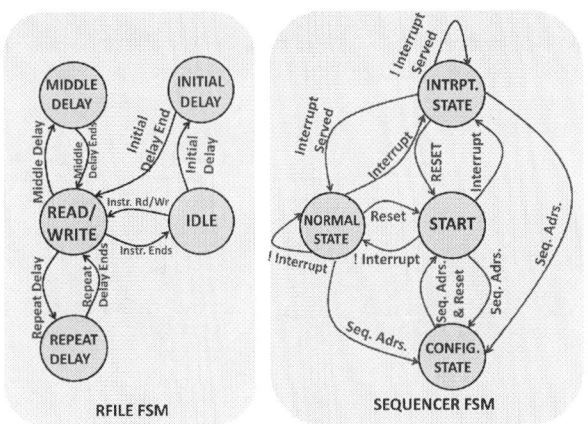

Figure 7. SEQUENCER & RFILE FSMs

Figure 7 shows the SEQUENCER [16] and RFILE FSM [12] (responsible for AGUs) in DRRA. SEQUENCERs can implement any arbitrary FSM. Sequencers Addresses are pointing towards the sequencer memory (orange boxes in Figure 6) and instructions are issued sequentially. Instruction 2 in SEQUENCER 0 streams data from RFILE 0 (consider Figure 6) and instruction 3 in SEQUENCER 2 stores that data back in RFILE 2 after being processed by mDPU. These two instructions initiate two RFILE FSMs and start executing the data in stream. VESYLA synthesizes and synchronizes these multiple FSMs to implement the correct behavior.

V. SYNCHRONIZATION ALGORITHMS & METHODOLOGY

A. Synchronization Methods Available in DRRA

There are multiple ways for synchronizing distributed FSMs in DRRA and VESYLA uses the most appropriate feature for this purpose. The techniques used by VESYLA are : (a) Use of the delays (Initial, Middle, Repetition) provided by the AGU of RFILE. (b) Using dynamic delays for conditional branches and (c) Hierarchical Control mechanism [16]. This paper will only provide details about using methodology (a).

B. Synchronization using Initial Delays

Figure 8 shows the detail data path (by a single SIMD thread) generated by the C code presented in Figure 6. There are two SIMD threads working in parallel with each other and with in each thread we have set of interacting FSMs. It can be observed that Read FSMs must be initiated together as they are providing the data and Write FSM must wait until the output of arithmetic operation is available at it's input.

These FSMs must be synchronized with each other during various phases while executing the algorithm. We have to synchronize the FSMs in such away that X_0 & X_1 should start executing together and after 4 cycles (Delay of a Multiplier) Y_0 should begin it's execution (Remember that SEQUENCER issues instructions sequentially, which means that instruction 'n' will start executing k cycle before instruction 'n+k'). The synchronization can be achieved by using Initial Delays available in AGU (discussed in Section III). This feature would help in delaying the start of each FSM with 'k' number of cycles and VESYLA will calculate the exact values of these delays for adjusting these FSMs together. Algorithm for balancing these operations is described in Figure 9.

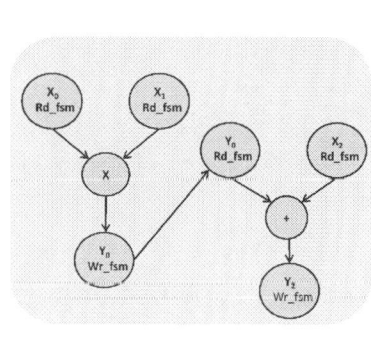

SEQUENCER 0 (Row 0, Col 0)		SEQUENCER 1 (Row 1, Col 0)	
Instr. Num.	Instr.	Instr. Num.	Instr.
0	Mul; prod	0	Add; sum
1	Read X0	1	Read X2
2	Read X1	2	Write Y2
3	Write Y0		
4	Read Y0		

SEQUENCER 2 (Row 0, Col 1)		SEQUENCER 3 (Row 1, Col 1)	
Instr. Num.	Instr.	Instr. Num.	Instr.
0	Mul; prod	0	Add; sum
1	Read X0	1	Read X2
2	Read X1	2	Write Y2
3	Write Y0		
4	Read Y0		

Figure 8. Datapath and state of sequencers for a single SIMD Threads

1) STEP 1 : The algorithm starts with examining Read FSMs in every statement and for each Read FSM it calculates the start exe. cycle (clock cycle at which that Read FSM will start its execution). This calculation is based on (a) The FSM's own initial delay (which is equal to 0 at the beginning) and (b) FSM's instruction number in the SEQUENCER memory. In this case start exe. cycle for X_0 is 1 (initial delay + instr.num = 0 + 1) and start exe. cycle for X_1 is 2 (initial delay + instr.num = 0 + 2). Read after write dependency don't exists for these FSMs, so these delays are not adjusted further.

2) STEP 2: Next step is to synchronize these Read FSMs. It can be observed that start exe. cycle for X_1 > start exe. cycle for X_0 (i.e. 2 > 1), which means that X_0 must be delayed by 1 cycle ($X_1 - X_0 = 2 - 1$) in order to to be synchronized with X_1. This value (1) is added to the initial delay of Read FSM X_0.

3) STEP 3: In the next step, algorithm calculates the time for performing the operation which is equal to 6 (max. start exe. cycle + num. of cycles taken by multiplier to generate the output = 2 + 4).

4) STEP 4: Lastly we have to synchronize the Write FSMs w.r.t. operator completion time. FSM Y_0 will get the execution time as 3 by the method shown in STEP 1(initial delay + instr.num = 0 + 3) and the execution time for completing the multiply operation is 6, which implies that FSM Y_0 should be delayed by 3 more cycles. All the FSMs in statement 1 are synchronized with each other after this value (3) is added into the initial delay of FSM Y_0.

Algorithm 1
(a) Initialize all variables to Zero
(b) Start Exe. Time represents the time/cycle when these FSMs should start their execution

For N number of threads
  For S number of statements
    For every node in each statement
      For each read node
        calculate start exe. cycle(for Read FSMs)
        If (Read After Write dependency) Then
          Adjust start exe. cycle for That read node
      End For

    get max. start exe. cycle
    (i.e. get the value of highest Start Execution Time among all read nodes)

    Set Initial Delays for each read node
    (w.r.t. Maximum Start Execution Time)

    Calculate Op. completion Time
    (Add delay of the operator in Max. Start Exe. Time For Adder This delay is 1 & For Multiplier it is 4)

    For each write node
      calculate start exe. cycle (for Write FSMs)
      If (Op. completion Time > Write start exe. cycle)
        Adjust Initial Delay For Write node
        (w.r.t. Op. CompletionTime)
      Else
        Go back to respective read nodes &
        Adjust Initial Delays accordingly
    End For

SEQUENCER 0 (Row 0, Col 0)		
Instr. Num.	Instr.	Init Delay
0	Mul; prod	-
1	Read X0	1
2	Read X1	1
3	Write Y0	3
4	Read Y0	3

SEQUENCER 1 (Row 1, Col 0)		
Instr. Num.	Instr.	Init Delay
0	Add; sum	-
1	Read X2	6

SEQUENCER 2 (Row 0, Col 1)		
Instr. Num.	Instr.	Init Delay
0	Mul; prod	-
1	Read X0	1
2	Read X1	0
3	Write Y0	3
4	Read Y0	3

SEQUENCER 3 (Row 1, Col 1)		
Instr. Num.	Instr.	Init Delay
0	Add; sum	-
1	Read X2	6

Figure 9. Algorithm For Calculating Initial Delays

In statement 2 we have Read FSM X_2 with execution time as 1 (initial delay + instr.num = 0 + 1) and Read FSM Y_2 with execution time as 4 (initial delay + instr.num = 0 + 4). Read FSM Y_2 has a read after write dependency, so the algorithm will re-calculate execution time for this FSM. Read FSM Y_0 have to wait at least 7 cycles (initial Delay Write FSM + instrnum Write FSM + num. of cycles for completing Write operation = 3 + 3 + 1) before starting execution. Initial delays for Read FSMs X_2 & Y_0 are calculated w.r.t to Maximum start Execution time (which is 7 now).

C. Synchronization using Middle Delays

Initial delays are used to synchronize the FSMs at the start of their execution i.e. they are used to delay the start of address execution process. For each FSM, they can only be used once, but it would often be the case to synchronize the FSMs at different stages of their execution. The main reason is that these FSMs might be working independently while operating on same/different resource with variable data consumption/production rates. For example we may need to halt an FSM at a particular stage in order to synchronize it with a FSM that will start its execution after 'n' cycles.

Consider the code presented in Figure 10 and the datapath generated for the specified values of generics. This is the same code as presented in the last example but now (Read FSM) X_0 is used as an operand in both statements (18 & 20), moreover statement 20 is dependent upon the completion of statement 18. This implies that X_0 (Read FSM) should be synchronized with X_1(Read FSM - statement 1) & Y_0(Read FSM - statement 2) respectively during the execution of this algorithm.

This type of synchronization can be achieved by using

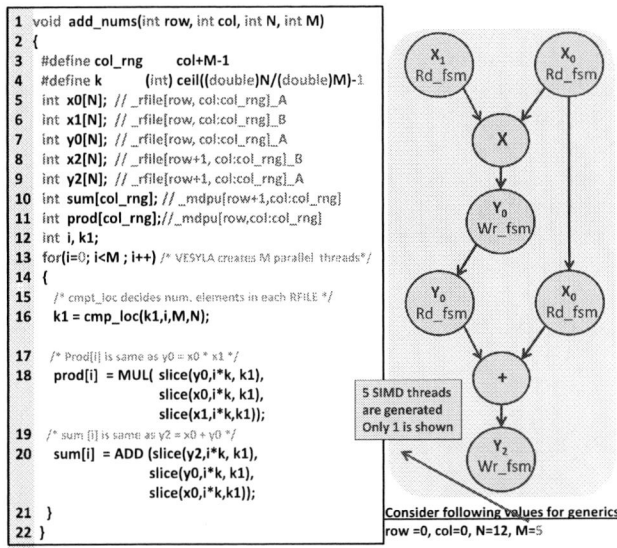

```
1  void add_nums(int row, int col, int N, int M)
2  {
3      #define col_rng      col+M-1
4      #define k            (int) ceil((double)N/(double)M)-1
5      int x0[N]; // _rfile[row, col:col_rng]_A
6      int x1[N]; // _rfile[row, col:col_rng]_B
7      int y0[N]; // _rfile[row, col:col_rng]_A
8      int x2[N]; // _rfile[row+1, col:col_rng]_B
9      int y2[N]; // _rfile[row+1, col:col_rng]_A
10     int sum[col_rng]; // _mdpu[row+1,col:col_rng]
11     int prod[col_rng];//_mdpu[row,col:col_rng]
12     int i, k1;
13     for(i=0; i<M ; i++) /* VESYLA creates M parallel threads*/
14     {
15         /* cmpt_loc decides num, elements in each RFILE */
16         k1 = cmp_loc(k1,i,M,N);

17         /* Prod[i] is same as y0 = x0 * x1 */
18         prod[i] = MUL( slice(y0,i*k, k1),
                          slice(x0,i*k, k1),
                          slice(x1,i*k,k1));
19         /* sum [i] is same as y2 = x0 + y0 */
20         sum[i] = ADD (slice(y2,i*k, k1),
                         slice(y0,i*k, k1),
                         slice(x0,i*k,k1));
21     }
22  }
```

Figure 10.   C code for Middle Delays

the Middle delays available in the AGUs. Algorithm for calculating middle delays is represented in Figure 11. We have modified algorithm 1 by keeping track of resources being read/write for multiple times and clock step in which that read/write operation was performed (Remember that X_0 is providing data from address 0 - 8 in streaming mode and after every 1 cycle it increments the address).

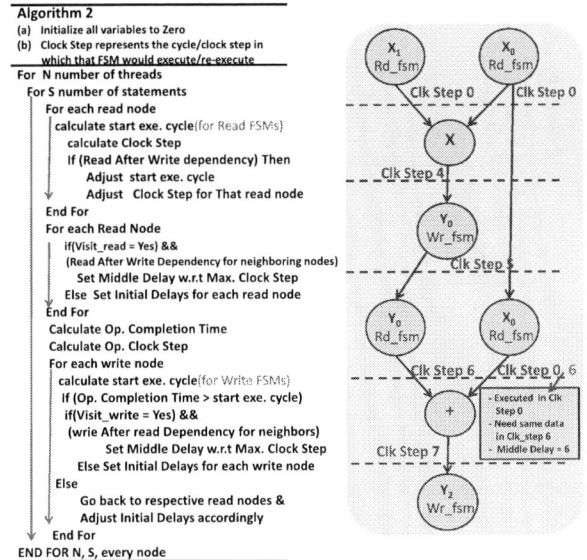

Figure 11.   Algorithm for computing Middle Delays

The algorithms starts by analyzing statement 1 and it computes initial delays for FSMs X_0, X_1 and Y_0 by following STEPS 1 - 4. After computing initial delays it marks the FSMs as being visited (visit = YES). This variable helps in keeping

track if the resource is read/write multiple times.

Initial Values of clock steps for both Read FSMs are equal to 0. Value of clock step is updated to 4 after performing the Multiply operation (multiplication takes 4 cycles) as shown in Figure 11 and it gets incremented, once initial delay for Write FSM (Y_1) is computed. .After analysis on Statement 18 is completed, it starts computing start execution times for Read FSMs (Y_0 & X_1) and tries to compute initial delays, but X_1 has already started its execution in clock step 0 and is providing a new value after every clock cycle (using address 0 - 8). It is not possible to shift execution of Y_0 in clock step 0, because it has a read after write dependency and at earliest it can be scheduled in clock step 6 (as shown in Figure 11). The only solution is that X_1 must delay the process of address generation, in order to be synchronized with both Read FSMs X_0 & Y_0.

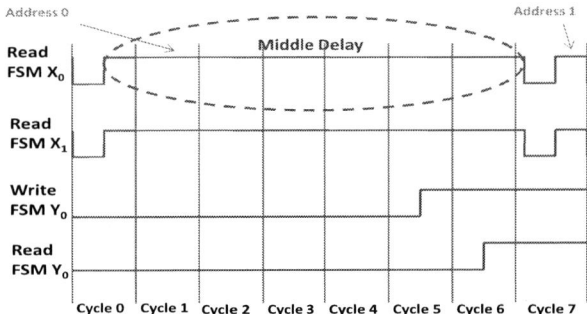

Figure 12.   Waveform showing Middle Delays

This synchronization is achieved by setting the value of middle delay as 6 for X_1. A new address is generated after every 6 cycles now. Since all the dependent FSMs are affected by this change, the value is assigned to all FSMs in that path. Figure 12 shows the exact usage of middle delays between successive address generation of X_0. At cycle 6, Read FSM Y_0 starts executing and at the same cycle value form X_0 is still available. At this point these all FSMs are synchronized with each other and in the next cycle (cycle 7) X_0 generates the next address.

D. Synchronization using Repetition Delays

Various FSMs need to repeat the execution of their states and while reaping the states they might need to be synchronized with another FSM that has not yet started its execution. There is also a possibility that it needs to be synchronized with another FSM which will also repeat its states but is working on slower data consumption/production rate. These types of synchronization is achieved by using repetition delays.

Consider the code presented in Figure 13 and data path generated by giving appropriate values to generics. The RFILE in Figure 13 represents a 3 x 3 matrix where data at consecutive address represents the rows. Matrix B represents the transpose of Matrix A. We have to read data from rows and store it in the form of columns (or vice versa) which means that Read FSM has to read from address 0,1,2 and then repeat the execution for

978-1-4577-0671-4/11 $26.00 © 2011 IEEE

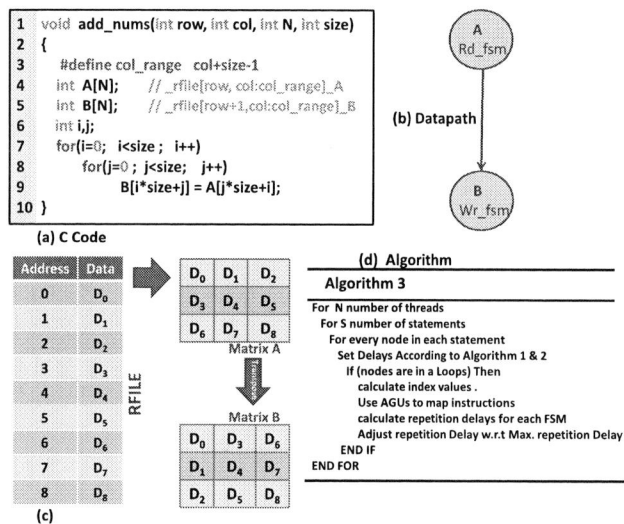

Figure 13. An Example of Matrix Transpose

threads, therefore we have chosen a fully serial FIR filter for demonstration by using the generic values 1, 1, 0, 0 for N (Number of Taps), M(Degree of parallelism), r and c respectively.

The _rfile pragma (st. 6) specifies that x (delay line) is placed at row r, and will occupy columns c to c+M-1. [1]. As this is a fully serial solution, so the delay line is occupying the (single) RFILE present at row 0, column 0 in DRRA and the _mdpu pragma (st. 9) allocates the number of mDPUs required for performing computations. The co-efficients are also using a single RFILE (row 1, col 0) as shown in Figure 15 .

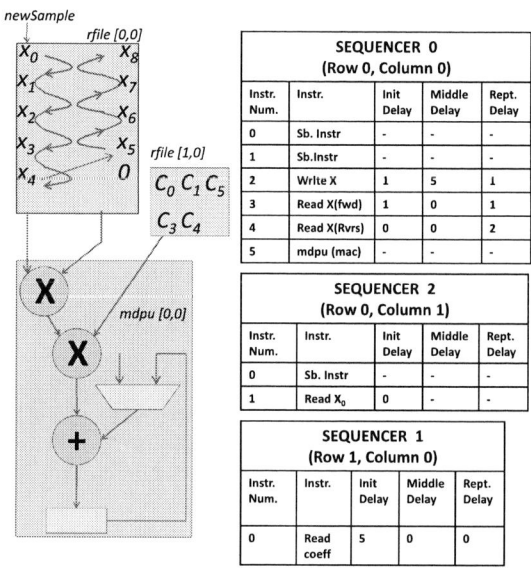

Figure 15. Datapath generated by VESYLA for symmetric FIR Filter for N=9, M=1, r=0 and c=0

addresses 3,4,5/7,6,8. Similarly the Write FSM needs to repeat its execution for some particular address patterns as well. We can use the repetition delays to synchronize the repetition of FSMs.

VI. EXPERIMENTAL RESULTS

In this section we will apply the methodology for synchronizing FSMs on a DSP algorithm. This synchronization methodology is related to distributed control schemed implemented in DRRA, so we cannot compare our results in this section with other architectures and programming solutions.

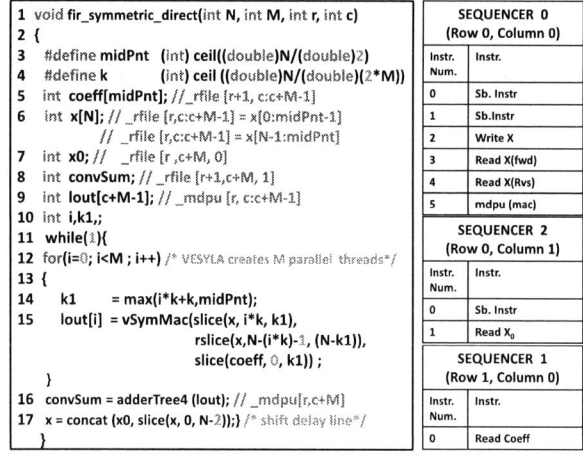

Figure 14. C Code of N Taps Symmetric FIR Filter

Code presented in Figure 14 can generate a N Tap FIR Filter, from fully serial to full parallel and anything in between (serial-parallel). The synchronization methodology works correctly for every type of filter. We will not be able to show all the waveforms and computations for parallel SIMD

Statement 15, 16 and 17 are implementing the functionality of FIR filter. Statement 15 is implementing the symmetric MAC operation and VsymMAC is a pre-defined function that computes MAC operations on vectors of different sizes obtained from the pre-defined slice functions [15], [1]. Statement 16 implements a Shift Delay line. This single statement implements both forward and reverse delay lines for shifting the samples before the arrival of new sample. All RFILEs are initialized by 0s (i.e. 0 are stored at location 0 - 63). X_0 represents the new sample.

The algorithm starts analysis at the arrival of new sample X_0. Read FSM X_0 and Write FSM x must be synchronized with each other by having 0 & 1 as their initial delays by using steps 1 - 4 described in algorithm 1. For reading values in circular buffer the start execution cycle for X(Fwd_line) is 3 (initial delay + instr.num = 0 + 3), start exe. cycle for X (RVRs_line) is 4 (initial delay + instr.num = 0 + 4) and Start Execution Cycle for coeff is 0 (initial delay + instr.num = 0 + 0). Since Read FSM x has a dependency on Write FSM x, so the start execution cycle of Read FSM X must be delayed by until the write operaion is completed. It is evident that start exe. cycle for X(Rvrs_line) > X(Fwd_line) > coeff,

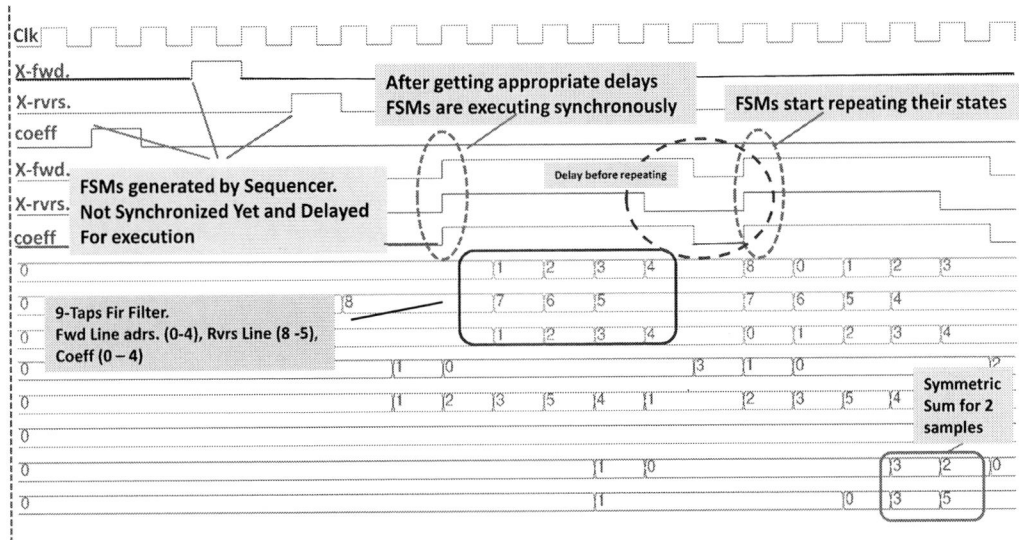

Figure 16. Waveform generated From Modelsim

which means that X(Fwd_line) & coeff must be delayed by 1 (X(Rvrs_line) - X(Fwd_line) = 4 - 3) and 4 (X(Rvrs_line) - coeff = 4 - 0) cycles respectively in order to to be synchronized with X(Fwd_line). These values are added to the initial delay of both Read FSM.

This algorithm is working in streaming mode, and the outer most loop indicates to VESYLA that these FSMs must repeat their executions. It takes one cycle to clear the internal accumulator of mDPU, so before repeating their states, they must wait for at least 1 cycle. Reverse Line is operating on 4 addresses as compared to forward line which is operating on 5 addresses, therefore Reverse line completes its execution 1 cycle before forward line. This 1 cycle is added into the repetition delay value for Reverse line. Figure 16 shows the waveform for these FSMs which are working in fully synchronized manner now. The circular buffer must read all the addresses before accumulating the next sample value. In this case both fwd and rvrs streams are working in parallel, so the write x FSM must wait for at least 5 cycles before accepting the next sample (which is also the middle delay according to Algorithm 2 - condition -> Write after read Dependency).

VII. CONCLUSION & FUTURE WORK

In this paper we have presented the methodology for synchronizing distributed FSMs generated during mapping an application on DRRA. VESYLA is capable enough to synchronize FSMs generated in case of Adaptive Algorithms and memory architecture as well. The hierarchical control scheme [16] still needs to be implemented in VESYLA and is the part of our future work.

ACKNOWLEDGMENT

This work is supported by Vinnova funded CREST (Coarse Grain Reconfigurable Embedded Systems Technolo-

gies) project with Diarienr 2010-01453 and HEC (Higher Education Commission) of PAKISTAN.

REFERENCES

[1] O. Malik and et al, "A library framwork for a coarse grain reconfigurable architecture," in *IEEE 24th International Conference on VLSI Design*.

[2] G. Lee and et al, "Automatic mapping of application to coarse-grained reconfigurable architecture based on high-level synthesis techniques," in *IEEE SOC Design Conference*, 2008, pp. I–395 – I–398.

[3] D. Shin and et al, "An interactive design environment for c-based high-level synthesis of rtl processors," in *IEEE Transaction on VLSI Systems*, vol. 16, 2008, pp. 466–475.

[4] "http://www-labsticc.univ-ubs.fr/www-gaut/."

[5] B. Thomas and et al, *Catapult Synthesis: A Practical Introduction to Interactive C Synthesis: From Algorithm to Digital Circuit, Springer*.

[6] P. Coussy and et al, "An introduction to high-level synthesis," in *IEEE Design and Test*, 2009.

[7] B. Mei and et al, "Dresc: A retargetable compiler for coarse-grained reconfigurable architectures," in *International Conference on Field Programmable Technology*, 2002.

[8] Z. A. Ye and et al, "A c compiler for a processor/fpga architecture," in *Proceedings of the 8th ACM Internation Symposium on Field-Programmable Gate Arrays*, 2000.

[9] Alle, Myhthri, and et al, "Compiling techniques for coarse grained runtime reconfigurable architectures," in *5th International Workshop on Applied Reconfigurable Computing*, 2009.

[10] S. Chalamalasetti and et al, "Mora - an architecture and programming model for a resource efficient coarse grained reconfigurable processor," in *AHS 2009. NASA/ESA Conference*, pp. 389 –396.

[11] M. A. Shami and et al, "Morphable dpu: Smart and efficient data path for signal processing applications," in *SIPD*, 2009.

[12] M. A. Shami and A. Hemani, "Address generation scheme for a coarse grain reconfigurable architecture," in *IEEE 22nd conference on ASAP*, 2011.

[13] M. A. Shami and et al, "An improved self-reconfigurable interconnection scheme for a coarse grain reconfigurable architecture," in *IEEE 28th NORCHIP Conference*, 2010.

[14] F. Jafari and et al, "Optimal selection of function implementation in a hierarchical configware synthesis method for a coarse grain reconfigurable architecture," in *DSD*, 2011.

[15] O. Malik and et al, "A high level synthesis framework for a coarse grain reconfigurable architecture," in *IEEE 28th NORCHIP Conference*, 2010.

[16] M. A. Shami and et al, "Control scheme for a cgra," in *22nd International Symposium on Computer Architecture and High Performance Computing*, 2010, pp. 17 – 24.

978-1-4577-0671-4/11 $26.00 © 2011 IEEE

Automatic Calibration of Streaming Applications for Software Mapping Exploration

Weihua Sheng, Stefan Schürmans, Maximilian Odendahl, Rainer Leupers, Gerd Ascheid

Institute for Communication Technologies and Embedded Systems

RWTH Aachen University, Germany

E-mail: sheng@ice.rwth-aachen.de

Abstract— Streaming models have lately gained a lot of interest in embedded software design as they closely resemble computation of signal processing applications typically found in wireless and multimedia domains. To map streaming applications onto MPSoCs (Multi-Processor System-on-Chips) efficiently, programmers need not only to validate software but also to estimate the performance of their software accurately. Therefore, fast MPSoC virtual platforms which support fully functional execution of software with good timing accuracy are required. In this paper, we propose a tool-flow to construct such MPSoC virtual platforms. The key idea is to annotate timing of sequential execution of streaming applications automatically by calibration in a configurable abstract MPSoC virtual platform. A case study of applying the tool-flow to a real-life heterogeneous MPSoC, TI's OMAP, has been conducted to prove the tool-flow's feasibility and show good accuracy of the calibrated virtual platform for software mapping exploration.

I. INTRODUCTION

Heterogeneous MPSoCs are widely used in modern embedded systems. Both complexity and cost of nowadays MPSoC designs increase so rapidly that only a few semiconductor houses can afford new designs. Therefore, software design becomes the differentiator of embedded systems. Streaming or dataflow programming models such as KPN (Kahn Process Networks) [1] and SDF (Synchronous Dataflow) [2] have gained acceptance recently, as they closely resemble computation of signal processing applications typically found in wireless and multimedia domains. However, software mapping exploration of streaming applications onto MPSoCs is yet a difficult and daunting task for MPSoC programmers.

Virtual platform technologies ([3], [4]) have recently become main-stream of system simulation for ESL (Electronic System Level) designs. In order to map parallel software onto MPSoCs, programmers need not only to validate software but also to get accurate performance estimation of their software. That requires that MPSoC virtual platforms support fully *functional* execution of software with good timing *accuracy*. In addition, simulation speed of virtual platforms also needs to be *fast* so that software developers can explore the huge software mapping space efficiently.

Targeted for architecture exploration, detailed virtual platforms built with hardware models of low abstraction levels (instruction or cycle accurate) provide accurate simulation results. However, they are usually too slow and thus inefficient for software mapping exploration.

Most of the previous work ([5], [6], [7]) on generation of fast and accurate performance models for MPSoCs running streaming applications is also rooted in the scenario of architecture exploration. The common approach is to obtain program traces from software functional simulation and use them in a separate trace-driven system simulator to evaluate different architecture options. The separation of software functionality and hardware performance models allows fast architecture exploration. However, it causes inconvenience for programmers to do software mapping exploration, which demands both functionality and accuracy.

In this work, we investigate how to construct fast, accurate MPSoC virtual platforms which also support fully software functionality for the scenario of software mapping exploration. The key concepts are listed as follows:

- **Speed**: A configurable SystemC-based MPSoC virtual platform is used in our work as a base system. An abstract processor modeling technology called VPU (Virtual Processing Unit) [8] is used for modeling PEs (Processing Elements) instead of low abstraction level models. This enables fast simulation speed which is necessary for software mapping exploration.

- **Accuracy**: We annotate the timing for sequential execution parts of streaming applications in the VPUs by pre-calibrated data. Prior to mapping exploration, a calibration database to contain such data is generated. This can be done e.g. by using software profilers or measurements on the hardware (in this work calibration based on hardware measurements is used). Communication delays caused by data exchange between software tasks, which are highly dynamic depending on mapping, are simulated using detailed models such as buses and memories in virtual platforms. Therefore, accuracy is ensured.

- **Functional**: Unlike previous approaches, streaming programs are compiled and directly run in the VPUs. Fixed inputs like application traces are not needed in the mapping exploration. Programmers can fully debug software as in traditional simulators with good accuracy and speed.

In this paper, we present a tool-flow to fully automate calibration of abstract MPSoC virtual platforms mentioned above for mapping streaming applications. The rest of this paper is organized as follows: Background information about

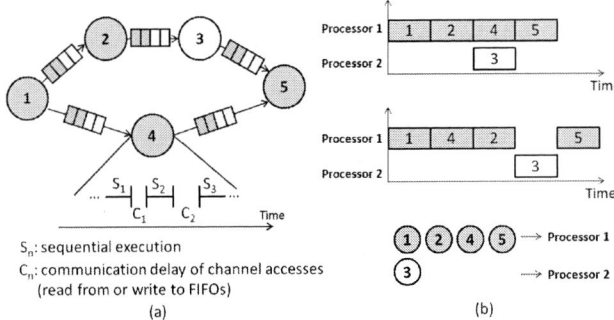

S_n: sequential execution
C_n: communication delay of channel accesses
(read from or write to FIFOs)

(a)

(b)

Fig. 1: *(a) A five-process KPN example with a zoomed view of internal process execution (b) Different scheduler decisions result in different execution times: A dual-processor architecture is assumed. Processes 1,2,4,5 which are marked in gray are mapped to one processor while process 3 runs on the other processor.*

the streaming model we use is given in section II, followed by a summary of the related work in section III. Section IV introduces the key idea of calibrating streaming applications automatically. Architecture modeling using VPUs is explained in section V. The implemented tool-flow is described in section VI. Section VII presents our case study of applying the tool-flow to TI's OMAP3530 [9] in detail and the results are shown in section VIII. The paper is concluded by section IX, which also gives a brief outlook into the future work.

II. Streaming Model - KPN

We use KPN as the streaming model for applications in our work. KPN [1] is a widespread MoC (Model of Computation) for parallel programming especially for signal processing applications. A KPN is represented as a graph $G = (V, E)$ which is built of nodes representing autonomous processes of computation. The edges between the nodes represent unbounded unidirectional FIFO channels that transmit data items, though in reality FIFO channels are bounded. Reading from an empty FIFO channel results in the process being blocked until data to read become available. A KPN is determinate, i.e. the channel history of tokens from process computation is independent of the order of process execution (scheduling). A five-process KPN example is shown in Fig. 1(a).

Execution timing of KPN applications is decided by two factors as follows.

1) **Execution of each KPN process**: This can be further broken down into *sequential code execution in processes* and *inter-process communication by channel accesses*. Fig. 1(a) has a zoomed view of process 4's internal execution as an example.

2) **Scheduling of KPN processes**: This is usually determined by the scheduler of the processor to which KPN processes are mapped. An example is demonstrated in Fig. 1(b). In both top and bottom scenarios, process 2 and 4 are both ready to run after process 1 finishes. The top scenario picks process 2 over process 4 while the

bottom one does it the other way. The different scheduling decisions result in different performance results.

In summary, accurately modeling both factors is key to good performance estimation of KPN applications.

III. Related Works

Trace-driven performance estimation approaches ([5], [6], [7], [10], [11]) are popular for architectural exploration. They require *a-priori* application traces which are commonly obtained from software functional simulation firstly. Trace-driven simulators (usually not with full software functionality) are then used to replay traces to estimate the system performance and evaluate different architecture options. We target the scenario of *software mapping exploration* which asks for system simulation being both accurate and software functional to handle different behaviors of the subsystems depending on system setup, configuration and inputs. Differing from trace-driven work, in this work we have integrated functionality and accuracy in one model.

Accurate and functional simulation for software development has been addressed in many research works. Cheung et al. have proposed automatic generation of structural models for software performance estimation in [12]. The annotation granularity is at basic block level. However, complex processor features like superscalar pipelines and branch prediction cannot be handled well and significant work needs to be done in the compiler backend to enable the tool-flow. HVP [13] is a high-level virtual platform to support software development in early system design. The granularity of timing annotation is at compiler IR (Intermediate Representation) statement level. While it is relatively easy to set up functional simulators using HVP, the weaknesses lie in the simplistic performance estimation technique and the platform not being easily extendable. Our approach uses pre-calibration to ensure accuracy. It annotates at a much *coarser* granularity, i.e., sequential execution segments between communication events. For the streaming applications we target, this gives us the best trade-off between timing annotation granularity and simulation speed.

A VPU-based MPSoC simulation framework for C application mapping has been proposed in [8] and later enhanced with a fine-grained C source code instrumentation technique at compiler IR level, which is suitable for RISC-like processors to obtain accurate software performance [14]. NXP has lately developed the AMM (Application Modeling and Mapping) methodology based on the VPU for abstract system simulation and exploration [15]. The applications are written in dataflow models in SystemC with some pre-defined primitives and task execution times are measured and annotated manually. Unlike those, the calibration process in our work is *automated* by a tool-flow which eliminates the manual annotation and the method is generic and not bound to any specific processor type.

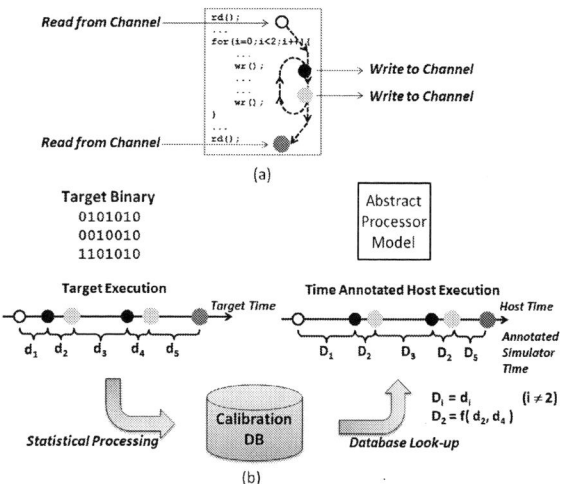

Fig. 2: *Key Concept (a) An example KPN application with marked channel accesses (b) Mechanism of calibration*

IV. AUTOMATIC PLATFORM CALIBRATION

As mentioned in section II, timing of each KPN process is determined by sequential code execution in processes and communication by channel accesses. In this section, we discuss how those two parts are calibrated and modeled.

We introduce a number of definitions firstly to characterize execution of a KPN process. As the computation is local to each process, the behavior of a process P on PE U can be described as a finite[1] sequence of channel access events $E^{P,U} = (e_1^{P,U}, \ldots, e_n^{P,U})$. Fig. 2(a) shows the execution trajectory of a small example process P_{ex} and the events: there are two channel accesses (read) before and after the loop while there are two channel accesses (write) inside the loop, marked with circles of different colors that represent the events $e_1^{P_{ex},U}, \ldots, e_4^{P_{ex},U}$.

The program execution trajectory *between* the channel accesses results in computation time, whereas the channel access events themselves also take some time on a real platform, e.g. due to reading from or writing to a memory location via a bus, leading to the following definitions:

$T_{\mathrm{B}}(e)$ wall clock time at which the event begins

$T_{\mathrm{E}}(e)$ wall clock time at which the event ends

The wall clock times of the events $e^{P,U}$ are influenced by the scheduling of the process P and its mapping to a PE U. Therefore, we define the following for an event $e^{P,U}$ of process P on PE U:

$t_{\mathrm{B}}(e^{P,U})$ total execution time of P running on processor U at wall clock time $T_{\mathrm{B}}(e^{P,U})$

$t_{\mathrm{E}}(e^{P,U})$ total execution time of P running on processor U at wall clock time $T_{\mathrm{E}}(e^{P,U})$

[1] A PN process in reality will either terminate, deadlock or be aborted by the user.

Then, the duration of the computation between two events e_1, e_2 can be calculated as follows:

$$d(e_1, e_2) := t_{\mathrm{B}}(e_2) - t_{\mathrm{E}}(e_1)$$

To model the computation and communication times on the abstract processor model, we propose:

- for sequential computation between two channel accesses e_1, e_2, the time $d(e_1, e_2)$ is annotated by pre-calibrated values (either by profiling or measurement), as signal processing algorithms mostly possess static computation patterns.
- for communication time spent in a channel access e, it is simulated as part of the MPSoC platform simulation, as it is highly dynamic depending on mapping where analytical or estimation approaches fall short.

Fig. 2(b) demonstrates the mechanism of our proposed calibration for sequential computation. The application is firstly instrumented to report timing durations d of sequential execution between channel accesses. After compilation and execution on the target, the timestamps of $E^{P,U}$ are collected per code segment and per processor. We define a property $l(e)$ of an event e as the code location where the event occurs. Then, all observed durations of code segments between locations l_1 and l_2 on processor U can be obtained by finding two events e_1 and e_2 that occur at those locations and appear sequentially in the sequence $E^{P,U}$:

$$\mathbb{D}^U(l_1, l_2) := \{\, d(e_1, e_2) \,|\, \forall e_1, e_2 : l(e_1) = l_1, l(e_2) = l_2, \\ \exists P : E^{P,U} = (\ldots, e_1, e_2, \ldots) \,\}$$

As code execution times do not depend on the processor instance U, but only on the processor type C which U belongs to, the durations can be merged:

$$\mathbb{D}^C(l_1, l_2) := \bigcup\nolimits_{U \text{ is of PE type } C} \mathbb{D}^U(l_1, l_2),$$

A statistical post-processing utility application using an averaging function f is used to compute the so-called *calibration database*:

$$D^C(l_1, l_2) := f\left(\mathbb{D}^C(l_1, l_2)\right)$$

When the simulation of the same application takes place on the host, the host time for executing the same sequential segments will not match the target execution time correspondingly (see the different placement of the events on the time axis in Fig. 2(b)). As a matter of fact, the host time *does not* play a role here, because the simulator time is only advanced by *explicitly* calling a simulator-specific function with a parameter of time increment. So, the time annotated host execution between the channel accesses at the code locations l_1 and l_2 will look up the pre-calibrated time $D^C(l_1, l_2)$ in the calibration database to annotate the sequential execution segments. In the example shown in Fig. 2(b), all annotated D_i values are equivalent to d_i except for D_2. The sequential segment between both channel accesses in the loop is executed twice in the calibration phase (d_2 and d_4). Its annotation D_2 is therefore calculated by averaging: $D_2 = f(d_2, d_4)$.

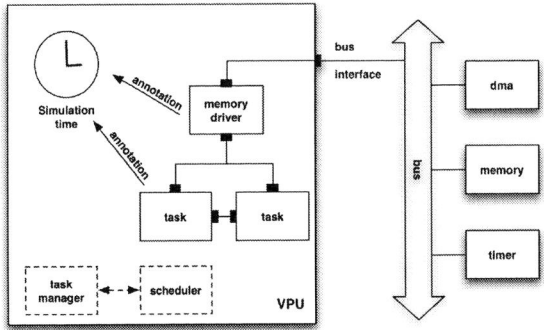

Fig. 3: *Virtual Processing Unit (VPU)*

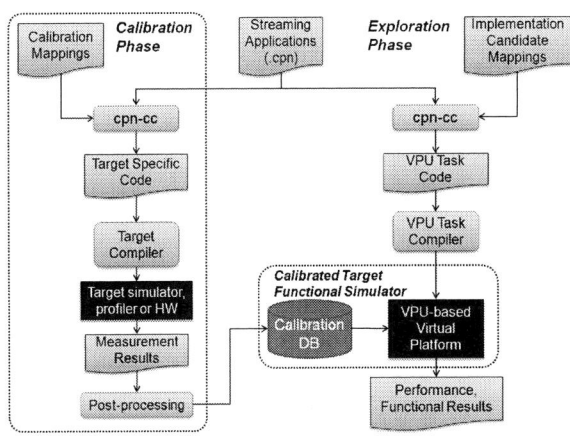

Fig. 4: *System Overview - Software Mapping Exploration of Streaming Applications*

For heterogeneous platforms having L PEs of N types, each sequential code segment needs to have an entry per PE type in the calibration database to be able to simulate all mappings on the abstract virtual platform. That means the calibration phase requires a minimum of N profiling runs (e.g. all processes being mapped to one PE type per run). After this small one-time effort, the calibrated host simulator can be used to explore as many as L^M ($L^M \geq N^M$ as $L \geq N$) different software mapping possibilities of streaming applications (M being the number of processes).

V. ARCHITECTURE MODELING

SystemC is chosen for modeling architectures in our work. To model various kinds of PEs of MPSoC platforms, an abstract processor modeling technology called VPU has been used. The VPU [8], shown schematically in Fig. 3, is a configurable abstract processor model for SystemC-based virtual platforms, which is able to simulate execution of software tasks under control of an abstract operating system. The tasks are compiled for the simulation host to have full software functionality and only consume simulation time via explicit timing annotations. Communication can be done explicitly between tasks within the VPU or between a task and the hardware layer outside the VPU using a *driver*, which can also contain timing annotations. The abstract OS is simulated by the *task manager* implicitly present inside the VPU. It controls the execution state of all tasks and supports preemption and time-slicing. A pre-defined or user-specific *scheduler* is used to implement the actual task selection algorithm.

For our work, the explicit timing annotation feature that VPU supports is used for sequential execution parts of KPN process tasks. As discussed in section II, another key factor in determining the timing behavior of KPN applications is how processes are scheduled to run. This is usually controlled by the OS on a processor. As schedulers of VPUs are highly customizable, we are able to model different OS scheduling and task selection policies for different PEs which exist in target MPSoC architectures.

VI. SYSTEM OVERVIEW

In this section we present an overview of the system we implemented to perform software mapping exploration of streaming applications enabled by the ideas mentioned in section IV and V. An overview is shown in Fig. 4. There are two phases, namely *calibration phase* and *exploration phase*, which are described below. The streaming programming model that we used and its compiler are also explained at the end of this section for completeness.

A. Calibration Phase

A tool-flow was implemented to fully automatically calibrate VPU-based MPSoC virtual platforms in the calibration phase. Streaming applications are written using a lightweight C extension programming model CPN (C for Process Networks). A retargetable source-to-source compiler (*cpn-cc*) has been developed to take inputs of CPN code together with mapping descriptions (process-to-PE) and generate code for target architectures. For each calibration mapping, *cpn-cc* generates target C code which can be further compiled by the target compiler tool-chains. Instrumentation is done to output the profiles of processes' computation times. A post-processing utility application transforms the results into a database using statistical analysis, which the VPU-based virtual platform will use for look-up of timing annotation. Shown in the left part of Fig. 4, the calibration phase is done once per application (a minimum of N iterations, N being the number of PE types).

B. Exploration Phase

After the calibration phase is done, programmers are able to explore an arbitrary software mapping using the calibrated MPSoC virtual platforms in the exploration phase, shown in the right part of Fig. 4. With small effort, an abstract VPU-based MPSoC virtual platform which resembles the target platform is set up. Thanks to the retargetability of *cpn-cc*, it translates the same streaming application into VPU task code that can be compiled and run on the VPUs. Together with

```
1   extern unsigned int zigzag[64];
2   void f_idct(short coeff[64], short pix[64]);
3
4   __PNkpn Unzigzag __PNin(short val)
5                    __PNout(short coeff) {
6   unsigned int i;
7   while(1) {
8      __PNout(coeff:64) {
9         memset(coeff, 0, 64 * sizeof(int));
10        for (i = 0; i < 64; ++i) {
11           __PNin(val) {
12              if (val == END_OF_BLOCK) break;
13              coeff[zigzag[i]] = val;
14   } } } } }
15
16  __PNsdf Idct __PNin(short coeff:64)
17              __PNout(short pix:64) {
18     __PNloop {
19        f_idct(coeff, pix);
20     } }
21
22  __PNchannel short cVal, cCoeff, cPix;
23  __PNprocess pUzz = Unzigzag __PNin(cVal)
24                          __PNout(cCoeff);
25  __PNprocess pIdct = Idct __PNin(cCoeff)
26                          __PNout(cPix);
```

Listing 1: *Excerpts from the CPN code of a JPEG decoder*

querying timing annotations from the calibration database, the virtual platform can simulate any software mapping with high accuracy. Therefore, efficiency of mapping space exploration is greatly improved.

C. CPN and its Compiler

Our streaming programming model is the CPN language based on ANSI C extended with a limited number of new syntactic elements for processes and first-in first-out (FIFO) channels. The new keywords begin with the common prefix __PN to avoid name clashes.

Lst. 1 shows excerpts from a JPEG decoder application written in CPN. A KPN-like processes template (l. 4-14) can contain arbitrary control flow and accesses the channels explicitly (l. 8, 11) using __PNin and __PNout. The SDF-like template in lines 16-20 contains an infinite loop (__PNloop, l. 18) implicitly accessing each channel once per iteration. The access windows to the channels can be used like C variables (l. 12) or arrays (l. 19) inside the channel access blocks. The remaining part of the code declares three channels containing tokens of type short (l. 22) and instantiates processes from templates (l. 23-26) while connecting them to channels.

We implemented a retargetable CPN to C compiler, called *cpn-cc*, based on Clang [16]. With an additional input specifying the mapping of the processes to the PEs and the capacities of the channels, *cpn-cc* is able to compile CPN source to plain C source code containing target-specific APIs which e.g. realize instantiation of KPN processes and inter-process communication. The output C code is intended to be compiled to binaries by the native compilers of the PEs on the target or the host compiler in case of an abstract simulator target. As shown earlier in this section, *cpn-cc* is a key enabler in our system allowing to automate the calibration and mapping space exploration.

VII. CASE STUDY - OMAP

A case-study of applying our tools for a commercial heterogeneous MPSoC, TI's OMAP3530, has been carried out. A hardware evaluation board of OMAP is used in our study, which allows us to obtain calibration data and compare the calibrated virtual platform simulation and real hardware run. It is noted that our approach can work with or without existing hardware, e.g. getting calibration data from cycle accurate virtual platforms or even from PE profilers is also possible.

A. OMAP

The OMAP3530 [9] is a heterogeneous MPSoC which features an ARM Cortex A8 processor running at 550 MHz and a TI C64x+ DSP (Digital Signal Processor) clocked at 400 MHz. The hardware evaluation board possesses 128 Mbyte DDR SDRAM at 166MHz shared between the ARM and the DSP. Software-wise, on the ARM side, Linux is used as OS, while on the DSP side a lightweight multitasking operating system, called DSP/BIOS [17] from TI, is used. To run CPN applications on the OMAP, a shared memory based FIFO implementation has been developed.

B. Calibration Measurements and Post-processing

As we perform measurements in the calibration phase on the hardware board, logging of events is intrusive and modifies the application behavior. In order to reduce this effect, special care was taken to implement logging on both the DSP and the ARM processor with minimal overhead. For instance atomic ARM instructions are used instead of expensive locks for ensuring log data consistency. The time-stamps are collected by the Linux call clock_gettime() on the ARM and the hardware performance counter on the DSP. The raw log-file is post-processed afterwards to establish a database containing the computation times needed by the abstract processor model. For every pair of successive channel accesses in the logged trace, there might be multiple logged time durations. Different statistical processing methods can be applied here to produce an estimate for the sequential execution code. We used the *arithmetic mean* in our tool-flow to calculate the entry in the calibrated database.

C. Abstract OMAP Virtual Platform

The abstract virtual platform of the OMAP system was created using Synopsys Platform Architect [18] version 2010.09_SP1. The ARM and DSP processors were modeled using the VPU block. Both were connected to the shared memory using a TLM 2.0 bus. As the abstract OMAP virtual platform consists of less than 10 SystemC blocks, it was set up within a few hours.

On the software level inside the VPU blocks, custom scheduler implementations are used to approximately model the scheduling behavior of Linux on the ARM and of DSP/BIOS on the DSP. The FIFO communication is modeled explicitly via the shared memory using VPU driver blocks with a behavior close to the FIFO implementation on the OMAP. They contain fixed timing parameters to model channel accesses.

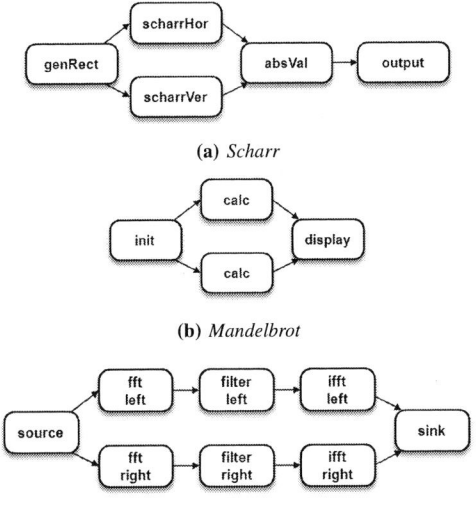

(a) *Scharr*

(b) *Mandelbrot*

(c) *Band-pass Filter*

Fig. 5: *Benchmark Applications*

VIII. RESULTS

Three benchmarks have been chosen to evaluate our proposed approach. *Scharr* is an edge detection algorithm used in image processing. The algorithm uses a 3x3 filter which is convolved with the original image and calculates an approximation of the gradient in horizontal and vertical directions. *Mandelbrot* computes the Mandelbrot fractal which is a mathematical set of complex values. *Band-pass Filter* implements fast convolution filtering using FFT (Fast Fourier Transform) and inverse FFT. It performs low-pass filtering on the left and right channel of a stereo audio signal in parallel. The graphical representations of the process networks of the benchmark applications are shown in Fig. 5.

As the OMAP has two PE types (ARM and DSP), the calibration phase is carried out by running two mappings on the OMAP board per application, namely all-on-ARM and all-on-DSP[2]. The profiling results are processed statistically and packed into a calibration database. Then in the exploration phase, we run different mappings of the streaming applications on the calibrated OMAP virtual platform (see section VII-C).

We firstly examined the *accuracy* of the calibrated virtual platform. In our experiments a number of possible mappings of process-to-PE (here, ARM and DSP) were simulated using the calibrated OMAP virtual platform. To compare with the real execution time on the OMAP board, the same mappings ran on the hardware. The relative error is used as the measure for accuracy. The results are shown in Table I. Overall, the calibrated VPU-based virtual platform for OMAP has shown good accuracy for the three benchmarks (average error 2.51%, 4.53% and 3.59% respectively). The relative errors are mainly due to the scheduler modeling in the VPU for ARM not fully

---

[2]For the OMAP case study, sink and source nodes must be on the ARM, as ARM controls the peripherals for input and output.

---

matching a full-fledged Linux scheduler and instrumentation overhead to get measurement data. We will improve those in our future work.

We also measured the simulation *speed* of the calibrated OMAP virtual platform. The measurements were done using Synopsys Platform Architect 2010.09_SP1 on an AMD Athlon 64 X2 6000+ machine with 6GB memory. To compare with virtual platforms of low abstraction levels, we used ARM926EJS (instruction and cycle accurate) models as reference (a fully detailed OMAP virtual platform is not available to us). We took the all-on-ARM mapping of Band-pass Filter in Table I(c). The instruction-accurate simulator used 36.768s and the cycle-accurate one used 30.5 minutes. Our calibrated OMAP virtual platform just took 1.028s, showing speed-up of 36x and 1778x respectively.

The calibrated OMAP virtual platform also supports fully software *functional* execution. Fig. 6 shows an example. Waveforms can be generated during simulation to visualize run-time behavior. Both waveforms shown in Fig. 6 were taken from benchmark Scharr. Fig. 6 (a) shows the situation when all processes are mapped to ARM. Scharr horizontal computation (scharrHor) started only after the image is generated (genRect), because both reside on ARM. Fig. 6 (b) demonstrated the behavior of another mapping where scharrHor has been moved to DSP. The computation could immediately start as soon as first parts of the image arrive from genRect to scharrHor. The generated waveforms also have source code locations of channel accesses to ease further analysis. It is straightforward for programmers to see the timing change from one mapping to another, while at the same time debugging through lines of source code is also possible.

IX. CONCLUSIONS

We have presented an automatic tool-flow to calibrate VPU-based MPSoC virtual platforms for mapping of streaming applications. A complete case-study on OMAP for software mapping exploration has shown promising results of the flow's feasibility and the calibrated virtual platform's accuracy. Our future work includes applying the tool-flow to more MPSoC architectures. The tool-flow can also be coupled with intelligent algorithms to generate mapping decisions, further helping MPSoC programmers in the software development.

ACKNOWLEDGMENT

This work has been supported by the UMIC Research Centre, RWTH Aachen University.

REFERENCES

[1] Gilles Kahn, "The Semantics of a Simple Language for Parallel Programming," in *IFIP Congress 74*, North Holland, Amsterdam, 1974, pp. 471–475.

[2] Edward A. Lee and David G. Messerschmitt, "Synchronous Data Flow," in *Proceeding of the IEEE*, vol. 75, no. 9, 1987.

[3] "Synopsys Virtual Prototyping Tools," [Online] Available http://www.synopsys.com/Systems/VirtualPrototyping/Pages/VP-Tools.aspx (accessed 11/2010).

[4] "Wind River Simics," [Online] Available http://www.windriver.com/products/simics/ (accessed 11/2010).

TABLE I: *Results* $\left(Error = \left|\frac{Simulation\ Estimated\ Time - Real\ Execution\ Time}{Real\ Execution\ Time}\right|\right)$: *Processes with white background run on the ARM while processes with black background run on the DSP.*

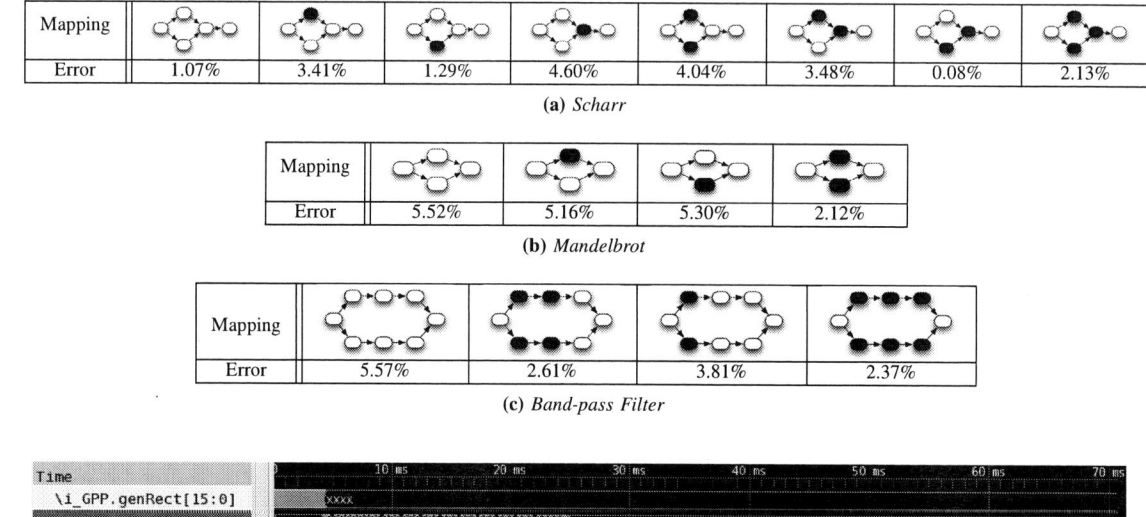

(a) *Scharr*

(b) *Mandelbrot*

(c) *Band-pass Filter*

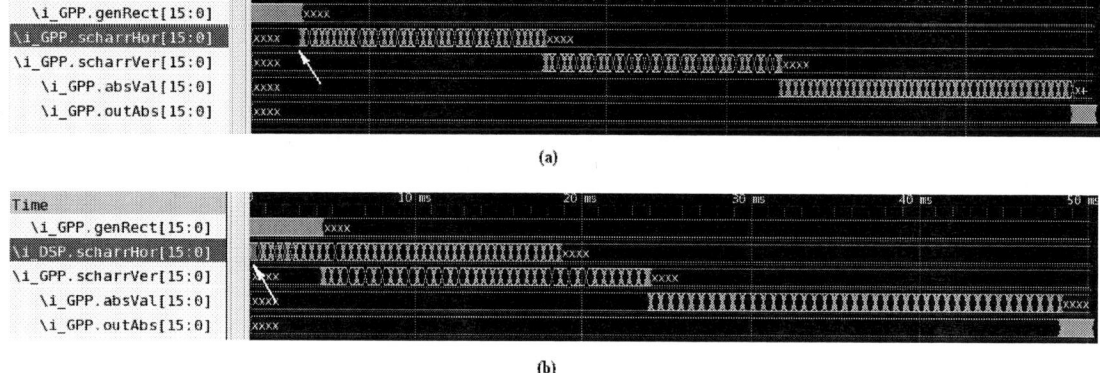

Fig. 6: *Calibrated OMAP Simulation Waveforms - Benchmark Scharr (a) all-on-ARM mapping (b) all-on-ARM except scharr horizontal computation on DSP (Task execution or accessing channels is recorded while task being idle is annotated with xxxx.)*

[5] A. D. Pimentel, C. Erbas, and S. Polstra, "A Systematic Approach to Exploring Embedded System Architectures at Multiple Abstraction Levels," *IEEE Trans. Comput.*, vol. 55, February 2006.

[6] A. D. Pimentel, M. Thompson, S. Polstra, and C. Erbas, "Calibration of abstract performance models for system-level design space exploration," *J. Signal Process. Syst.*, vol. 50, February 2008.

[7] K. Huang, I. Bacivarov, J. Liu, and W. Haid, "A modular fast simulation framework for stream-oriented MPSoC," in *Industrial Embedded Systems, 2009. SIES '09*, 2009, pp. 74 –81.

[8] T. Kempf, M. Doerper, R. Leupers, G. Ascheid, H. Meyr, T. Kogel, and B. Vanthournout, "A Modular Simulation Framework for Spatial and Temporal Task Mapping onto Multi-processor SoC Platforms," *Date '05*, 2005.

[9] Texas Instruments, "OMAP35x Product Bulletin," [Online] Available http://www.ti.com/lit/sprt457 (accessed 11/2010).

[10] T. Isshiki, D. Li, H. Kunieda, T. Isomura, and K. Satou, "Trace-driven workload simulation method for Multiprocessor System-On-Chips," in *Proceedings of the 46th Annual Design Automation Conference*, ser. DAC '09, 2009.

[11] R. Plyaskin and A. Herkersdorf, "A Method for Accurate High-Level Performance Evaluation of MPSoC Architectures Using Fine-Grained Generated Traces," in *ARCS*, 2010.

[12] E. Cheung, H. Hsieh, and F. Balarin, "Fast and accurate performance simulation of embedded software for MPSoC," in *Proceedings of the 2009 Asia and South Pacific Design Automation Conference*, 2009.

[13] J. Ceng, W. Sheng, J. Castrillon, A. Stulova, R. Leupers, G. Ascheid, and H. Meyr, "A high-level virtual platform for early MPSoC software development," in *Proceedings of the 7th IEEE/ACM international conference on Hardware/software codesign and system synthesis*, 2009.

[14] T. Kempf, K. Karuri, S. Wallentowitz, G. Ascheid, R. Leupers, and H. Meyr, "A SW performance estimation framework for early system-level-design using fine-grained instrumentation," in *DATE '06*, 2006.

[15] R. van den Berg, W. Tibboel, R. Wieringa, and M. Klompstra, "Using the application modeling and mapping methodology for system-level performance analysis," [Online] Avaiable, http://www.eetimes.com/General/DisplayPrintViewContent?contentItemId=4208820 (accessed 11/2010), September 2010.

[16] "clang: a C language family frontend for LLVM," [Online] Available http://clang.llvm.org/ (accessed 11/2010).

[17] D. Dart, "DSP/BIOS Kernel Technical Overview," *Texas Instruments Application Report*, no. SPRA780, August 2001.

[18] "Synopsys Platform Architect," [Online] Available http://www.synopsys.com/Tools/SLD/VirtualPrototyping/Pages/PlatformArchitect.aspx (accessed 11/2010).

Building a RTOS for MPSoC Dataflow Programming

Yaset Oliva, Maxime Pelcat, Jean-Francois Nezan,
Jean-Christophe Prevotet
IETR, INSA Rennes, CNRS UMR 6164, UEB
20, Av. des Buttes de Coesmes, 35708 Rennes
Email: yaset.oliva, maxime.pelcat, jean-francois.nezan,
jean-christophe.prevotet@insa-rennes.fr

Slaheddine Aridhi
Texas Instruments
06271 Villeneuve Loubet, France
Email: saridhi@ti.com

Abstract—Multiprocessor Systems-on-Chip (MPSoC) are becoming the standard high performance Digital Signal Processing (DSP) systems. Hardware complexity abstraction is needed to enable efficient MPSoC programming. A major challenge of MPSoC programming is efficiently handling the combination of new features necessary in a MPSoC operating system: load balancing and efficient use of the parallel resources, with the more traditional features of Real-Time Operating Systems (RTOS): resource sharing between applications, task priorities and reactivity to events. This paper presents a method to combine dataflow methods and RTOS features. The resulting system prototypes an RTOS for symmetric multiprocessing MPSoCs whose inputs are dataflow graphs of applications. The prototype is built on the μC/OS-II RTOS. Experimental results are given on a 3GPP Long Term Evolution algorithm executed on a 4-core MPSoC.

I. INTRODUCTION

In [1], Edward Lee shows that programming with threads is an error-prone operation and proposes several alternatives, including using process networks, to make software behavior more predictable. The dataflow process network [2] Model of Computation (MoC) models an algorithm by concurrent and independent modules known as actors which communicate ordered tokens (data quanta) through First-In First-Out channels. A set of firing rules defines when an actor executes. Dataflow models have been shown to favor parallel algorithm description as they favor data locality and reduce multi-core scheduling constraints to data dependencies [3]. Thus, dataflow models are well suited for use with signal processing algorithms

MPSoC systems used for signal processing are increasingly complex to program. Tools that ease MPSoC programming are thus more and more needed. This paper outlines a method of using dataflow graphs instead of thread declarations as inputs to an RTOS. In this case, the RTOS is then able to dispatch the actors which compose the dataflow graph to the cores of an MPSoC. These actors are directly executed by the MPSoC RTOS, instead of requiring the programmer to program primitives for task migration and synchronization. This method ensures the synchronization of actors and the management of actor input and output data.

This paper presents experiments performed with a MPSoC RTOS prototype incorporating dataflow model management

into the μC/OS-II kernel. The first target application is the uplink data decoding algorithm executed in base stations supporting the 3GPP Long Term Evolution (LTE) telecommunication standard.

Section II presents related works on multi-core runtime management and dataflow MoCs. Section III introduces the MPSoC RTOS structure. Section IV explains the dataflow management part and Section V considers the multi-core scheduling part. Experimental results of the prototype are shown in Section VI.

II. RELATED WORKS

A. RTOS and Runtime Management Systems

In [4], Nollet, et al. present an overview of runtime management systems for MPSoCs. The overview covers both industrial and academic systems. Commonalities between systems are identified: their structure can be divided into two parts, the quality manager and the resource manager. The quality manager tries to optimize the Quality of Service (QoS) of the system, i.e. to find the best application configuration; the resource manager offers mechanisms to allocate cores, communication media and memory.

The runtime systems reviewed in [4] consider only sequential applications which share cores of an MPSoC. The MPSoC RTOS presented in this paper differs, as it partitions each application between available cores. Parallelism of each application is explicitly stated using a dataflow graph. Such partitioning aims at ensuring good load balancing between cores and at reducing the computation latency of each application. In this way, the MPSoC RTOS may be considered to be equivalent to a quality manager of low granularity. The StreamIt [5] runtime system has similar goals than the MPSoC RTOS. However, it processes a specific streaming language while the MPSoC RTOS reuses C/C++ legacy code for the actors and combines it with a parameterized dataflow coordination language.

The MPSoC RTOS prototype is built on an existing kernel: μC/OS-II [6]. This choice was made through consideration of the small footprint, the simplicity and the available source code for this kernel. Since the original μC/OS-II doesn't implement

978-1-4577-0671-4/11 $26.00 © 2011 IEEE

any multiprocessor mechanism, its sources are modified to manage MPSoCs.

B. Modeling Applications with a Parameterized Dataflow Model

Many dataflow MoCs have been introduced in the literature, each offering a tradeoff between compile-time predictability and capacity to model dynamic run-time variability. It may be seen that the most obvious difference is their firing rules. Signal processing applications, such as telecommunication or video processing, are based on a loop which repeats a pattern of execution on a sequence of input data. Describing the repeated pattern with a dataflow MoC consists of dividing the computation into actors that exchange data only through input and output data queues without sharing any state.

In this paper, applications are described using parameterized dataflow modeling [7]. The chosen model of Parameterized Cyclo-Static Directed Acyclic Graph (PCSDAG) states that the graph can be totally reconfigured once before starting a new execution. This model is a parameterized and acyclic version of the Cyclo-Static Dataflow (CSDF) model [8]. The reconfiguration enables dynamic behavior of the algorithm, i.e. computation strongly depending on the input data. The PCSDAG model has been shown to be suitable for describing uplink and downlink data processing algorithms of 3GPP LTE base stations [9].

III. STRUCTURE OF THE MPSoC MODEL-BASED RTOS

The RTOS obtained is divided into two modules: dataflow management and RTOS scheduling (Figure 1). The dataflow graph is described with C++ objects and compiled within the dataflow management module. At each OS clock tick, the dataflow graph is parameterized, i.e. new parameter values are retrieved, and a temporary graph of execution is obtained. For each actor within the temporary graph, an RTOS task is created. The task corresponding to an actor must manage actor data and call actor code. Flags are automatically generated to synchronize actors based on their dependencies. RTOS scheduling is completed by assigning each actor to a core.

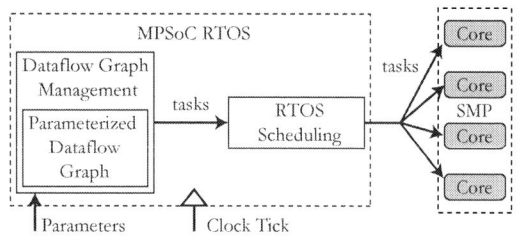

Fig. 1.   Structure of the MPSoC Model-Based RTOS

Next sections detail the dataflow management and RTOS scheduling modules.

IV. DATAFLOW GRAPH MANAGEMENT

The test application used is the uplink decoding of 3GPP LTE base stations. The PCSDAG model was specifically designed for this application but is likely to be also suitable for video encoding and decoding algorithms. Managing the graph consists of performing an expansion. This is the process where the dataflow graph is transformed into the temporary graph in which each actor is executed only once in a graph iteration. The temporary graph depends strongly on parameter values that fluctuate between consecutive iterations.

Figure 2 illustrates the application and the expansion. Uplink decoding consists of retrieving data sent simultaneously by nb_{user} users. This data is divided into Code Blocks (CB) of variable size. Both the number of users nb_{user} and the number of code blocks nb_{CB} for each user vary every millisecond. nb_{CB} is different for each user and may be represented by a cyclic pattern $nb_{CB} = nb_{CB_{user0}}, nb_{CB_{user1}}, \ldots$. The computation is divided into four phases: Multiple Input Multiple Output (MIMO) decoding, symbol to bit conversion, bit processing, and Cyclic Redundancy Check (CRC) [9]. Each millisecond, the graph is parameterized and transformed into a single rate directed acyclic temporary graph (srDAG) in which production and consumption rates of each FIFO queue are equal. Figure 2 shows 4 possible configurations. The expansion phase has been demonstrated to be executable in real-time while respecting the 3GPP LTE constraints in [9].

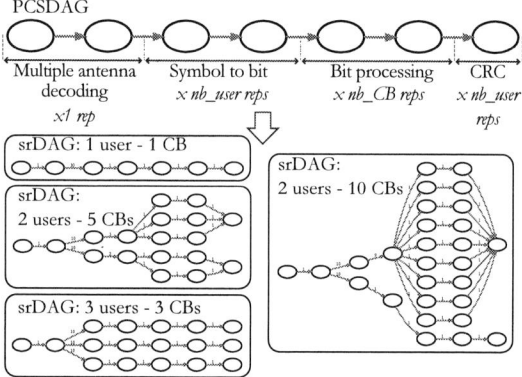

Fig. 2.   Expanding the input dataflow model.

After the expansion phase, an RTOS task is created for each actor in the srDAG graph. These tasks are then scheduled on the multiple cores of the MPSoC architecture. It may be noted that an alternate algorithm model to PCSDAG can be chosen provided that the srDAG graph can still be produced.

V. MULTI-CORE SCHEDULING

RTOS scheduling is divided into two phases: the master/slave phase and the symmetric phase. In the master/slave phase, the core designated as master executes the dataflow graph management and posts actors to the slave cores and to itself. When the scheduling enters the symmetric phase, each core is part of a pair and calls successively the common μC/OS-II scheduler to execute the highest priority task. The objective of this division is to limit the master/slave phase that naturally puts the master in the position of the bottleneck. The symmetric phase also enables preemptions and passive wait of events.

A. Master/Slave Phase

The master/slave phase is illustrated in Figure 3. Each core has a 1-place queue to receive an order of task execution. After an OS timer tick, the previously described dataflow graph management is performed and one task per actor is created. The master processor then dispatches the ready tasks with highest priority to the available slave processors. The dispatching process is performed by adding a message into the 1-place queue of the selected slave. The message contains the address of the shared memory location where the code of the task has been placed as well as input/output buffer management information. The dispatching process finishes when all ready tasks have been mapped or when there are no more available slave processors. When all slave processors are busy and there are still ready tasks, then the master processor executes the next task itself.

At the same time as the master executing, each slave processor waits for a message to be placed into its 1-place queue. When the message arrives, the slave processor places its program counter to the designed address and executes the task. After this first task dispatching is done, the symmetric phase starts and all cores become pairs.

B. Symmetric Phase

During the symmetric phase (Figure 4), each core can access the schedule function. This common access requires the use of mutex semaphores provided within the architecture support library. As in a typical RTOS, the scheduler function consists of identifying the highest-priority task that is ready and running it. The scheduler is called if either the task execution is preempted by a higher execution task or if the task execution is completed.

After the execution of the schedule function, and if no task is ready to be executed by the core, the scheduler saves the current task's context state and gets returned to its private memory. The slave is now ready to receive another message from the master processor.

Using RTOS scheduling, several independent applications can share the MPSoC with each actor with its own priority. Moreover, passive wait of events is allowed. This is very important in the case of systems with co-processors. In the cases where a programmer wishes to offload a costly operation such as turbo-coding on a co-processor, an actor can wait for turbo-coding completion passively on a core and can be preempted by another ready actor while turbo-coding is running.

The next section gives experimental results of the prototype.

VI. EXPERIMENTAL RESULTS

The experimental prototype is executed on an Altera Cyclone II FPGA integrating 4 NIOS cores [10] and a shared memory. The memory footprint of the system is shown in Figure 5. As can be seen, the operating system data occupies

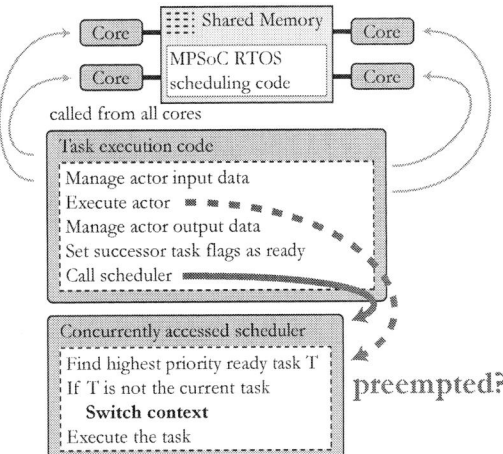

Fig. 4. Symmetric Scheduling Phase: Self-Organizing Execution

more than 50% of the 510kb of total system memory. This expensive cost is largely due to the size and number of task stacks statically allocated by the kernel (64 in this case).

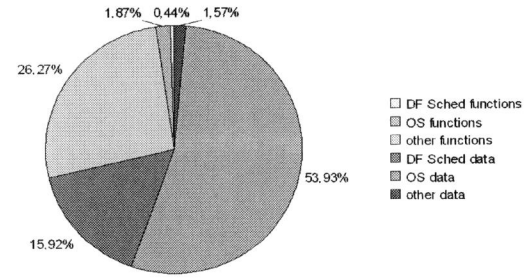

Fig. 5. Memory Usage

The prototype has been run for six iterations. At the end of each iteration, the application parameter values are modified to generate a different number of actors, as shown in Table I.

Iterations	1	2	3	4	5	6
nb_{user}	10	3	6	8	10	2
$max\ nb_{CB}$ per user	5	4	3	1	2	5
$max\ nb_{CB}$	14	10	13	5	8	14
nb_{actors}	39	31	43	27	30	36

TABLE I

ITERATION PARAMETERS

Figure 6 shows the system performance in terms of execution time, against the number of cores. The iterations correspond to the ones presented in Table I. These curves confirm that the system performances improve as the number of cores increases. However, the concurrent shared memory accesses also increase and become a bottleneck. This is a limitation of the current system. The performance improvement naturally depends on the shape of the srDAG because the exposed parallelism depends on the graph parameters.

978-1-4577-0671-4/11 $26.00 © 2011 IEEE   145

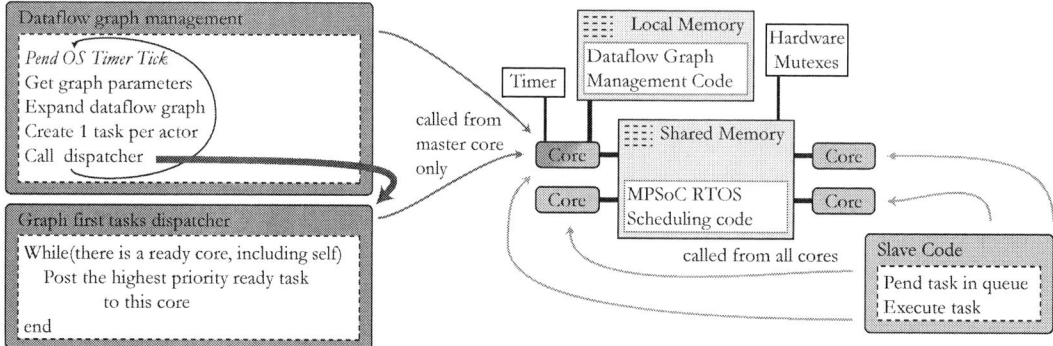

Fig. 3. Master/Slave Scheduling Phase: Expanding Dataflow Graph and Launching Slave Cores.

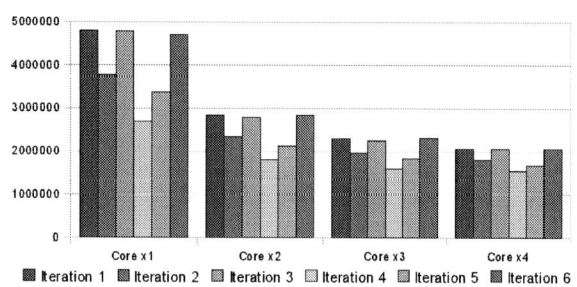

Fig. 6. Execution time performances

VII. FUTURE WORKS

The current prototype may be improved both in terms of latency and in terms of memory. Speed can be gained by exploiting quiescent points [11] of the single rate DAG. Quiescent points are points between two actors' execution when no context needs to be maintained during rescheduling. This case corresponds to the scheduler call at the end of the "Task execution code" in Figure 4. This call does not require context switch and the task stack can even be flushed.

From a memory viewpoint, the task stacks can be significantly reduced. μC/OS-II associates independent stacks to each task. Having a pool of tasks and reusing tasks associated to finished actors will automatically reduce task creation time and stack memory. It is planned to test the code on a Texas Instruments TMS320TCI6486, a 6-core SMP DSP with a 500MHz clock. Each core of this platform also has a private memory. Thus, application data and code can be removed from the shared memory and managed automatically using dataflow graph information. The TMS320TCI6486 enables research towards a RTOS for non-SMP MPSoC architectures.

VIII. CONCLUSION

In this paper, we detailed an RTOS for Symmetric Multiprocessing MPSoC combining ideas from dataflow models and from commonly-used RTOS. The advantages of such a system are numerous. Dataflow graphs bring automatic parallelization and ease of application description. The traditional difficulty of

manually synchronizing tasks disappears. Moreover, the actors themselves can be written in C or C++ code, so legacy code can be easily reused. The system obtained is portable to SMP architecture with considerable number of cores.

The MPSoC RTOS obtained goes successively through a master/slave phase that parameterizes the dataflow graph and a symmetric phase that permits any core to access the RTOS scheduler. This allows the reactivity to events of an RTOS to be combined with the automatic partitioning and portable execution of an algorithm described in a parameterized dataflow graph. Experimental results are shown on a 3GPP Long Term Evolution algorithm running on a 4-core MPSoC. The MPSoC RTOS prototype is built on the μC/OS-II RTOS.

Based on the present prototype of the MPSoC RTOS, further research will be conducted to identify the dataflow models best suited to signal processing applications. Additional scheduling methods and architectures will also be tested to exploit a prior knowledge on graph execution.

REFERENCES

[1] E. A. Lee, "The problem with threads," *Computer*, vol. 39, no. 5, p. 33–42, 2006.

[2] E. A. Lee and T. M. Parks, "Dataflow process networks," *Proceedings of the IEEE*, vol. 83, no. 5, p. 773–801, 1995.

[3] W. B. Ackerman, "Data flow languages," *Computer*, vol. 2, 1982.

[4] V. Nollet, D. Verkest, and H. Corporaal, "A safari through the MPSoC Run-Time management jungle," *Journal of Signal Processing Systems*, vol. 60, no. 2, p. 251–268, 2010.

[5] W. Thies, M. Karczmarek, M. Gordon, D. Maze, J. Wong, H. Hoffmann, M. Brown, and S. Amarasinghe, "StreamIt: a compiler for streaming applications," *Technical Report MIT , Cambridge, MA*, 2001.

[6] J. J. Labrosse, *MicroC/OS-II: the real-time kernel*. Newnes, 2002.

[7] B. Bhattacharya and S. Bhattacharyya, "Consistency analysis of reconfigurable dataflow specifications," in *Embedded processor design challenges*, 2002, p. 308–311.

[8] G. Bilsen, M. Engels, R. Lauwereins, and J. A. Peperstraete, "Cyclostatic data flow," in *icassp*, 1995, p. 3255–3258.

[9] M. Pelcat, J. F. Nezan, and S. Aridhi, "Adaptive multicore scheduling for the LTE uplink," in *Adaptive Hardware and Systems (AHS), 2010 NASA/ESA Conference on*, 2010, p. 36–43.

[10] Altera, "Literature on NIOSII software embedded processor," http://www.altera.com/devices/processor/nios2/ni2-index.html.

[11] S. Neuendorffer and E. Lee, "Hierarchical reconfiguration of dataflow models," in *Formal Methods and Models for Co-Design, 2004.*, 2004.

Author Index

Eero Aho	62
Slaheddine Aridhi	143
Gerd Ascheid	136
Armando Astarloa	96
Rabie Ben Atitallah	56
Mohammad Badawi	102
Jürgen Becker	92
Luca Benini	9, 34
Yolande Berbers	48
Heikki Berg	62
Davide Bertozzi	9, 122
Zubair Wadood Bhatti	48
Daniele Bortolotti	34
Claudio Brunelli	62
Henk Corporaal	14
Cristian Cuna	108
Issam W. Damaj	72
Jean-Luc Dekeyser	56
Georg Ellguth	112
Otto Esko	29
Xin Fan	9
Marco Ferraresi	122
Martin Gag	116
Marc Geilen	14
Giuseppina Gobbo	122
Eckhard Grass	9
Vladimír Guzma	82
Christoph Heer	9
Ahmed Hemani	102, 128
Timo D. Hämäläinen	78, 86
Marko Hännikäinen	86
Sebastian Höppner	112
Tomoki Ikegaya	22
Amir-Hossein Jahangir	68
Jaime Jimeńez	96
Pekka Jääskeläinen	29
M.R. Kakoee	9
Subayal Khan	1
Ilya Klotchkov	42
Uli Kretzschmar	96
Miloš Krstić	9
Abdallah Lakhtel	92
Jesús Laźaro	96
Lasse Lehtonen	78
Rainer Leupers	136
Daniele Ludovici	122
Mikko Majanen	1
Omer Malik	128

Andrea Marongiu	34
Hiroshi Matsuo	22
Narasinga Rao Miniskar	48
Mehdi Modarressi	68
Mikhail Moiseev	42
Orlando Moreira	14
Yasuhiko Nakashima	22
Rick Nas	14
Jean-Francois Nezan	143
Seyyed Hossein Nikounia	68
Jari Nurmi	1
Ryosuke Oda	22
Maximilian Odendahl	136
Yaset Oliva	143
Nicolae Olteanu	108
Francesco Paterna	34
Esko Pekkarinen	78
Maxime Pelcat	143
Christian Pinto	34
Teemu Pitkänen	82
Davy Preuveneers	48
Jean-Christophe Prevotet	143
Santhosh Kumar Rethinagiri	56
Martino Ruggiero	34
Jukka Saastamoinen	1
Sergey Salishev	42
Erno Salminen	29, 78, 86
Birgit Sanders	9
Christoph Schmutzler	92
René Schüffny	112
Stefan Schürmans	136
Weihua Sheng	136
Martin Simons	92
Firew Siyoum	14
Alessandro Strano	9
Jarmo Takala	29, 82
Dirk Timmermann	116
Teodor Tite	108
Tomoaki Tsumura	22
Adelina Vig	108
Dennis Walter	112
Tim Wegner	116
Roel Wuyts	48
Tatsuhiro Yamada	22
Alexey Zakharov	42
Aitzol Zuloaga	96

9781457706714